THE EMERALD PLANET

David Beerling is the Sorby Professor of Natural Sciences and Director of the Leverhulme Centre for Climate Change Mitigation at the University of Sheffield. He has published numerous articles in learned academic journals. *The Emerald Planet* is his first popular science book and formed the basis of a major three-part BBC Two television series, How to Grow a Planet. He was elected to the Fellowship of the Royal Society, London, in 2014.

Praise for *The Emerald Planet*

'My favorite nonfiction book this year. A minutely-argued but highly readable history of the last half-billion years on earth. The story Beerling tells could not have been put together even ten years ago, for it depends upon the latest insights from palaeontology, climate science, genetics, molecular biology, and chemistry, all brilliantly and beautifully integrated together. I got a special deep, quiet pleasure from reading *The Emerald Planet*—the sort of pleasure one gets from reading Darwin.'
Oliver Sacks, Books of the Year, *Observer*

'Within these pages is one of the greatest stories ever told: the story of the way plants have shaped our planet and how they will shape its future as the climate changes more rapidly than ever before. It is as fascinating as it is important.'
Stephanie Pain, *New Scientist*

'A beautifully detailed account of the puzzles of reconstructing Earth's climatic history...a gorgeous book about plants and the fantastically complex way they can be used to deduce past climatic events.'
Steven Poole, *The Guardian*

'Here at last is David Beerling as the Green Knight, revealing the extraordinary story of the construction of our emerald planet. Rigorous science joins hands with an enthusiastic delivery to re-awaken our fascination in plants, while engaging anecdotes provide a thrilling background to an extraordinary story of climate change and our current environmental crisis.'
Professor Simon Conway-Morris, author of *Life's Solution*, University of Cambridge

'A fascinating, ambitious and well-written book [on] how the pursuit of plants promises to unlock greater riches from their fossil record...a fresh form of critical analysis on some of the major issues of Earth's history. Beerling makes a compelling case for the power of plants and for research on living systems as a way of unlocking the potential of the fossil record.'
Paul Kenrick, Natural History Museum, London

'David Beerling's fascinating new book offers a new global perspective on the evolution of our planet... [a] vivid account...The environmental legacy of the plant kingdom upon our world can only be better appreciated after reading this book.'
Louis Ronse De Craene, Book of the Month, *Geographical*

'Beerling uses evidence from the plant fossil record to reconstruct past climates and to help explain mass extinctions. Too often this evidence has been disregarded, but Beerling gives it its due, and then some. [He] introduces us to the scientists of the past and their contributions to today's hypotheses...and successfully conveys the incremental nature of science and that new hypotheses often emerge from a combination of observations and syntheses of previous work.'
Pamela Soltis, Florida Museum of Natural History

'David Beerling tells two stories in parallel. Both are eloquently and engagingly merged in a scholarly, yet generally accessible book...Beerling provides for the reader a fascinating history of the discovery of fossils and the inferences drawn from them...this book is a wonderful example of the nascent field of Earth systems science.'
Paul Falkowski, Nature

'Beerling gives us the big picture of how plants have changed our planet—and poses the key question of how we will manage the emerald planet to ensure the kind of future we desire.'
Professor Sir Peter Crane, University of Chicago

'The Emerald Planet is a serious talking to about why plants must not be ignored. Everyone should appreciate this...and also know, Beerling argues, how plants fit into the global picture.'
Jonathan Silvertown, Times Literary Supplement

'[A] fascinating overview of green evolution.'
Karl Dallas, Morning Star

'Reads like a novel...but is constructed solidly on a foundation of diverse scientific literature. The stars are the scientists from the past, the present and the near future, pushing back ignorance, racing against each other and sometimes against impending natural threats...The coverage is sweeping. An engaging historical narrative in which science seems to be a fast-paced enterprise. Provocative, well written and well researched.'
William A. DiMichele, Smithsonian Institution, Washington DC

'David Beerling is to be congratulated on producing a highly readable account of an underappreciated aspect of Earth history—the role of plants in shaping today's planet, and our views of it. Palaeobotany has probably never been so topical.'
Nigel Chaffey, Annals of Botany

'Of great value and relevance to all interested in plants, climate and, equally, the future of our "emerald planet".'
John MacLeod, Royal Horticultural Society

'Refreshingly novel...a thought-provoking book...by better understanding the role that plants played during extreme episodes in Earth history we are better equipped to understand the changes that might occur in response to global warming.'
Howard Falcon-Lang, University of Bristol

'His account interweaves hard scientific facts with rich anecdotes about the scientists who have pieced together the evolutionary record over time. The result is a book that is fascinating and exciting to read. *The Emerald Planet* is beautifully written, fresh and provocative.'
Jennifer McElwain, *American Scientist*

THE
EMERALD
PLANET

*How plants changed
Earth's history*

DAVID BEERLING

OXFORD
UNIVERSITY PRESS

OXFORD

UNIVERSITY PRESS

Great Clarendon Street, Oxford, OX2 6DP,
United Kingdom

Oxford University Press is a department of the University of Oxford.
It furthers the University's objective of excellence in research, scholarship,
and education by publishing worldwide. Oxford is a registered trade mark of
Oxford University Press in the UK and in certain other countries

© David Beerling 2007
First published 2007
First published in paperback, 2008
Revised impression, as Oxford Landmark Science 2017

The moral rights of the author have been asserted

First Edition published in 2017

Published in the United States of America by Oxford University Press
198 Madison Avenue, New York, NY 10016, United States of America

British Library Cataloguing in Publication Data
Data available

Library of Congress Control Number: 2016955210

ISBN 978-0-19-879832-3

Printed and bound by
CPI Group (UK) Ltd, Croydon, CR0 4YY

For Juliette

FOREWORD

The Emerald Planet is one of those books that makes you think differently about the world you live in. Or at least it certainly did that for me. I had been used to thinking of how the Earth worked in terms of those big physical processes you can see—shuddering earthquakes, belching volcanoes, grinding glaciers. But what unfolds in the pages of this book is the story of a silent force that has shaped our planet and almost everything living on it, one that is hidden in plain sight: plants.

I'm not sure what your feeling about plants is, but mine had always been rather ambivalent. Plants seemed... well... dull. Even a brief childhood phase of force feeding insects to my Venus Fly Trap simply highlighted how little plants did. Animals were far more exciting. Even rocks were more impressive, courtesy of the spectacular landscapes they helped sculpt. True, plants could be beautiful too, even extravagant, but they were passive—bit players in the workings of a natural world that elsewhere was bristling with shock and awe. Then, I read *The Emerald Planet*.

I came to the book just as I was finishing a television series called 'Earth: The Power of the Planet', which celebrated the wonderfully intricate interplay between our land, ocean, ice and atmosphere. It showcased the new geological perspective of Earth System Science, which brought together insights from all aspects of science to understand how the planet we call 'home' came to be. For me, the revelation of *The Emerald Planet* was that when you are interested in those planetary scale machinations, and when you have hundreds of thousands or millions of years to play with, that's when plants come to the fore. That's when they become a planetary force in their own right - orchestrating the evolution of our climate, our landscapes and our animal life. It is true that plant power is a slow and almost imperceptible force, but set against a backdrop of creeping continents or the measured rise and fall of mountains, the actions of the plant world become dramatic and fundamental. The result is an epic story that deserved to be told, and so the seeds of a television series were sown: *How to grow a planet*.

The Emerald Planet opens with a humorous but telling remark from the book *Science made stupid* which states 'The history of plants is an important chapter in the story of life. Unfortunately it's a pretty dull chapter, so we'll skip it.' David Beerling's achievement with this engaging book is that no reader will ever think that way ever again.

Iain Stewart

PREFACE

The great evolutionary biologist J.B.S. Haldane (1892–1964), on being asked by a cleric what biology could say about the Creator, entertainingly replied, 'I'm really not sure, except that the Creator, if he exists, must have an inordinate fondness of beetles.' Haldane was referring to the fact that approximately 400 000 species of beetles make up roughly 25% of all known animal species. Current estimates for the total number of species of flowering plants in the world (300 000–400 000), had they been available to him at the time, may have given Haldane pause for thought about his riposte.

Plants and beetles may be tied, stem and thorax, in the global biodiversity stakes but when it comes to capturing our own fascination, plants are way ahead, clear winners in the popularity stakes. We have been collecting, classifying, and cultivating floras worldwide for centuries. Not only do plants provide us with fuel, food, shelter, and medicines that sustain the human way of life, but they also uplift and inspire us. Irrespective of the season, we flock to fine gardens, elegantly sculpted landscapes, botanical gardens, and arboreta to pay homage to the plants and trees.

But how many of us have stopped to wonder how remarkable plants are, how profoundly they have altered the history of life on Earth, and how critically they are involved in shaping its climate? Only now are we unlocking vital information about the history of the planet trapped within fossil plants. My aim in writing this book has been to provide a glimpse of these exciting new discoveries because they offer us a new way of looking and thinking about plant life. It recognizes—indeed emphasizes—

that plants are an active component of our planet, Earth. At the global scale, forests and grasslands regulate the cycling of carbon dioxide and water, influence the rate at which rocks erode, adjust the chemical composition of the atmosphere, and affect how the landscape absorbs or reflects sunlight. In this book, I reveal how plant activities like these have added up over the immensity of geological time to change the course of Earth history. Never mind the dinosaurs, here is a revisionist take on Earth history that puts plants centre stage.

My hope is that the book will further stimulate readers' natural fascination with plants—both the living and the long dead—by revealing their activities in this new light. Each chapter leads the reader through a scientific detective story describing a puzzle from Earth history in which plants have played a starring role. Occasional linkages with themes from other chapters are pointed out as they arise. This format allows individual chapters to stand alone or be read in sequence. I provide a short summary at the start of each chapter to help readers quickly grasp the nature of the puzzle and glimpse the scientific excitement ahead. In writing a popular science book like this, it is true that, in Mark Twain's words, I have got 'wholesale returns of conjecture out of ... a trifling investment in fact'. All sources of the 'facts' taken from the published scientific literature are given in the notes, and where my ideas and conjecture are more speculative, I hope I have clearly signposted them as such. I have made every effort to keep the text free of scientific jargon, but admit that the odd word or term has proved indispensable. These are defined or explained where they occasionally crop up.

PREFACE

He had been eight years upon a project for extracting sunbeams out of cucumbers, which were to be put into vials hermetically sealed, and let out to warm the air in raw inclement summers.

Jonathan Swift (1726), *Gulliver's travels*

Humankind continues to take liberties with our planet, although not, of course, in the gentle manner Jonathan Swift described in *Gulliver's travels*. By consuming fossil fuels and destroying tropical rainforests, we are undertaking a global uncontrolled experiment guaranteed to alter the climate for future generations. Plants and vegetation are major actors in the environmental drama of global warming now as they have been in the recent and more distant past. This book focuses on the distant past, Earth history from millions of years ago. As we shall see, though, this investigation of the past has much to teach us about our present predicament. It offers us cautionary lessons about the current mismanagement of our planet's resources we would be wise to heed.

July 2006, Sheffield D.B.

ACKNOWLEDGEMENTS

This book had its genesis in discussions with colleagues over a beer in a sushi bar in San Francisco, in December 2002. San Francisco is home of the fall meeting of the American Geophysical Union, an annual gathering of several thousand scientists from a host of disciplines who congregate for a science feast. At the 2002 meeting, I had the prospect of delivering a belated inaugural lecture the following spring hanging over me, and was searching for an effective way to present some of the findings of my research group over the past decade in an engaging way to a lay audience. One idea was to present them as a series of short stories, each beginning with a seemingly straightforward question, an approach used to good effect by Paul Colinvaux in his admirable 1980 book *Why big fierce animals are rare* (Penguin, London). The basic concept of individual stories, each with plants playing the starring role, worked well on the night, and I subsequently adopted that format here, although in all but one case the inclusion of a question-in-the-title has been abandoned.

Many people have been instrumental in helping to put this book together. I extend warm thanks to Bill Chaloner (University of London) and Colin Osborne (University of Sheffield) for patiently and critically reviewing earlier drafts of the text. Many other colleagues also kindly gave of their time to critically read and comment on various chapters, provide data, ideas, and images, and engage in detailed discussions about the different scientific issues and queries raised during the writing process. I have benefited greatly from their input and special thanks must go to Paul Kenrick (Natural History Museum, London), Karl

Niklas (Cornell University), Charles Wellman, Doug Ibrahim, Barry Lomax, Peter Mitchell, Andrew Fleming, and Ian Woodward (University of Sheffield), Robert Berner (Yale University), Jon Harrison (Arizona State University), Robert Dudley (University of California, Berkeley), Don Canfield (Odense University, Denmark), Henk Visscher (Utrecht University), Dana Royer (Wesleyan University), Charles Cockell (Open University), Kevin Newsham and Jonathan Shanklin (British Antarctic Survey), Virginia Walbot (Stanford University), Sheila McCormick (University of California, Berkeley), Lee Kump (Pennsylvania State University), Michael Benton and Paul Valdes (University of Bristol), John Pyle and Michael Harfoot (University of Cambridge), Tim Lenton (University of East Anglia), Paul Wignall, Jane Francis and Jon Lloyd (Leeds University), Gavin Schmidt (NASA/Goddard Institute for Space Studies, New York), Barry Osmond (Australian National University) and Govindjee (University of Illinois). The corrective feedback of all of these individuals trapped numerous errors of interpretation, and crucial omissions. Any remaining errors and over-enthusiastic interpretations of datasets and published papers remain my own responsibility.

The groundwork for my thinking about plants as a geological force of nature was laid in large part during my tenure of a Royal Society University Research Fellowship held between 1994 and 2001. I am extremely grateful to the Royal Society for funding my research through this mechanism. These fellowships continue to offer unsurpassed opportunities to young scientists by giving them the most valuable commodity in their armoury—time to think, free from the usual burdens of administration and teaching that normally accompany academic life. I am also grateful to the Leverhulme Trust and the Natural Environment Research Council, UK for their financial support of my research.

Popular science writing requires a step change in style from the more turgid prose used in writing scientific papers. Francis Crick (1916–2004), the British molecular biologist and co-discoverer of the structure of DNA, commented in his 1990 book *What mad pursuit: a personal view of scientific discovery* (Penguin, London) that 'there is no form of prose more difficult to understand and more tedious to read than the average scientific paper'. I am extremely grateful to my editor, Latha Menon, for her wise counsel and suggestions on earlier drafts that have eased the transition, and which have been instrumental in shaping the current direction of the book. Whether I have been successful in this endeavour or not is another matter; any failings remain my own. I also thank the production team at Oxford University Press for efficiently shepherding me through the production process, especially Michael Tiernan the copy-editor and Sandra Assersohn for efficiently sourcing some delightful images.

Finally I thank my partner Juliette for her forbearance far above and beyond the call of duty. The time that writing this book has stolen from us over the past three years astonished me as well.

CONTENTS

ILLUSTRATIONS

PLATES

Plate 1 A fossil of *Cooksonia*.
(© The Natural History Museum, London. Reproduced with permission.)

Plate 2 The leafless and the leafy.
(*Upper image*: from Osborne, C.P., Beerling, D.J., Lomax, B.H., and Chaloner, W.G. (2004) Biophysical constraints on the origin of leaves inferred from the fossil record. *Proceedings of the National Academy of Sciences, USA*, **101**, 10360–2. *Lower image*: courtesy of Colin Osborne. Both photos reproduced with permission.)

Plate 3 Antoine Lavoisier.
(© Getty Images. Reproduced with permission.)

Plate 4 Robert Berner.
(Photo © Robert Berner. Reproduced with permission.)

Plate 5 Fossil charcoal of gymnosperm woods from wildfire in Nova Scotia.
(From Falcon-Lang, H.J. and Scott, A.C. (2000) Upland ecology of some Late Carboniferous cordaitalean trees from Nova Scotia and England. *Palaeogeography, Palaeoclimatology, Palaeoecology*, **156**, 225–42. Reproduced with permission.)

Plate 6 Robert Strutt.
(National Portrait Gallery, London. Reproduced with permission.)

Plate 7 Mutated fossil plant spores dating to 251 million years ago.
(From Visscher, H., Looy, C.V., Collinson, M.E. *et al.* (2004) Environmental mutagenesis during the end-Permian ecological crisis. *Proceedings of the National Academy of Sciences, USA*, **101**, 12952–6. Reproduced with permission.)

Plate 8 William Buckland.

Plate 9 Buckland's table of polished coprolites.
(Lyme Regis Museum. Reproduced with permission.)

Plate 10 Above: solid methane hydrate brought up from the depths of the ocean. Below: small fragments of icy hydrate burning in air.

(*Upper image*: Leibniz Institute of Marein Sciences (IFM-GEOMAR). Reproduced with permission. *Lower image*: courtesy of Tom Pantages.)

Plate 11 Scott's party at the South Pole.

(Scott Polar Research Institute. Reproduced with permission.)

Plate 12 Albert Seward.

(National Portrait Gallery, London. Reproduced with permission.)

Plate 13 Fossil remains of polar forests discovered in Axel Heiberg Island in the Canadian High Arctic and on the Antarctic Peninsula (top left). The tree stump (top left) is thought to be of dawn redwood (*Metasequoia*), a deciduous species with feathery leaflets still widely planted today (top right). The substantial fossil tree trunk discovered on Antarctica (bottom left) belongs to the southern beech (*Nothofagus*) family, the relatives of which form extensive natural forests in New Zealand (bottom right).

(*Top left*: Eocene fossil stump, courtesy of Jane Francis, University of Leeds. *Top right*: Ming Li/Photolibrary. *Bottom left*: courtesy of Jane Francis, University of Leeds. *Bottom right*: courtesy of Ian Woodward, University of Sheffield. All photos reproduced with permission.)

Plate 14 John Tyndall.

(© Getty Images. Reproduced with permission.)

Plate 15 Martin Kamen and Samuel Ruben.

(*Kamen image*: AIP Emilio Segre Visual Archives, Segre Collection. *Ruben image*: Ernest Orlando Lawrence Berkeley National Laboratory. Both photos reproduced with permission.)

Plate 16 The complex web of feedbacks between biology and the climate system, linked by fire, which might have accelerated the global expansion of C_4 savannas some 8 million years ago.

(Photo courtesy of Doug Ibrahim, University of Sheffield. Reproduced with permission.)

1

Introduction

This book tells for the first time stories of the evolution of plants. It illuminates the exciting role fossil plants are playing in unravelling the history of our planet. The illumination is made possible thanks to the emergence of an exhilarating new discipline, one that integrates unprecedented knowledge of plants as living organisms with their fossil record and the role they play in driving global environmental change. As we do so, we can see clearly that plants are not 'silent witnesses to the passage of time' but dynamic components of our world that shape and are, in turn, shaped by the environment. The power of the new science is that it brings to life the plant fossil record in previously hidden ways to offer a deeper understanding of Earth's history and pointers to our climatic future.

The evolution of plants is an important chapter in the history of life. However, it's a pretty dull chapter, so we'll skip it.

Tom Weller (1985), *Science made stupid*

CHARLES Darwin (1809–82), the greatest naturalist of all, was fascinated by them, Richard Dawkins all but ignored them.[1] The world, it seems, is divided about the charms of the plant kingdom. The opening quotation of this chapter is from the American popular science author Tom Weller's witty and provocative 1985 book *Science made stupid*, and sums up the malaise afflicting those on one side of the great divide. To these folk, plants have an unexceptional evolutionary trajectory leading up to the emergence of our modern floras and play no appreciable role in unravelling Earth's history. Too often, this view is reiterated, reinforced, in Earth science textbooks, where it is palmed off on the unwary reader as received wisdom. Many such scholarly tomes devote a few pages to Earth's first green spring, that decisive moment of our past when terrestrial plants turned the continents green. A few graciously give more space— an entire chapter, perhaps—to the progression of plants up the evolutionary ladder from their earliest beginnings through to the appearance of the first forests, the emergence of seed plants, and the blooming of the Earth with the rise of flowering plants. Fewer still recognize plants as important players in the game of life.[2]

In this book I argue that Weller's viewpoint, and the conventional view of textbooks, is now outdated, redundant even, and misguided. The scientific investigation of fossil plants is on the threshold of an exciting new era, a grand synthesis illuminating new chapters in the inseparable stories of plant evolution and Earth's environmental history. This book is about that new science. It is an endeavour that has emerged unnoticed in the last two decades but which is proving a powerful tool for clearing a path through the dense, sterile thicket of entrenched orthodoxy. It advocates fossils not as the disarticulated remains of ancient plant life gathering dust deep within the basements of museums, but as exciting, dynamic entities brought to life in new ways by the scientific investigation of their living counterparts. The *Emerald planet* is not a textbook, nor an attempt at describing, blow-by-blow, the detailed evolutionary history of plant life over the ages in a manner accessible to the general reader. Neither will the reader find a classical treatment of the detailed history of the Earth, with its shifting continents, the opening and closing of ocean gateways, and the changing climate of the past 4.5 billion years. To be sure, plant evolution, global climate change, and the theory of plate tectonics are all elements that form a crucial part of what the new science is about. But the argument is that we must marry these traditional elements of geology with a focus on plants as living organisms to mount a frontal attack on the citadels of received wisdom and orthodoxy and reach a deeper understanding of Earth history.

The endless fascination of reaching for this deeper understanding of Earth history is that it has already happened. It establishes the sparkling intellectual adventure of unravelling the what, why, and how of it all. Ancient fossils and rocks document it, and by decoding the different languages they are written in we find that they often betray the processes involved in shaping Earth's history. The grand challenge is piecing it all together from a fragmentary record of events. Unlike the science of the future, the science of the past holds out the ultimate reward—the exciting prospect of understanding the causes of things to better comprehend how the world works. Projections of future climates and ecology, like the retreat of mountain glaciers and the polar ice caps, the migration of forests, and so on, are really just proposals, made in spite of real ignorance about the critical physical and biological processes involved, and the difficulty of actually evaluating them.[3]

The key to it all lies in recognizing the urgent need to understand how the environment shapes plants, and how plants shape the environment, over the immensity of geological time. My intention is to show that with this recognition come two new ideas. First, that plants exquisitely record previously hidden features of Earth history, and second, that plants are a geological force of nature, one to be added to the pantheon of mighty forces traditionally thought to have moulded and recycled the Earth's landscape and climate throughout its 4.5 billion years. Yet the underlying rocks of our familiar modern world, weathered by the action of climate, so obviously govern the character of the landscape around us, and influence the formation of soils and the nature of agriculture and natural vegetation, that it seems an impossible task to think of the reverse situation.

But for this scientists have a trick up their sleeve. It has been likened in significance to the Copernican revolution, the seminal moment in history that properly put Earth, and the other planets in our solar system, in orbit around the Sun some five hundred years ago. The second 'Copernican' revolution is emerging in the form of a general class of mathematical models, grandly dubbed 'Earth system' models.[4] Earth system models vary enormously in complexity, forming a dynamic hierarchy that ranges from those that run in seconds on desktop computers to state-of-the-art examples demanding weeks of processing time on the world's fastest supercomputers. It is axiomatic that even the most sophisticated models are incomplete; their value lies in their capacity to simulate how the biological and physical components of our planet—the atmosphere, oceans, and biosphere—interact with each other across a very wide range of timescales, from days to millions of years. When the newly discovered activities of plants are included in such models, we glimpse their capacity to shape the global environment of our planet.

Before we embrace these new ideas, it is perhaps time to say something about the thorny issue of the Gaia hypothesis of James Lovelock and colleagues. The Gaia hypothesis originally stated that the Earth's environment is regulated 'at a state comfortable for life by and for the biosphere.'[5] Many scientists were understandably aghast at such an extravagant claim and took issue with the implied teleological suggestion that life could consciously bend the climate to its collective will to improve its lot. Indeed, less than a decade later Lovelock abandoned the idea, writing 'It is important to recognize that the Gaia hypothesis so stated is wrong'.[6] In its place, rising phoenix-like from the ashes, is a revised concept called 'Gaia theory', in which 'active feedback processes operate automatically and solar energy sustains comfortable conditions for life. The conditions are only constant in the short-term and evolve in synchrony with the changing needs of the biota as it evolves'.[7] Again the language hints at the uneasy notion that living organisms regulate the environment to maintain conditions comfortable for themselves. I show in several chapters that this is often not the case at all, but there are many other examples.[8]

The problem is that if we look around us, life seems to be supremely adapted to its environment and this simple observation tempts us to the false conclusion that organisms orchestrated things this way. Yet the logic is flawed by the obvious fact that, as Darwin observed, natural selection ruthlessly weeds out those life forms that are poorly adapted to their environments. Douglas Adams (1952–2001), author of the *Hitchhiker's guide to the galaxy*, commented on the Gaia hypothesis with characteristic flair, 'imagine a puddle waking up one morning and thinking, "This is an interesting world I find myself in—an interesting *hole* I find myself in—fits me rather neatly, doesn't it? In fact it fits me staggeringly well, must have been made to have me in it!"' Suspended uncomfortably between tainted metaphor, fact, and false science, I prefer to leave Gaia firmly in the background.[9]

In the chapters that follow, I show how plants are painting a vivid and revealing picture of the dramas in Earth's history. The timeframe for this ambitious venture is the last 540 million years, a thick slice of Earth history known as the Phanerozoic eon, characterized by the evolution of complex plants and animals that define our modern world. The chapters documenting the lifting of our 'veils of ignorance' are organized along a timeline, from the oldest events discussed in Chapter 2 to the youngest in Chapter 8. Figure 1 outlines where each chapter slots into the geological timescale.[10] Although I have tried whenever possible to keep the use of geological names to a minimum, a passing familiarity with the different eras and periods will be helpful (Fig. 1).

My other intention in writing this book, besides casting the spotlight sharply on plants' proper place in Earth history, is to place these stories in their proper historical context by highlighting the brilliant achievements of the generations of scientific pioneers and adventurers who have shaped scientific thought. I have attempted to do this by bringing to life key figures with biographical sketches, and by occasionally outlining historical scientific developments and events. These are not intended in any way to be complete but rather to give the reader a flavour of the personalities of the pioneers and the excitement

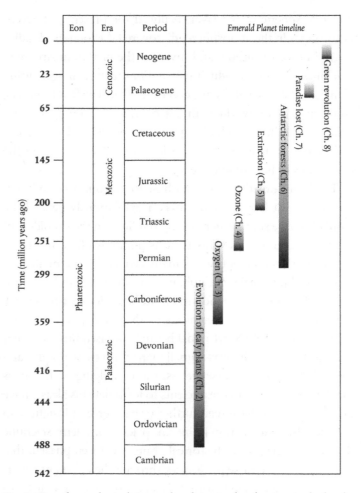

Eon	Era	Period	Emerald Planet timeline

Time (million years ago)

0
23
65
145
200
251
299
359
416
444
488
542

Phanerozoic

Cenozoic — Neogene, Palaeogene

Mesozoic — Cretaceous, Jurassic, Triassic

Palaeozoic — Permian, Carboniferous, Devonian, Silurian, Ordovician, Cambrian

Green revolution (Ch. 8)

Paradise lost (Ch. 7)

Antarctic forests (Ch. 6)

Extinction (Ch. 5)

Ozone (Ch. 4)

Oxygen (Ch. 3)

Evolution of leafy plants (Ch. 2)

Fig. 1 How the geological timescale relates to the chapters in the book.

of their discoveries on which a particular story builds. Some
may be familiar while others, I hope, will be less so. We will
learn of the contributions of pioneering chemists and physicists
who laid the foundations of modern chemistry, discovered

the stratospheric ozone layer, deduced the presence of green-house gases in the atmosphere, discovered the long-lived radio-active isotope of carbon, and invented the first atom-smashing machines, cyclotrons, which ushered in the nuclear age. Sitting alongside these scientists are eccentric Victorian fossil hunters, who amazed the world with discoveries of giant prehistoric animals and the weird-looking remains of the early plant life, and heroic polar explorers who lost their lives extending the boundaries of human knowledge.

The English mathematician, physicist, astronomer, and one-time alchemist Isaac Newton (1643–1727) famously penned the words 'If I have seen farther, it is by standing on the shoulders of Giants.' The meaning of this famous phrase is often misunder-stood and the comment was actually coaxed from him after he was cajoled into making a public reconciliation with his sworn enemy, the formidable polymath Robert Hooke (1635–1703), following several years' acrimonious dispute between the two men. It seems likely Newton deliberately phrased this comment as a dig at Hooke, who was a small man with a twisted spine, and certainly no giant.[11] Nevertheless, the underlying sentiment is that he borrowed from the ancients to formulate his ideas. I have no pretensions to have seen further or have greater insight than anyone else; rather my point in placing modern scientific debates in their proper historical context is to emphasize that the scientific enterprise progresses through the efforts of gener-ations who have gone before. It has become almost a cliché to point out that scientific progress is an incremental affair, a journey not a destination, characterized by being wrong as often as being right. Too often the historical flesh of discovery documenting this progress is filleted from the textbooks, and yet clearly those involved, either by luck, judgement, or special insight, at significant moments deserve proper credit.

The stories I describe selectively illustrating the new science can be classified into three broad non-exclusive categories. First, there are those in which fossil plants contribute to the debate as we come to appreciate that they record previously unrealized facets of Earth history (Chapters 4 and 5). In this category, I introduce the idea that fossil leaves can 'breathalyse' the ancient atmosphere for its carbon dioxide content. Here we will also find the contentious notion that mutated fossil spores, which suddenly appear in rocks dating to the 'mother of mass extinctions' towards the end of the Permian, may be signalling significant disruption to the ozone chemistry of the atmosphere. Second, there is a group of four chapters (2, 3, 7, and 8) that reveal plants to be powerful agents of global environmental change. These chapters describe how the evolution and spread of plants inexorably altered the composition of the atmosphere with, in some cases, dramatic consequences for their own ecological success, that of the animals, and the Earth's climate. Finally, a third group document remarkable stories about the evolutionary history of a particular vegetation type and its surprising interaction with the environment (Chapters 6 and 8). In these chapters, I revive the flagging fortunes of several forgotten heroes of palaeobotany and photosynthesis research whose pioneering efforts paved the way for a deeper understanding of the forests that once clothed the polar regions millions of years ago and of the dramatic appearance of our modern savannas onto the evolutionary stage.

Several chapters can be collected into yet another important category, one offering lessons from the past for our own climatic future (Chapters 5, 6, and 7). We live in an age when the escalating influence of humankind on the environment is only too apparent. In fact, so dramatic is the overprint of human society on the environment that a new term has been assigned for our present human-dominated geological epoch—the anthropocene.[12] Bubbles of air trapped in ice cores have revealed that the global carbon dioxide concentration began increasing late in the eighteenth century, around the time the Scottish inventor James Watt (1736–1819) designed the steam engine, and this is considered the start of the anthropocene. Over the past three centuries, industrial and agricultural expansion, driven by the rapidly growing global human population, have drastically increased emissions of greenhouse gases, especially methane and carbon dioxide, in concert with the continued destruction of the tropical rainforests. It is now beyond doubt that a serious consequence of all this will be a warmer climate.[13] Quite how much warmer is uncertain. Chapters 5, 6, and 7 highlight the dangerous game we are playing with the global climate system by showing that the consequences could be more far reaching and surprising than we might anticipate. The lesson Earth history teaches us is that by causing global warming, we are in danger of entraining unstable feedbacks in the Earth system which could propel us towards a far warmer world.[14] We skip plants as 'an important chapter of the history of life' at our peril.

2

Leaves, genes, and greenhouse gases

Leaves are remarkable structures that allow plants to conduct the daily business of photosynthesis and ensure the continuity of generations. Yet it took 40 million years for this seemingly simple evolutionary innovation to appear and spread throughout the plant kingdom. By comparison, humans evolved from primates in a tenth of the time. Quite why it took plants such an inordinately long time to evolve leaves puzzled scientists for nearly a century. Now a radical shift in our thinking points towards plummeting carbon dioxide levels as the key. By removing an environmental barrier to the evolution of leafy plants, falling carbon dioxide may have released the genetic potential of plants to fashion the blueprint for our modern terrestrial floras. In doing so, plant diversification transformed global climate and accelerated the evolution of terrestrial animals.

Any one whose disposition leads him to attach more weight to unexplained difficulties than to the explanation of a certain number of facts will certainly reject my theory.

Charles Darwin (1859), *The origin of the species*

THE *Galileo* spacecraft, named after the Italian astronomer Galileo Galilei (1564–1642), who launched modern astronomy with his observations of the heavens in 1610, plunged to oblivion in Jupiter's crushing atmosphere on 21 September 2003. Launched in 1989, it left behind a historic legacy that changed the way we view the solar system. *Galileo*'s mission was to study the planetary giant Jupiter and its satellites, four of which Galileo himself observed, to his surprise, moving as 'stars' around the planet from his garden in Pardu, Italy. En route, the spacecraft captured the first close-up images of an asteroid (Gaspra) and made direct observations of fragments of the comet Shoemaker–Levy 9 smashing into Jupiter. Most remarkable of all were the startling images of icebergs on the surface of Europa beamed backed in April 1997, after nearly eight years of solar system exploration. Icebergs suggested the existence of an extraterrestrial ocean, liquid water. To the rapt attention of the world's press, NASA's mission scientists commented that liquid water plus organic compounds already present on Europa, gave you 'life within a billion years'. Whether this is the case is a moot point; water is essential for life on Earth as we know it, but this is no guarantee it is needed for life elsewhere in the Universe.[1] Oceans may also exist beneath the barren rocky crusts of two other Galilean satellites, Callisto and Ganymede. Callisto and Ganymede probably maintain a liquid ocean thanks to the heat produced by natural radioactivity of their rocky interiors.

Europa, though, lies much closer to Jupiter, and any liquid water could be maintained by heating due to gravitational forces that stretch and squeeze the planet in much the same way as Earth's moon influences our tides.

To reach Jupiter, *Galileo* required two slingshots (gravitational assists) around Earth and Venus. Gravitational assists accelerate the speed and adjust the trajectory of the spacecraft without it expending fuel. The planets doing the assisting pay the price with an imperceptible slowing in their speed of rotation. In *Galileo*'s case, the procedure fortuitously permitted close observations of Earth from space, allowing a control experiment in the search for extraterrestrial life, never before attempted. Could we detect life on Earth with a modern planetary probe? Reporting in the journal *Nature*, Carl Sagan (1934–96) at Cornell University and his colleagues found that, on its December 1990 fly-by of Earth, *Galileo* detected abundant gaseous oxygen, large amounts of snow, ice, and extensive oceans, and amplitude-modulated radio transmissions of a type perhaps uniquely attributable to intelligence.[2] An oxygen-rich atmosphere is a suspicious pointer to life on Earth. One source of oxygen is that added to the atmosphere in small amounts when water molecules are broken apart by ultraviolet rays from the Sun. The hydrogen atoms from water escape into space as hydrogen gas while the heavier oxygen atoms are dragged back to Earth by gravity. Slowly our oceans are being lost to space, as astronauts on the *Apollo 16* mission observed from the surface of the Moon with a remarkable telescope. As the hydrogen escapes it gives off a fluorescent glow (Lyman alpha radiation) not visible on Earth because of the absorption properties of the atmosphere. The Moon provides an ideal vantage point, however, and when briefly stationed there the *Apollo 16* crew captured stunning images of hydrogen gas escaping Earth, revealed as a magical aura smeared towards the direction of the Sun. But the oxygen donated to the atmosphere by this route occurs far too slowly to account for the quantity detected by *Galileo*. Only biology can accomplish that feat by harnessing enzyme systems to split water and release prodigious amounts of oxygen.

The *Galileo* spacecraft also detected 140 times more methane in the atmosphere than expected from non-living considerations alone, and this disparity is, like oxygen, another indicator deeply suggestive of life. Methane is quickly converted to water vapour and carbon dioxide in the atmosphere, so very little is expected in the atmosphere of a lifeless planet. For methane to accumulate in the atmosphere, something has to be pumping it out at a rate faster than it is being destroyed. On Earth, the anaerobic microbial inhabitants of our swamplands and the great swathes of rice paddies in Southeast Asia perform the task, churning out over 200 million tonnes of the gas each year.

Onboard instruments measuring the composition of light reflected from Earth's surface provided further clues to life and the origins of the oxygen-rich atmosphere. Typical *Galileo* images of Earth revealed vast tracts of land mysteriously absorbing visible red light. The anomalous absorption fingerprint is unlike that of any common igneous or sedimentary rocks or soil surfaces seen elsewhere in the solar system and at least raises the possibility that some kind of light-harvesting pigment was responsible. And, indeed, chlorophyll molecules within green leaves absorb more than 85% of the incoming visible red light, which has sufficient energy to drive the splitting of water during photosynthesis. Large areas of Earth's land surface are characterized by the unusual absorption of visible light, allowing us to deduce that plant life is correspondingly widespread, an observation offering an explanation for the large amounts of oxygen in the atmosphere.

Galileo's fly-by of Earth convinced the scientific community it had a reasonable chance of successfully detecting photosynthetic life from space on other 'terrestrial' satellites orbiting nearby stars, assuming they exist and we can locate them. Even at an early stage of life's evolution, this should be possible because clues to Earth's photosynthetic biosphere were ripe for discovery by a similar fly-by with an alien spacecraft over two billion years ago, given primitive photosynthetic cyanobacterial crusts and algal mats instead of modern plant life. Whether or not plants might arise in a recognizable form is another matter, although there is a strong case to be made for expecting chlorophylls of some sort or another to evolve on a planet with an atmosphere offering the basis for photosynthesis.[3]

As we might expect, Galileo's mission foreshadowed better things to come from space technology. Today, observing the Earth from space is revolutionizing our understanding of the planet. Satellites now constitute our global macroscope. Just as Robert Hooke's newly invented microscope offered a fresh perspective on the natural world in the seventeenth century, so the macroscope operating at the opposite spatial scale provides a new means of observing natural and human influences on Earth's condition. Unfortunately, much of what we are learning is alarming. We are quickly coming to realize, for instance, that the portion of Earth's surface covered in snow, ice, and glaciers, the cryosphere (from the Greek word kryo meaning frost or icy cold), is shrinking alarmingly. Arctic sea ice is disappearing rapidly, with half a million square kilometres being lost every decade.[4] Elsewhere, ice grounded on land in western Antarctica and Greenland is on the move, threatening to raise the sea-level on the timescale of human economies. Ice shelves that once buttressed glaciers are melting, allowing them to surge spectacularly forward and discharge their icy cargo into the world's oceans.[5] On the Antarctic Peninsula, the disintegration of ice shelves larger than Hawaii is coincident with warming in the last half-century of 2–4 °C.[6] The distinctive fingerprints of the extraordinary influence on the Earth system of human activity are recognized all too easily from space.

Besides opening our eyes to a changing planet, powerful space-age technology is also creating an exciting new intellectual moment for the twenty-first century. The French mathematician and physicist Henri Poincaré (1854–1912) once remarked that 'Science is facts. Just as houses are made of stones, so science is made of facts. But a pile of stones is not a house and a collection of facts is not necessarily science.' In Poincaré's day, and indeed throughout the nineteenth and twentieth centuries, science revolved around observing the natural world and employing inductive and deductive reasoning to frame and test hypotheses.[7] Earth-observing systems are ushering in an unprecedented data-rich era—scientists are obtaining enormous datasets ('facts') about our planet—and this brings with it the twenty-first century challenge of making sense of the world. To do that requires theory, a step further on from reasoning that offers scientists an opportunity to draw on their inspiration. For Poincaré, inspiration struck when he was out on a geological excursion. His breakthrough in solving a difficult mathematical problem that had stumped him for some time, came 'at the moment when I put my foot on the step, the idea came to me, without anything in my former thoughts seeming to have paved the way for it'.[8] Poincaré's case is typical. Inspirational flashes leading to dramatic breakthroughs often seem to appear conjured out of nowhere. Einstein noted that there is 'no logical path' to connecting theoretical concepts with observations, noting in 1952 'the always problematical connection between the world of ideas and that which can be experienced'.[9] Increasingly sophisticated satellite technology is, then, opening new doors to the intellectual endeavour of devising theories to explain planetary biology, physics, and chemistry.

NASA's Compton Tucker led the way in understanding terrestrial vegetation in the 1980s when he developed the theory for exploiting the capability of satellite-borne instruments to detect the chlorophyll pigments in leaves.[10] The satellite sensors detect, in the absence of cloud and other atmospheric interferences, the contrasting strong absorption of visible red light and the weak absorption of infrared light by the green leaves of plants. The infrared 'glow' combined with a deficit of red light is a characteristic signature of terrestrial vegetation. The activity of plants is now routinely monitored by making these and other measurements with instruments on-board polar orbiting satellites that scan Earth's surface daily. The snapshot of green continents recorded by *Galileo* has been converted into the movie in which we can watch the seasonal springtime 'greening' of the northern hemisphere landmasses from space. As with the cryosphere, here, too, the influence of human activities is being revealed as forests respond to carbon dioxide fertilization. In 1989, for example, a team of researchers analysed satellite data to reveal a dramatic stimulation of forest growth throughout the northern high latitudes during one of the warmest periods of the last 200 years (1981 and 1991).[11]

Scanning the surface of the oceans and the land, the satellites revealed that terrestrial and marine plants synthesize a staggering 105 billion tonnes of biomass each year from carbon dioxide extracted from the atmosphere.[12] Remarkably, phytoplankton, the single-celled photosynthetic organisms drifting in the currents of marine and freshwaters, are responsible for about half of this productivity, yet account for less than 1% of Earth's photosynthetic biomass. Terrestrial plant life, on the other hand, contributes the remainder and constitutes over 90% of the world's biomass, a figure reflecting their true dominance in the biosphere.

Detecting future changes in the activity of the world's forests from space will be important. They could signal when nature's brake on global warming is about to be released, for forests are a major natural sponge soaking up some of our excess carbon dioxide. By burning fossil fuels and clearing tropical forests, humans are adding about 7 billion tonnes a year of the greenhouse gas carbon dioxide to our atmosphere. About half of this amount remains in the atmosphere, causing the current inexorable rise in the atmospheric concentration. The remainder is mopped up, in roughly equal proportions, by the oceans and terrestrial vegetation.[13] How much longer nature's carbon sinks will continue to buffer our carbon excesses remains uncertain. Forests could lose this capacity within the next fifty years as their ability to absorb carbon dioxide and synthesize biomass saturates is overtaken by the release of carbon dioxide by respiration in a hotter, drier, future climate.[14] If, or rather when this happens, for the models are quite consistent about this prediction, it will accelerate the accumulation of carbon dioxide in the atmosphere and climate change.[15] Indeed, the rate at which the carbon dioxide content of the atmosphere is rising hit an unprecedented high in 2002, and again in 2003, prompting some scientists to speculate, perhaps prematurely, that anthropogenic climate change is already causing forests to release rather than absorb carbon.[16]

All the features of terrestrial plant life that I have just outlined, its worldwide dominance on land, our capacity to observe and monitor it from space, and its influence on our own climatic future, pivot around a single remarkable organ—the leaf.[17]

Acting as innumerable solar arrays, leaves house the cellular and biochemical machinery necessary for plants to harvest sunlight and conduct the daily business of photosynthesis. Today, these pervasive photosynthetic structures cover 75% of the Earth's land surface, and prove their extraordinary versatility by enduring climatic extremes that range from the freezing temperatures of −56 °C in Siberia,[18] the terrestrial biosphere's coldest region, to over 40 °C in deserts. Only truly inhospitable deserts, the ice fields of Antarctica, and the highest altitudes of the world's mountains remain bare. Virtually all of the estimated quarter of a million or so species of flowering plants depend on leaves for capturing light to power photosynthesis and manufacture biomass. Numerous other non-flowering plant species depend on leaves to ensure growth, reproduction, and the continuity of future generations. In spite of the fantastic variety of shapes and sizes shown by leaves in nature, all conform to the same structural blueprint of cantilevered blade, and for good reason. The design elegantly solves the engineering dilemma facing plants needing flat photosynthetic surfaces. The surface should be sufficiently stiff to resist the tug of gravity, and yet at the same time sufficiently flexible to minimize damage on windy days.

On the evidence of the world's modern floras, leaves have an evolutionary inevitability about them. It seems unthinkable that plants needing to conduct the business of photosynthesis could do without them. Yet surprisingly, when plants began the great saga of colonizing the land around 465 million years ago, they did so without leaves. Initially, 'primitive' leafless plants conducted the frontal assault on land. The early terrestrial pioneers for this pivotal moment in Earth history evolved from a small group of predominantly freshwater green algae (Charophyceae) and left behind fragmentary fossil remains of reproductive structures closely resembling modern bryophytes (liverworts, hornworts, and mosses).[19] The fossils, and relationships based on the genetic make-up of living plants with different evolutionary histories, have revealed that terrestrial colonization by plants is essentially a story of evolutionary transition from green algae to bryophytes. Strangely, green algae occupied the oceans and shorelines for nearly half a billion years before a terrestrial existence proper beckoned for their descendants. Why plant life seemingly 'hesitated' on the strand line for so long remains mysterious.

From these simple photosynthetic organisms, it was to be another 40 million years (the dating is frustratingly imperfect and controversial) before the ancestors of our modern vascular floras finally arrived on the scene, and they too were leafless. When fossils from this important act in the drama were found, the early pioneers of the scientific study of fossil plants nearly missed their significance completely. Until the early nineteenth century, plant fossil hunters were preoccupied with the numerous leafy plants of the Carboniferous coal deposits, which held great commercial importance. But in 1859 the eccentric Canadian William Dawson (1820–99) collected fossil plant specimens completely different from anything found before from shoreline exposures around the Gaspé Peninsula, lying below where the St Lawrence River carves into the north-east Canadian coastline. Dawson's finds startled the palaeobotanical world and challenged botanists of the day to explain them. He firmly believed that these strange fragmentary fossils, lacking leaves and with a simple branched structure, predated those of the Carboniferous by tens of millions of years.[20] Few at the time agreed; his claims were greeted with fierce scepticism. Some thought the fossils represented the stems of ferns, roots, or even algae.

What finally turned the tide, some fifty years later, were reports by two botanists, Robert Kidston (1852–1924) and

William Lang (1874–1960), between 1917 and 1921 of near-complete early land plants from the small, picturesque village of Rhynie in Aberdeenshire, Scotland. Whereas before the discovery of the fossil plants in Rhynie descriptions of early land plants were based mainly on fragmentary fossil materials, afterwards the picture changed completely. The Rhynie plants are preserved by silica-rich volcanic fluids infiltrating the tissues of the dead plants and crystallizing out in the gaps between the organic matter. The process results in superbly preserved fossils that reveal exquisite cellular detail and provide a glimpse of how early land plants were put together. Strange, simple vascular land plants, photosynthesizing as naked stems without leaves, had once existed after all with a complexity lying somewhere between the mosses and true vascular plants.

Some years after the spectacular fossil finds in Rhynie, Lang topped that triumph when he excavated disarticulated fragments of the remains of the earliest vascular plant yet discovered. Rather charmingly, he named it *Cooksonia*, after one of his long-term collaborators, the Australian palaeobotanist Isabel Cookson (1893–1973). A description soon followed, and in a seminal 1937 paper,[21] he reported details of *Cooksonia* fossils squashed flat with simple vascular axes (primitive stems) from 417-million-year-old rocks of the Welsh Borderland. Lang was convinced the axes belonged to *Cooksonia*, but mere association made a far from compelling case and the possibility that it was indeed a true vascular plant remained uncertain. Such consideration may merely seem a nicety, but what is at stake here is the claim for the earliest vascular plant, and extraordinary claims require extraordinary evidence. It was not until fifty years later that more trustworthy evidence in the form of better-preserved specimens was unearthed from the older rocks of South Wales, dating to 425 million years ago. Here, at last, was the fossil evidence vindicating Lang's belief of *Cooksonia*'s vascular status by crucially showing specialized water-conducting tissues within the fossils themselves.[22] Subsequently, numerous fossilized remains of early vascular land plants have come to light, all showing a similar basic body plan of simple or branched stems, without leaves.[23] Frail and leafless, *Cooksonia* makes an unlikely herald foreshadowing the dawning of terrestrial floras proper (see Plate 1). Carpeting the floodplains of rivers meandering through the ancient terrestrial landscape, it seemingly added little to the green veneer of early photosynthesizers already present, but was to prove central to the assembly of terrestrial life as we know it.

Presaged by these humble beginnings, plant life began to flourish. Over the next 65 million years, between 425 and 360 million years ago, an unparalleled burst of evolutionary innovation and diversification followed.[24] In fact, the claim is that this chunk of geological time represents the botanical equivalent of the Cambrian 'explosion', a time when marine invertebrate animals went from being single-celled to complex multicellular organisms virtually in a geological instant, some 540 million years ago. In the botanical version, land plants became transformed, establishing in the process a blueprint for the present-day plant world. Extraordinarily complex body plans and sophisticated life cycles soon arose from a simple body plan of only a few cells. Yet in the midst of all this evolutionary excitement, leaves strangely became widespread at the last minute.

We know something of the evolutionary sequence leading up to the appearance and spread of leaves from the fossil record.[25] At first, knee-high trees and shrubs populated the landscape, maintaining their photosynthetic way of life by supporting bare branches and forking twigs for fully 30 million years without leaves. Then, gradually, over a period of some 10 million years, things started to change. Proper fossil leaf specimens begin to turn up borne on the earliest known modern trees belonging to the extinct genus *Archaeopteris*. These exciting fossils signal that plants had started to exploit the photosynthetic proficiency of a flat solar panel for capturing sunlight and powering photosynthesis. Leaves then originated independently in three other plant groups (the sphenopsids and pteridiosperms, ancestral forms of horsetails and ferns, respectively, and seed plants), as the plant kingdom seized the moment.[26] By the start of the Carboniferous, 360 million years ago, leafy plants were firmly established in the floras of the day.

If we start the clock ticking from the appearance of the first vascular plant *Cooksonia* and stop it when large leaves become widespread, we can see the whole affair is bracketed by a 40–50-million-year-thick slice of geological time, within the accepted dating uncertainties. It is genuinely puzzling why it took plants such an inordinately long time to come up with what, on the face of it, is a rather simple evolutionary innovation, and why when it did arrive it took an age to become widespread throughout the floras of the day. Consider, for example, that humans evolved from primates in a tenth of the time. Come to that, mammals sprang from being furtive bit-part players in the game of life to their present diversity and dominance in the 65 million years since the dinosaurs famously went extinct.

The mystery of the long-delayed appearance of leafy plants onto the evolutionary stage deepened with finds of fossilized marine algae in dolomitic rocks along the eastern shores of Lake Winnipeg, Canada, and discoveries of an enigmatic fossil plant unearthed from outcrops on mountain slopes in southeastern Yunnan, China. The fossil algae from Canada are noteworthy because their broad flattened fronds, several centimetres wide, pre-date the advent of large-leaved land plants by tens of

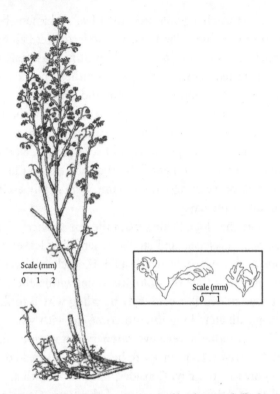

Fig. 2 The enigmatic early Devonian vascular plant *Eophyllophyton bellum*. It produced tiny leaves some 40 million years before they became widespread in the world's terrestrial floras.

millions of years.[27] Elsewhere, on the other side of the world, fossils of the enigmatic terrestrial vascular plant *Eophyllophyton bellum* turned up in Chinese rocks some 390 million years old (Fig. 2).[28] *Eophyllophyton* fossils are remarkable because they possess tiny (1–2-mm diameter) proper leaves distributed regularly along the stems and branches, as in modern leafy plants. What are we to make of these fossilized botanical oddities? I believe both the existence of broad algal fronds and *Eophyllophyton* hints at something rather important; they suggest that marine and terrestrial plants evolved the capacity to make flattened photosynthetic organs long before the idea took off.

We should be cautious, however, in supposing from evidence of this sort that all land plants had the genetic capability of producing leaves whenever in their evolutionary history it suited them. To reach beyond speculation of this sort, we need to turn to molecular developmental genetics, the study of genetic pathways used and reused to build organisms. Making a leaf, and much else besides, requires homeobox gene networks (also present in animals) to organize growth and development by ensuring cells take on the right form and function depending on where they are on the plant. In plants, the so-called knotted homeobox gene (KNOX) family plays a critical role in leaf formation[29] and is present in some green algae, mosses, ferns, conifers, and flowering plants.[30] It functions in a similar manner in different plant groups: when KNOX genes of a fern are put into a flowering plant and vice versa, they still work. In other words, plants with diverse evolutionary histories possess them and their function is highly conserved, exactly as we would expect if the genes are very old.

The first step towards producing leaves is to turn off the KNOX genes. This 'KNOX-off' state causes sideways outgrowths to protrude from the shoot, which then go on to develop into leaves. If KNOX genes are left switched on, the plant continues to grow its shoot as normal without pausing to initiate leaf formation. Very different plant groups have followed this same approach to making leaves quite independently of each other. Only recently, evolutionary biologists discovered that leaf formation is controlled in a primitive group of plants, the lycophytes, in much the same way as in higher plants (angiosperms).[31] More than likely, then, all plants irrespective of their evolutionary history share a common genetic mechanism for making leaves.

Developing a leaf also requires that plants 'know' how to assemble their upper and lower surfaces. The upper layers of a leaf are specialized for intercepting and processing energy from sunlight while the structure of the lower part is arranged to optimize absorption of carbon dioxide. These specializations have to be built into the leaf as it develops and here, too, we find that genes for regulating this dichotomy are very old, dating back over 400 million years.[32] There is even some suggestion that they were 'borrowed' from those that organize the vascular tissues,[33] which as we have seen with *Cooksonia*, appeared 50 million years before leaves. As the attention of the molecular geneticists, usually held by crops of commercial importance, switches to more fundamental questions about how plants evolved, exciting discoveries surely lie ahead. For now, we can note that although a proper understanding of the genetic mechanisms underlying leaf evolution is still some way off, it does seem as if the genetic 'tool-kit' required to assemble leaves was in place long before large leaves appear in the fossil floras of the world.

If the molecular geneticists are correct and plants did possess the genetic capacity to produce leaves very early in their evolutionary history, some crumbs of comfort for palaeontologists can be derived from the fact that it is the fossil record which affords us a glimpse of how it was released. Nearly 75 years ago, the German palaeobotanist Walter Zimmermann (1892–1980), at the Universität Tübingen, Germany, published the first detailed attempt at charting the evolutionary trajectory of leaves, from the initial axial structures of early land plants to the eventual appearance of trees with true leaves.[34] Zimmermann's work built on the foundations laid by the great German poet and philosopher Johann Wolfgang Goethe (1749–1832), who published his seminal essay *Versuch die Metamorphosis der Pflanzen zu erklaren* ('Metamorphosis of plants') in 1790. It was during Goethe's epic Italian journey from 1786 to 1788 that his thoughts on the possibility that plant organs developed by modification of a single organ, a leaf, or *blatt* in German, crystallized. Goethe proposed, for example, that petals are modified leaves.[35] Scientists from Charles Darwin to Zimmermann accepted Goethe's adventurous and highly original ideas, and Goethe has come to be widely regarded as the father of plant metamorphosis.

Zimmermann presented his scholarly synthesis of the fossil evidence and its integration with theories of plant morphology as the 'telome theory'. It describes how leaves arose along an evolutionary series with four main steps, each one representing a genuine evolutionary innovation and recognizable repeatedly in different groups of plant fossils of progressively younger ages. The transformation begins with the simple three-dimensional branching architecture of early land plant stems, as typified by the Rhynie Chert fossils. In the second step, the main stem bears dividing side branches without further branching of the central axis. Eventually, the side branches all divide in the same spatial plane, essentially giving the appearance of being flattened (planated). This intermediate form paves the way for the final stage in the evolution of the leaf, the development of 'webbing', which joins the segments of the flattened side branches with a sheet of photosynthetic cellular tissue. Because flat-bladed leaves evolved independently in several groups, it appears that three of these transformations, planation, webbing, and fusion, are steps in the evolution of leaves that have been recruited multiple times during the evolutionary history of land plants.[36]

Thanks to Zimmermann, by 1930 palaeobotanists had in place a theory describing the different steps in the evolution of leaves.

However, the problem with his telome theory is that it is not a 'theory' in the formal sense of the word at all, but rather a description of the 'how'. Zimmermann's valiant efforts neatly sidestep the thorny question of 'why' the whole business took so long and the mystery endures. It rightly received sharp criticism for this shortcoming: in effect it 'describes everything but explains nothing'.[37] How can it be that plants as complex as trees reproducing with sophisticated life cycles evolved smoothly and apparently without difficulty, yet leaves proved to be so difficult, even though plants seem to have been equipped with the genetic tool-kit for making them. The remarkable contrast is not easily shrugged aside, and alerts us to a new possibility: was some feature external to biology—the environment—holding back leaf evolution? New questions open new doors and suddenly cast a different light on the notion that the photosynthetic proficiency of leaves makes their evolution inevitable.

Crucial elements of the conceptual framework necessary for a radical rethink began to emerge, with reports showing that a remarkable change in the carbon dioxide content of the ancient atmosphere had taken place between 400 and 350 million years ago.[38] This is the same time slice covering plants' dramatic evolutionary diversification—the botanical version of the Cambrian explosion. The carbon dioxide record, based on chemical analyses of fossil soils, revealed that levels plummeted, falling by 90%. When *Cooksonia* and its cousins began colonizing the land, they enjoyed the benefits of 15 times more carbon dioxide atmosphere than today, and a very much warmer climate. By the time the first leafy forests of *Archaeopteris* spread across the ancient continents, 40 million years later, carbon dioxide levels had plummeted tenfold. As far as we know, the drop is unprecedented in the last half billion years of Earth's history and the inference from fossil soils is supported by the theorists concerned with defining Earth's carbon dioxide history from first principles.[39] The fall in carbon dioxide levels weakened the atmospheric greenhouse effect and led ultimately to a major ice age, triggering the formation and growth of massive glaciers that spread out from the South Pole and across the contiguous continents until they reached into the tropics.

One further piece of evidence required for establishing a connection between ancient carbon dioxide levels and leaf evolution was the discovery that the gas itself affected the number of microscopic pores (stomata) on the surface of leaves.[40] Stomatal pores act as miniature valves through which leaves absorb carbon dioxide, the raw material for photosynthesis, and lose water. This important discovery was reported in a landmark paper published in *Nature* in 1987 by Ian Woodward when at the University of Cambridge. Noted for his quick, often acerbic wit, a fondness of cricket, and a weakness for supporting Sheffield United Football Club, Woodward uncovered the remarkable effect of carbon dioxide on stomatal formation not with living plants but through examination of long-dead specimens of native British trees preserved in the University of Cambridge herbarium. As collections of pressed plant specimens, herbariums represent a fortuitous legacy of generations of naturalists who have collected and preserved plants since at least the eighteenth century. They were fuelled by the belief that recording patterns of variation within and between plant populations afforded a means for understanding the nature of a species, one of the big questions in the natural history of the day. For no lesser scientists than the Swedish naturalist Carl Linnaeus (1707–78) and Charles Darwin, herbariums formed a standard means of studying the natural history of plants. Linnaeus, who introduced the— for the time controversial—new system for naming plants (it was based on their sexual characteristics), built up a vast herbarium of many thousands of specimens.[41] Much of it still resides today in the vaults of the Society bearing his name in Burlington House, London. Darwin's familiarity with the organization of the dried plant collections of his mentor, the kindly John Henslow (1796–1861), when at Cambridge ensured that he carefully identified plants by date and location while touring the Galapagos Islands in 1865.[42] This careful attention to detail later proved crucial in showing the highly endemic nature (geographically restricted range) of the Galapagos flora, an issue never quite clinched for his famous finches because he had not labelled the birds with the same care.[43]

Little did the Victorians realize the importance their collections would hold for future generations. Thanks to their efforts, herbarium archives make it possible to go back in time to understand historical responses of plants to the rise in the carbon dioxide content of the atmosphere initiated at the onset of the industrial revolution in the Western world. And much to his surprise, Woodward found that the herbarium collections had borne witness to the silent revolution of the trees during our own industrial revolution. Trees had responded to the rising atmospheric level of carbon dioxide by producing fewer pores; the leaves of trees in southern England today have 40% less than 150 years ago. What finally clinched it, and won over the sceptics, was experimental evidence demonstrating that the carbon dioxide content of the air in which plants grew altered the numbers of stomata on leaves.[44] The experiments also revealed why—to conserve water. When carbon dioxide is plentiful, plants maintain their uptake for photosynthesis, but having fewer stomatal pores has the benefit of restricting the loss of water out of the leaf by transpiration.

Later work revealed that plants have an exquisite signalling system allowing mature leaves to communicate with newly developing leaves to inform them of the optimum number of pores to produce for the environmental conditions they will shortly be encountering.[45] Moreover, the whole business of regulating stomata numbers by carbon dioxide was found to have a genetic basis when Julie Gray at the University of Sheffield and her team identified the so-called *HIC* (*High Carbon Dioxide*) gene.[46] The *HIC* gene controls the formation of stomatal pores according to the level of carbon dioxide in the atmosphere. The switch is believed to operate by influencing the deposition of specific chemicals (fatty acids) in the specialized cells of the stomatal pore. When switched on by a carbon dioxide-rich atmosphere, fatty acids inhibit neighbouring cells from developing into more stomata. In mutant plants, with a defective *HIC* gene that cannot be switched on properly, fewer fatty acids are made and stomatal production spirals out of control. Working out how fatty acids prevent pore formation is the next step, and still under investigation.[47]

Conceivably, genes such as *HIC* are of considerable antiquity, programming plant responses to changes in the atmospheric carbon dioxide concentration. Fossils illuminate the debate by revealing trends consistent with our expectation from the leaves of herbarium specimens. Fragmentary fossilized remains of early land plants that originally enjoyed a carbon dioxide-rich atmosphere possess very few stomata, usually less than five per square millimetre.[48] Such low values are exceptional among modern plants, which typically have several hundred per square millimetre of leaf surface. Only desert succulents compare. Forty million years of dwindling carbon dioxide levels later, we find the earliest *Archaeopteris* leaves with six times as many pores as in their axial ancestors.[49] By the time leafy plants became widespread pore numbers had jumped tenfold, almost as if plants were gasping for air in the struggle to combat carbon dioxide starvation. Circumstantial though the evidence is, it is hard to avoid the conclusion that plants have coped with changes in carbon dioxide in the same way for an exceptionally long time, perhaps even their entire evolutionary history.

To recap, the evidence suggests very high carbon dioxide levels prevailed when early plants appeared, which then plummeted as the botanical version of the Cambrian explosion unfolded. As it did so, the fossil record shows that leaves gradually appear, tiny at first and then getting progressively larger, all the while with an increasing number of stomatal pores as the stress of carbon dioxide starvation forced their hand. The significance all this held for understanding the exceptionally long delay in the evolution of large leaves becomes obvious when we realize that the number of microscopic pores holds important consequences for the ability of plants to keep cool. As we have already seen, stomatal pores act as valves regulating the uptake of carbon dioxide and the costly, unavoidable, escape of water by transpiration, a process that cools the leaf. More pores confer an improved capacity to keep cool.[50] Keeping cool is also increasingly difficult as leaves get larger because they are much less effective at shedding heat on a cooling breeze than small ones. The reason for this is that as air flows across the leaf removing heat, it is slowed by the friction of the air molecules bumping against the leaf surface. Larger leaves exert greater friction because the air has to travel further, and this makes heat loss less efficient. So, the argument can be made that large leaves awaited a drop in the carbon dioxide content of the atmosphere, which permitted plants to produce more pores, allowing bigger leaves to stay cool.

We can easily see that elements of these ideas help explain the paradox presented by the presence of large and small-leaved plants in today's desert, savannah, and Mediterranean regions. Many desert species evolved small leaves as a strategy to reduce the interception of sunlight and keep cool while at the same time helping to limit water loss by transpiration. However, it is often the case that in Mediterranean-style floras, small-leaved species are found together with large-leaved species. How odd. This paradox was resolved by William Smith at the University of Wyoming investigating plants in the Coachella Valley area of the Sonoran Desert in southern California.[51] In the baking hot Sonoran Desert, Smith recorded leaf temperatures of large-leaved desert perennials 20 °C below scorching air temperatures of 40 °C. The difference is due to very high rates of transpiration cooling the leaves. The seemingly contradictory combination of a hot climate and high water losses is made possible by the climate of the region ensuring adequate rainfall throughout the year. In the Sonoran Desert, large-leaved species even evolved cooler optimum temperatures for photosynthetic productivity in tune with their cooler leaves, an alternative evolutionary solution to tolerating extreme heat.

There are some interesting parallels here with heat control in cold-blooded animals, organisms adept at exploiting sunlight to regulate their body temperatures. Lizards, for example, being unable to generate heat internally, rely on heat from the Sun to warm up. Cold and sluggish early in the morning, they tend to align themselves at right angles to the Sun's rays to maximize exposure and warm up quickly. By contrast, when in danger of getting too hot while out foraging during the day, they sporadically turn to face the sun head-on to minimize exposure to sunlight and stay cool. Under drastic circumstances plants can play the same game, wilting their leaves to reduce the surface area exposed to sunlight, but only as a short-term strategy for coping with the heat.

Drawing on principles governing heat regulation in plants, a team of scientists from Sheffield and London Universities attempted to put all these ideas together in an effort to shed light on the puzzle of leaf evolution.[52] They were able to show that long-extinct simple-stemmed early land plants stayed cool because their slender stems intercepted small quantities of sunlight and were adequately cooled by a limited transpiration stream escaping through sparse numbers of stomata. Probing the mysteries of leaf evolution a little further, the team then asked what might have happened if these small plants had evolved flat leaves, a situation that did not, as far as we know, actually happen. In this instance, the calculations revealed that leaf temperatures soared above the lethal limit of about 50 °C, and the leaves would have started to cook. The temperature of 50 °C seems to be the upper limit most plants and animals can stand. Much hotter and the proteins involved in essential metabolic processes begin to irreversibly break down. Any organisms experiencing this would soon be dead. Why the drastic outcome? The reason is that a simple flat leaf intercepted three times the amount of sunlight as a simple stem, and a severely restricted transpiration stream due to the scarcity of pores prevented effective cooling. It seems that the earliest land plants could not have evolved large leaves without violating the chemistry of how proteins are put together.

Of course, sceptics might take issue with this analysis. Recall that the limited number of stomata is thought to be due to the very high carbon dioxide content in the atmosphere 425 million years ago when *Cooksonia* evolved. Suppose, then, that for whatever reason the number of pores is not regulated by carbon dioxide. Then our mythical early land plants could evolve flat leaves with abundant pores and the benefit of efficient cooling. It would also allow plants to capture more sunlight, promoting photosynthesis in the carbon dioxide-rich atmosphere. Interesting as it may be, the idea doesn't hold up to closer scrutiny.

Theoretical calculations indicate that water exits from such a mythical leaf punctured by abundant microscopic pores ten times as fast as the primitive root and plumbing systems can supply it. So, early land plants simply could not have coped with evolutionary novelty of this sort, and any random suite of mutations leading to it by chance would probably have been lethal.

Forty million years after the landmark appearance of *Cooksonia*, and with the carbon dioxide content of the atmosphere plummeting, the economics governing heat loss from photosynthetic organs changed. Sophisticated root and shoot plumbing systems evolved. Root traces left behind in fossilized soils beneath *Archaeopteris* forests penetrated over a metre deep into the ground, providing much needed mechanical support and an effective means of extracting water and nutrients from the soil.[53] These structures are a far cry from the diminutive rootlets put down by the earliest land plants, with their limited capacity for absorbing water. Less carbon dioxide in the atmosphere allowed pore production to be stepped up and the generation of a high-cooling transpiration stream. With the appearance of sophisticated root and shoot systems, the uptake and delivery of water necessary to support these high rates of transpiration were in place. The moment when larger-leaved plants became a possibility had finally arrived. As we might imagine, it is no coincidence that the evolution of the root, shoot, and leaf—the holy trinity of the plant world—took place simultaneously.[54] As plant life became freed from its earlier strictures, leafy canopies began their ascendancy.

It is hard to know for certain whether things happened like this. Obliterated by the ravages of time, the fragmentary nature of the fossil record means that only vestiges of past events remain. Piecing together a narrative of what might have actually happened several hundred million years ago is a difficult task.

However, the arguments we have used are consistent with the fragments of fossil plants that do survive. They hinge on our understanding of how modern plants cope with changing greenhouse gas concentrations and combine this with the laws of physics dictating the conservation of energy and matter. So the case for carbon dioxide as the culprit delaying the appearance of leaf plants would seem to be compelling, but is it correct? A critical test of any hypothesis is its capacity to make specific predictions potentially falsifiable by observations. The prediction in this case is that as carbon dioxide levels dwindled and pore numbers increased, larger leaves evolved, better equipped to cope with dissipating the heating caused by intercepting more solar energy.

Evidence to test this prediction came from investigations of fossil plants held in museum collections. It is easy to forget that natural history museums are dynamic entities preserving not only the spectacular dinosaur skeletons adorning the magnificent display halls, but also preserving and cataloguing the dramatic story of how plants evolved. By mining the collection of fossil plants held in museums across Europe, from the Swedish Museum of Natural History in Stockholm across to the Senckenberg Institute in Frankfurt and down to the Royal Belgian Institute of Natural Sciences in Brussels, researchers have analysed three hundred specimens of fossil plants. The results revealed increased over time as carbon dioxide levels fell (see Plate 2).[55] The largest leaves (nearly 8 cm across) belonged to fronds of primitive seed ferns (pteridosperms) and only appeared after carbon dioxide levels had fallen to their lowest point. After the seed ferns, waves of plants quickly seized the opportunity to produce larger leaves for capturing sunlight and powering photosynthesis.[56] *Archaeopteris* was first, followed by the pteridosperms and then a group comprising the forerunners

of modern conifers—fossil evidence confirming a central prediction of the carbon dioxide hypothesis.

We are glimpsing a new framework for interpreting why it
took such an inordinately long time for large-leaved plants to
become widespread. It marks a radical departure from Zimmermann's theory seventy years earlier, and points towards a carbon
dioxide-rich atmosphere as an environmental 'barrier' to leafy
plants, the barrier at plants' evolutionary gateway only being
lifted after carbon dioxide levels dwindled. Interestingly, an
analogous barrier has been proposed for the Cambrian explosion
of marine invertebrate animals in the form of low atmospheric
oxygen levels.[57] In the case of marine animals, a step increase in
the oxygen content of the atmosphere seems to have coincided
with the geologically 'sudden' evolution of complex multicellular
animals. The idea appears to have some merit because physically
larger invertebrate animals require more oxygen to sustain their
metabolism. Evolutionary developmental biology has also indicated for certain that the genetic tool-kit needed for building
multicellular animals was in place, untapped, for millions of
years before large animals appeared on the scene.[58]

We should, however, resist the notion that the genetic tool-kit
for making leaves—or indeed complex invertebrate marine
animals—was, metaphorically speaking, 'ready and waiting' to
spring into action. Molecular biologists believe that evolutionary
novelty in animals was achieved through the evolution of
switches and gene networks that regulate how fundamental
genetic tool-kits for building plants and animals are deployed.[59]
Investigations into the genetic basis of key plant evolutionary
innovations are only just beginning, but early insights reveal
interesting parallels with animals.[60] We have already discovered
that flowering (eullophytes) and non-flowering plants (lycophytes) produce very different sorts of leaves by utilizing a

similar genetic pathway.[61] It has also recently been found that land plants contain a common very powerful regulator of gene expression that in mosses controls their life cycles but in higher plants controls flower formation.[62] The details of how this remarkable regulator works are still far from clear, but one possibility is that it influences a similar network of genes in both mosses and flowering plants. If leaves evolved in this way, the role of the environment is critical. Any proto-leaves arising by the evolution of new switches regulating a basic genetic tool-kit will be doomed to extinction in a hostile environment.

It is also important to recognize that neither the fall in carbon dioxide nor the rise in oxygen is really the 'trigger' for these dramatic evolutionary events in the plant and animal kingdoms. Ecological phenomena play the starring role in both dramas. Controversy abounds when it comes to the Cambrian explosion,[63] but it is increasingly clear that as larger, more diverse, marine animals evolved with structures like complex eyes, jointed appendages, and shells, they in turn drove the evolution of more complex animals better equipped for defence, evasion, or protection against predation. As one leading developmental biologist comments: 'Genes in the tool kit are important actors in this picture, but the tool kit itself represents only possibilities, not destiny. The drama of the Cambrian was driven by ecology on a grand scale.'[64] Ecology, too, played a decisive role in the transformation of terrestrial ecosystems as carbon dioxide levels fell, creating the ecological opportunities for land plants. Once the limitation on large leaves imposed by a high carbon dioxide concentration was removed, we must picture how competition intensified between neighbouring plants as their shoot and root systems evolved. The race to become taller as leaves got ever larger, and plants struggled to avoid being shaded out in the competition for light, is well documented in the fossil record.[65]

Tall leafy forest trees are a striking feature of terrestrial ecosystems worldwide by the start of the Carboniferous, 360 million years ago, an evolutionary crescendo signifying the end of a remarkable chapter in the history of plants.

When integrated with studies of the genetic make-up of organisms, we can see that the historical perspective of biology afforded by the study of fossils tells us a richer story about the evolution of life on Earth. It deepens our understanding of the nature of evolutionary processes. As we move towards a common framework for understanding leaf evolution, it appears increasingly to be the product of a chance combination of 'genetic potential and environmental opportunity'.[66]

An obvious question to ask from all this is what caused carbon dioxide levels to plummet in the first place? Surprisingly, it seems that the plants themselves are the answer. Through a myriad different ways, plants have a remarkable ability to upset the network regulating the exchange of carbon dioxide between the rocks, oceans, and the atmosphere—a network known as the long-term carbon cycle.[67] By long term I mean that it operates incredibly slowly, on a timescale of millions of years. Nevertheless, in spite of its stately pace, the cycle is hugely important for regulating the climate of Earth and indeed of other planets.[68] This is not the cycle familiar to most of us, whereby plants take up carbon dioxide by photosynthesis, only for it to be eventually released back to the atmosphere after they die and decompose. These natural processes control the carbon dioxide content of the atmosphere on a timescale of years to tens of thousands of years. The long-term carbon cycle is, as we shall see, completely different, dwarfing its short-term counterpart involving forests, grasslands, and soils of the world by cycling a hundred times as much carbon as that contained in the world's terrestrial ecosystems every million years.[69]

We can think about the cycling of carbon on Earth over millions of years by starting with volcanoes which, when they erupt, vent carbon dioxide from deep within the Earth's crust into the atmosphere. In the atmosphere, it dissolves in rainwater to form a weak acid (carbonic acid) that is still sufficiently strong to dissolve silicate rocks and flush bicarbonate and other ions into rivers and streams. After transport to the oceans, marine organisms use the bicarbonate ions to construct shells, a small fraction of which eventually become buried in sediments on the seafloor. The famous white cliffs of Dover in southern England are composed of the calcium carbonate shells of phytoplankton called coccoliths and were formed by these processes in the Cretaceous. Eventually, over millions of years, the dense ocean plates plunge beneath the less dense continental plates and carry the seafloor sediments to depths where they are cooked under high pressure until carbon dioxide reforms, ready to be released by volcanoes. If the Earth did not have a molten core to drive the recycling of the rocky crust, all of our planet's carbon dioxide would end up on the seafloor. Locked into carbonates or other minerals, the carbon dioxide greenhouse effect would collapse, converting Earth into a giant snowball planet.

The long-term cycling of carbon dioxide between the Earth's rocky crust, the oceans, and the atmosphere operates as a thermostat to prevent the climate becoming too hot or cold; analogous devices perform the same purpose in car engines and domestic central heating systems. The planetary thermostat operates even in the absence of life and works because the weathering of silicate rocks, which consumes carbon dioxide from the atmosphere, is strongly dependent on temperature.[70] Warmer climates accelerate weathering, lowering the carbon dioxide content of the atmosphere, while cooler climates decelerate it. Imagine a situation reminiscent of early Earth, when excessive volcanic activity liberated enormous quantities of carbon dioxide from the mantle and released it into the atmosphere. A carbon dioxide-rich atmosphere creates a warm greenhouse climate that promotes the weathering of rocks, consumes carbon dioxide from the air, and weakens the greenhouse effect. The net effect is a cooler climate. Because a warm climate can lead to a cooler one, and a cool climate to a warmer one, the feedback loop is said to be a negative or stabilizing feedback. It has likely prevented runaway planetary warming on Earth for the last billion years or so. Unfortunately, it takes from hundreds of thousands to millions of years to work and is far too slow to counteract human-induced global warming.

The thermostatic control of global climate by this slow cycling between the rocks, oceans, and atmosphere operates in the absence of life, and so can explain what has gone 'wrong' with the climates of other planets. Consider, for example, our neighbouring planets Venus and Mars.[71] Surface temperatures on Venus are in excess of 460 °C and on Mars typically −55 °C, occasionally dipping to −140 °C. Climate regulation on both planets has clearly failed. What went wrong? Why didn't their climate stabilize to be 'just right' like that on Earth? The reason appears to be that, because Venus is situated much closer to the Sun than Earth, whatever water it was originally endowed with simply boiled off. Obviously, no water means no rain, and no rain halts rock weathering and prevents the removal of carbon dioxide released by volcanoes. Consequently, Venus is a hot, dry planet with a carbon dioxide-rich atmosphere. The situation on Mars is quite different because it is further away from the Sun, with a surface temperature far below the freezing point of water. The answer to the question of what went wrong in this case is more speculative. The main cause seems to be Mars' small size; it is half the diameter of Earth and its interior cooled off much more quickly. Without being able to maintain a molten core, volcanoes on Mars soon went extinct, breaking the chain of carbon cycling as they ceased to act as conduits returning carbon dioxide into the atmosphere. Eventually most of the carbon dioxide became locked up in the crust of the Red Planet, putting it out of commission in the job of climate regulation.

Mars and Venus highlight the fragility of the carbon dioxide–weathering–climate thermostat. Returning to Earth, we find that as plants underwent their Palaeozoic 'big bang' they disrupted its previously smooth operation. For it turns out the activities of their roots—and their fungal partners—are effective at promoting the removal of carbon dioxide from the atmosphere by enhancing the rates at which continental silicate rocks undergo chemical weathering. Roots and their symbiotic fungal associates secrete organic acids that attack mineral particles in soils to liberate nutrients needed for growth. Roots also anchor soils, slowing erosion and giving the minerals more time to be dissolved by rainwater. Meanwhile, above-ground organic debris accumulates in a litter layer to form a continuous moist acidic environment that helps break down soil minerals, while higher up leafy forest canopies recycle precipitation. In Amazonia, rainforests recycle rain coming in off the Atlantic Ocean several times before it eventually returns to the sea, with the result that it repeatedly flushes through the soils to enhance weathering.[72] Plants, and their fungal partners below-ground, are evidently engaged in a conspiracy of silence as they gradually consume the rocks beneath our feet over the ages. Careful investigations have shown them to dissolve rocks five times faster than normal, irrespective of whether they are tropical rainforests in Hawaii or conifer forests in the Swiss Alps.[73]

Through these processes plants have been imperceptibly removing carbon dioxide from the atmosphere and regulating climate as the millennia ticked away. Initially, early in their evolutionary history, the small and leafless early land plants had rather insignificant effects on weathering rates. But as plant life began to spread across the continents and into upland areas, things changed. The 'greening of the land' and the rise of leafy forests with deeper rooting systems enhanced rock weathering on an unprecedented grand scale, stripping carbon dioxide out of the atmosphere.[74] Falling carbon dioxide levels allowed larger leaves and promoted competition for taller trees, further entraining the evolution of large rooting systems.[75] The rhythmic dance of plants, carbon dioxide, and climate reinforced itself, building in tempo over evolutionary time as carbon dioxide levels spiralled downwards, dragging the temperature of the Earth with them as the greenhouse effect weakened. As the Earth moved towards a catastrophic snowball state, threatening the global extinction of plant life, and probably much else besides, the thermostat kicked in. Rates of rock weathering slowed in the cooling climate, and the consumption of carbon dioxide from the atmosphere halted, stabilizing the climate and preventing the suffocation of plant life.

In the end, we can see that plant life was a major beneficiary of its own activities, falling carbon dioxide levels with each generation imperceptibly preparing the world for the next by facilitating the evolution of larger leaves. It was a progression of cause and effect, without foresight or planning. Any notion that plants 'pulled' carbon dioxide out of the atmosphere to permit the evolution of leaves is deeply flawed. Plants' modifications of their own environment were unconscious and undirected. Intriguingly, plant evolution generated changes in the global environment that persisted as a legacy to modify subsequent generations.[76] Falling carbon dioxide levels saw the evolution of leafy plants, which in turn accelerated the diversification of terrestrial animals and insects. Astonishingly, these far-reaching consequences for life on Earth stem from the interaction between greenhouse gases, genes, and geochemistry.[77]

3

Oxygen and the lost world of giants

Oxygen breathes life into all complex living organisms. For centuries, scientists peered keenly down the dimly lit half-billion-year-long corridor of Earth's past to discern whether the atmospheric content was set at the present-day level of 21% as complex life evolved. The dénouement to this mysterious puzzle emerged only recently, when a handful of pioneering scientists clarified the picture in the final decades of the twentieth century. We now believe that 300 million years ago, oxygen levels rose to a magnificent 30%, before sometime later falling to a lung-sapping 15%. Tropical swamplands played the starring role in oxygen's highs and lows and helped engender a remarkable evolutionary episode of gigantism in the animal world.

Deposition and denudation are processes inseparably connected, and what is true of the rate of one of them, must be true for the rate of the other.

Charles Lyell (1867), *The principles of geology*

OXYGEN, in its molecular form, is the second most abundant gas in our atmosphere but second to none in courting controversy. Its discovery is often credited to the great experimenter Joseph Priestley (1733–1804), who in 1774 showed that heating red calyx of mercury (mercuric oxide) in a glass vessel by focusing sunlight with a hand lens produced a colourless, tasteless, odourless gas. Mice placed in vessels of the new 'air' lived longer than normal and candles burned brighter than usual. As Priestley noted in 1775, 'on the 8th of this month I procured a mouse, and put it into a glass vessel containing two ounce measures of the air from my mercuric calcinations. Had it been common air, a full-grown mouse, as this was, would have lived in it about quarter of an hour. In this air, however, my mouse lived a full half hour.'[1] Later experiments revealed that mice actually lived about five times longer in the 'new air' than normal air, giving Priestley an early indication that air is about 20% oxygen.

About the same time, the Swedish chemist Carl Scheele (1742–86), working in Uppsala, showed that air contained a mixture of two gases, one promoting burning (oxygen) and one retarding it (nitrogen). Like Priestley, Scheele had prepared samples of the gas that encouraged burning ('fire air') by heating mercuric oxide, and also by reacting nitric acid with potash and distilling the residue with sulfuric acid. However, by the time his findings were published in a book entitled the *Chemical treatise on air and fire* in 1777, news of Priestley's discovery had already spread throughout Europe and the great English chemist lay claim to priority. Only later did it become clear from surviving notes and records that Scheele had beaten Priestley to it, producing oxygen at least two years earlier. The harsh lesson from history, which still rings true today, is that capitalizing on a new exciting discovery requires its expedient communication to your peers. The talented Scheele died at 43, his life shortened by working for much of the time with deadly poisons like gaseous hydrogen cyanide in poorly ventilated conditions. Nevertheless his achievements were considerable. He notched up the discovery of no less than five other gases besides oxygen that were new to science, laid the foundations for modern photography with his report of the action of light on silver salts, and was elected into the Swedish Royal Academy of Sciences in 1775.

Both Priestley and Scheele were tenacious devotees of the phlogiston theory, a concept hopelessly wide of the mark that had dominated chemistry for half a century. It was originally advanced by the German scholar and adventurer Johann Becher (1635–82), and shaped into the form in which Priestley and Scheele knew it by the German chemist George-Ernst Stahl (1660–1734). Stahl's theory maintained that air was necessary to absorb the imponderable fluid 'phlogiston' and without air the opportunity for absorption was lost—this was why things failed to burn in the absence of air. It held that burning a substance released phlogiston into the air in a form that manifested itself as a flame. After burning, the inert residue was said to have become 'dephlogisticated' while the gas driven off (oxygen) was designated 'dephlogisticated air'.

It fell to the founder of modern chemistry, Antoine Lavoisier (1743–94) (see Plate 3), to sort out the confusion. Lavoisier has been described as a man of 'conspicuous vanity, hauteur, and no little concupiscence',[2] who enjoyed the considerable advantage of starting out at the top, by being born into a wealthy French family. Lavoisier demolished the phlogiston theory in a masterly series of careful experiments showing that substances actually gained rather than lost mass after burning, stating in his *Mémoires* to the French Academy in 1786 that 'in every combustion there is an increase in weight in the body that is being burned, and this increase is exactly equal to the weight of the air that has been absorbed'. Priestley vainly attempted to repair the theory by proposing that phlogiston had negative mass, hence substances gained mass after burning, but there was no recovering from Lavoisier's awesome analytic evidence.

Exulted by the victory, Lavoisier organized one of his famous *soirées* in Paris to entertain society's elite in the form of a mock trial. The event is recorded as follows:

> He invited a large distinguished party, and enacted this trial in front of them. Lavoisier and a few others presided over the judicial bench, and the charge was read out by a handsome young man who presented himself under the name 'Oxygen'. Then the defendant, a very old and haggard man who was masked to look like Stahl, read out his plea. The court then gave its judgement and sentenced the phlogiston theory to death by burning, whereupon Lavoisier's wife, dressed in the white robe of a priestess, ceremonially threw Stahl's book on a bonfire.[3]

Not long after this charade on 8 May 1794, Lavoisier himself conspicuously failed to navigate the dangerous political undercurrents of the French revolution and was arrested, tried in less than a day, and tragically sentenced to death by guillotine the same afternoon. A contemporary of Lavoisier's, the mathematician Joseph-Louis Lagrange (1736–1813), famously observed at the time 'it needed but a moment to sever that head and perhaps a century will not be long enough to produce another like it'.[4]

Credit for the discovery of oxygen is often shared between Priestley, Scheele, and Lavoisier. Yet it is possible alchemists knew of the 'magical' properties of air at least 170 years before this dynamic trio came along. As early as 1604, the Polish alchemist Michael Sendivogius (1566–1636) reported that 'there is in the air a secret food of life' and may have explained how to produce it to the brilliant Dutch inventor Cornelius Drebbel (1572–1633).[5] If so, it could explain how Drebbel managed to keep the passengers alive in his new invention—the submarine. For in around 1621 Drebbel built three submarines, each bigger than its predecessor,[6] based on a design involving greased leather tightly stretched over a wooden frame. The largest vessel allegedly carried 16 people and was propelled by oarsmen whose oars projected out of the sides through ports sealed with tight-fitting leather flaps. Large pigskin bladders situated beneath the rowers' seats were filled with water through pipes and emptied by the crew squashing them flat to regulate buoyancy.

Drebbel's magnificent 16-man submarine, with six pairs of oars, was demonstrated to King James I on the banks of the River Thames with a return underwater voyage from Westminster to Greenwich (~22 km) lasting three hours. Fevered speculation surrounded how the sailors and passengers were kept alive for so long. Some said there were tubes to the surface and a set of bellows in the craft to circulate air. However, the famous chemist Robert Boyle reported that one of the passengers noted that a 'chemical liquor' had been used to replace the 'quintessence of air'.[7] We may never know for sure how it was achieved; Drebbel was notoriously secretive about his inventions, and never kept notes or diagrams of his work. The question is: did he use Sendivogius's method for producing oxygen? If he did, it was a brilliant demonstration that Drebbel understood our essential requirement for oxygen.

🌿

Nearly a century after Lavoisier brilliantly put paid to the phlogiston theory, his fellow Frenchman, the palaeontologist Charles Brongniart (1859–99), began exploring and excavating 300-million-year-old Carboniferous fossils from a quarry in Commentry, north-eastern France.[8] Charles Brongniart was the grandson of Adolphe Brongniart (1801–76), the renowned palaeobotanist and physician, who in turn was the son of Alexander (1770–1847), a chemist, mineralogist, and also, as it happens, palaeontologist. Evidently, a passion for fossils ran deeply in the Brongniart family. The quarry formed when the site was a narrow freshwater lake about 9 km long by 3 km wide. Streams and rivers from the surrounding mountains drained into the lake around its marshy shoreline, creating ideal environmental conditions for fossilizing plants and animals, especially, as it turned out, fossil insects.

In a remarkably productive collecting period between 1877 and 1894, Brongniart unearthed hundreds of previously unknown fossil insects and brought them to the attention of the scientific world. His most astonishing fossil find, announced to the world in a detailed paper published in 1894, was an ancient predatory dragonfly—*Meganeura*—with a giant wingspan of 63 cm.[9] Brongniart's *Meganeura* fossil resides today in the Muséum Nationale d'Historoire Naturelle, Paris.[10] Following the publication of this great work, other collectors soon began to turn up new specimens of giant insects in Carboniferous-aged rocks. To be sure, none were as spectacular as *Meganeura*, but some fossil mayflies rivalled the 16 cm wingspans of modern

Giant Andean Hummingbirds. Before the close of the century, spectacular finds like these made it obvious to palaeontologists that diverse groups of Carboniferous insects had reached giant proportions, dwarfing their living counterparts.

Two of the most celebrated giant insect fossils turned up in the coal seam of a colliery, long since closed, near the small medieval town of Bolsover, Derbyshire in 1978. The first had a wingspan of some 20 cm; the second specimen was over twice the size of the first, with a staggering wingspan of half a metre. Bigger than any living dragonflies, these fossilized flyers are the largest and oldest known examples of monster dragonflies; their discovery hit the headlines around the world. The first specimen, dubbed the 'beast of Bolsover', is not to be confused with Bolsover's longstanding and pugnacious Member of Parliament, Dennis Skinner. *The Daily Telegraph* proclaimed 'World's oldest dragonfly found in pit' and queues formed in Cromwell Road when the specimen was displayed in the Natural History Museum, London.

Giant winged insects were only one part of the story; the fossilized remains of Carboniferous floras and faunas demonstrated that gigantism was genuinely the order of the day. Some giant forms of plant life that today are characteristically small, like clubmosses and horsetails, colonized the ancient tropical coal swamps. Trees distantly related to our modern clubmosses reached heights of 40 m, while beneath them ancestral forms of horsetails inflated to majestic heights of 15 m or more towered above the luxuriant undergrowth of ferns. The evolution of huge plants created new habitats for animals to exploit for food and shelter, and fittingly insects, spiders, and millipedes also evolved into giants. The centipedes and millipedes inhabiting swamps reached lengths of a metre or more. Only recently, in 2005, a trackway was discovered in Scotland thought to have been made by a six-legged Carboniferous water scorpion some 1.5 m long.[11] Primitive amphibians, the biological intermediates between bony fish and reptiles that conquered the land, also attained gigantic proportions. The ancestors of our newts, resembling flattened crocodiles with large skulls, truncated snouts, and stocky limbs, reached several metres in length.[12]

The ecological community of giants must have been a spectacle to behold, with airborne insects swarming through the tropical forests, as beneath them the ground bristled with the activities of the giant swamp dwellers. The most extraordinary episode of gigantism that the world has ever witnessed was no brief evolutionary experiment, either. The world's gigantic fauna flourished for 50 million years from late in the Carboniferous, 300 million years ago, to the start of the subsequent Permian Period. After the Permian, the fossil evidence dries up. No giant animals have yet been unearthed in rocks younger than the Permian, 250 million years ago. The strange zoological oddities thrown up by the Carboniferous coal deposits are an intriguing puzzle demanding explanation.

In 1911 a French duo, Édouard and André Harlé, rose to the challenge and offered the world a controversial hypothesis about the giant insects: they argued that the flight of these giants might be explained if atmospheric pressure was higher in the Carboniferous relative to now.[13] Their rationale is straightforward. Higher atmospheric pressure increases air density (and resistance) on the wings and bodies of flying animals, allowing them to generate lift more easily. Ingenious though it was the idea had no precedent, and at the time no evidence to support it. Charles Lyell, the leading geologist of his day, for example, had earlier remarked in his 1865 book *The student's elements of geology*, that the raindrop prints he found on shale rock surfaces one day 'resemble in their average size those which now fall from the clouds. From such data we may presume that the atmosphere of the Carboniferous period corresponded in density with that now investing the globe'. After Lyell, few others followed up such observations and the questions concerning higher atmospheric pressure and the giant insects raised by the French duo remained firmly in the background.

If atmospheric pressure was higher 300 million years ago compared to now, then it can only have been brought about by a higher concentration of either or both of the two dominant gases making up the atmosphere, nitrogen and oxygen. Given that nitrogen and oxygen comprise the bulk (almost 99%) of our atmosphere between them, improbably large additions of the other constituents are required to significantly alter total atmospheric pressure. Argon, the leading player among the minor constituents, is just 1% by volume, carbon dioxide 0.04%, and methane only 0.0017%. Nitrogen is chemically far more inert than oxygen[14] and consequently its concentration has probably remained the same since reaching its current level early in Earth history. Oxygen, on the other hand, is quite reactive, making it a plausible contender for increasing atmospheric pressure.

What the question really amounts to is whether the oxygen content of our atmosphere has been fixed at 21% as the diversity of life on land and in the oceans blossomed over the past half billion years.[15] Although a challenging problem, even today, considerable scientific insight was brought to the matter as early as 1845 by the Frenchman Jacques Ebelmen (1814–52).[16] Ebelmen set out for the first time the fundamental concepts needed to understand the history of oxygen and carbon dioxide in our atmosphere in a perceptive paper in 1845. He recognized how the burial and oxidation of plant organic matter and sulfur compounds in rocks and sediments added or removed oxygen from the atmosphere over geological time. The rationale for invoking these processes will, I hope, become clear shortly. Unfortunately, as is often the case with radical ideas, the scientific world was not ready for the Frenchman's pioneering vision. In spite of the best efforts of others to draw it to the attention of a broader audience, his seminal contribution to our thinking languished unnoticed in an obscure scientific journal for over 170 years.[17]

This is not to say others hadn't given the problem of oxygen thought, too. John Ball (1818–89), Irish botanist and alpine explorer, was very struck by the significance the Carboniferous coal measures held for the oxygen content of the atmosphere. This was evidently the burial of plant matter on a grand scale, and he commented in his lecture to the Royal Geographical Society in 1879 'In the history of the earth, regarded as the scene of organic life, there is one event of transcendent importance, to which I think sufficient attention is not given. I allude to the deposition of the coal measures.'[18] Ball presciently grasped the significance of coal formation in the Carboniferous for the oxygen content of the atmosphere: 'In forming this amount of coal, the plants of that period must have set free more than 45 billions of tons of oxygen gas, increasing the quantity previously existing in the atmosphere by about 4 per cent.' With these comments echoing from the past, we cannot help but be struck by the near miss that occurred in the annals of scientific history. Between them, the trio of Ebelmen, Ball, and Brongniart, all scientific contemporaries, held the key pieces of a tantalizing scientific puzzle at their fingertips—massive carbon burial, atmospheric oxygen, and giant insects—yet fate decreed that the pieces were not yet ready to fall into place.

Earlier I casually dismissed nitrogen as a player in raising atmospheric pressure because its chemistry is too dull to cause large variations in its atmospheric content. But given that the organic debris of plants also contains small quantities of nitrogen, originally derived from the atmosphere by microbes, could the formation of the Carboniferous coal measures have altered the nitrogen content of the atmosphere significantly? Calculations suggest not. The flux into the Carboniferous coal measures is far too small relative to the very large mass in the atmosphere. Our best estimate is that the nitrogen content of the atmosphere dropped by less than one-hundredth of 1%—a trivial amount.[19]

From the very promising foundations laid by Ebelmen, the baton in the race that unfolded in the following century, to answer how Earth's oxygen content had varied as life evolved, was not so much dropped as misplaced. Lloyd Berkner and Lauriston Marshall at the Southwest Center for Advanced Study, Texas (among others) finally retrieved it in 1965 languishing in the intellectual dust, and picked it up and ran with it. At a symposium on the 'Evolution of Earth's atmosphere' convened by the National Academy of Sciences of the United States,[20] they argued that oxygen produced by land plants during photosynthesis was the primary source of oxygen in the atmosphere.[21] In other words, as plant life evolved and became more abundant leading up to the Carboniferous, 300 million years ago, it added increasing amounts of oxygen to the atmosphere. Oxygen levels were only restored to some sort of equilibrium by the subsequent decay of organic materials, a process that consumes oxygen.[22]

Here is an apparently plausible, if rather speculative, mechanism for driving up Earth's oxygen content during the rise of giant insects. Two decades later, greater scientific rigour was brought to bear on the matter by the pioneering geochemist Robert Garrels (1916–88).[23] Garrels was the son of a chemical engineer who had been an outstanding athlete: competing in the 1908 Olympics, his father came second in the 110 m hurdles and third in the shot-put. Garrels himself, though destined for a scientific career, maintained a keen interest in athletics and established an informal gang of like-minded scientists interested in swimming, rowing, and running known as the Bermuda Biological Station Athletic Club (BBSAC). We gain a glimpse of his contributions to science in the words of his protégé Robert Berner, who wrote of him as being 'among the handful of persons that over the past half century truly altered the course of geochemistry, which was his speciality, as well as that of earth science in general. Hidden within this modest, affable, kind, and considerate man was the soul of a revolutionary.'[24] Berner is no slouch when it comes to geochemistry or athletics. An outstanding BBSACer and swimmer, he is also an innovator of scientific thought and was elected to the National Academy of Sciences in North America for his accomplishments at a young age. As we shall see, Berner plays a pivotal role in discovering Earth's oxygen history (see Plate 4).

Garrels' geochemical brilliance was to recognize not only that Earth's atmospheric oxygen content is intimately tied to the evolution of plant life,[25] but also that it is intimately tied to the recycling of Earth's rocky crust. The day-to-day activities of oxygen-producing plants and microbes are linked with the very slow processes of sedimentation and carbon burial in rocks and on the ocean floor. The chain of reasoning begins with photosynthesizing plants adding oxygen to the atmosphere as they manufacture biomass. Eventually when the forests (marine or terrestrial) die, they offer rich pickings for animals, bacteria, and fungi that break down their remains, consuming oxygen in the process. Decomposition of organic matter reverses the effects of photosynthesis, reclaiming the oxygen originally released during its synthesis and returning carbon dioxide back to the atmosphere. However, a small fraction of plant biomass, less than 1%, escapes complete decomposition.

We see one expression of this today in the high Arctic, where the activity of nature's decomposers is slowed by the cold climate and waterlogged conditions to promote the development of peatlands. Organic carbon also accumulates on the continental shelves that receive water discharged from the world's rivers; these are Earth's carbon superhighways transporting huge quantities of floating organic debris and dissolved organic carbon compounds out to sea. The Amazon, for example, by far the largest river system on Earth, exports a staggering 70 million tonnes of carbon to the coast every year.[26] On being flushed into the sea, much of it becomes buried in near-shore sediments containing little or no oxygen and, with rates of decomposition slowed in such circumstances, thick deposits of organic matter accumulate in the sediments of continental margins.

The gradual and continual burial of the fragmentary remains of plants on land, and in the sea, means that a fraction of the oxygen produced during its synthesis cannot be reclaimed by chemical or biological processes. Instead, it is free to accumulate, adding tiny amounts of oxygen to our atmosphere from year to year. Over millions of years, these marginal increments accrue. Unlikely though it seems, the swamplands and continental shelves are the lungs of the Earth on a geological timescale of millions of years, slowly exhaling oxygen into the atmosphere.

Garrels and colleagues' further revelatory insight was to recognize that the natural cycling of sulfur through the oceans, atmosphere, and crust by chemical and biological processes adds oxygen to, and removes it from, our atmosphere. Heterotrophic bacteria are unable to manufacture biomass from simple chemical compounds and instead consume other organisms—living or dead. Some of these bugs are strict anaerobes that specialize in metabolizing sulfur compounds. They obtain energy by feasting on dead plant material and dissolved sulfur (sulfate) to make hydrogen sulfide. Under the right conditions, the hydrogen sulfide precipitates out as the common mineral known as iron pyrite or fool's gold.[27] The microbial conversion of organic matter to pyrite takes place in the oxygen-free inter-tidal mudflats around the coastlines of the world. Because the organic matter consumed was synthesized by photosynthesis, the net effect is to release oxygen.[28] Via this route, the burial of pyrite at sea also slowly adds oxygen to the atmosphere.

Eventually, over the eons, sediments containing the obliter-ated remains of plants, and pyrite, are heated and compressed until their transformation into sedimentary rocks is complete. These newborn rocks sit tucked within the Earth's crust until tectonic upheavals or retreating seas expose them as mountain ranges or the coastal cliffs of newly formed continents. Weath-ered by wind and rain, they slowly erode, giving up their cargo of organic matter and pyrite, which come under chemical attack.[29] Organic matter is oxidized with oxygen back to carbon dioxide and water and iron pyrite is oxidized back to iron oxide and sulfate. Slowly, the cycle of burial and oxygen production is reversed by exposure and oxygen consumption. In a surprise recent discovery, scientists have found that microbes, too, par-ticipate in the leg of the cycle that converts organic matter back to carbon dioxide while consuming oxygen. Laboratory experi-ments revealed super-bugs capable of digesting 365-million-year-old organic matter present in shale, a common sedimentary rock.[30] It is too early to say how important shale-eating microbes are for accelerating oxygen consumption; for now we can marvel at their amazing metabolic versatility.

We can see now that the daily activities of plants and microbes are locked in a very slow dance to the rhythmic cycling of the Earth's crust. The burial of organic matter and pyrite in rocks adds oxygen to the atmosphere, while the uplift, exposure, and weathering of those same rocks removes it. Garrels realized that the astonishing significance of this usually unhurried dance is that it controls the oxygen content of the atmospheric over geological time. In elucidating all this, he rediscovered what Ebelmen had glimpsed 130 years earlier. Rarely is anything new in science. Today, humans are artificially accelerating the 'weathering' of organic matter by burning our fossil fuel reserves of coal and gas. Like the natural weathering processes, our combustion of fossil fuels is consuming atmospheric oxygen.[31] We are, however, in no danger of even denting our oxygen reserves, let alone consuming them and driving ourselves to extinction. At the present glacial-paced rate of oxygen consumption it would take 70 000 years for that disastrous scenario to unfold, and our fossil fuel reserves are estimated to last only the next 1000 years or so.[32]

With a solid theoretical foundation for unravelling the evolution of Earth's oxygen content through the ages in place, the next task was to retrieve it from the rock record. That aim required drawing up an inventory of changes in the abundance of sedimentary rocks rich in organic matter and pyrite over geological time. The technique, pioneered in the late 1980s, is called the 'rock abundance' approach and involves determining the cycling—the birth and death—of rocks over millions of years. Robert Berner and his then-graduate student Donald Canfield were the architects of the approach.[33] A glimpse of Canfield's unusually modest personality surfaced in an anecdote after he moved to the NASA Ames Research Center in California. While at the NASA Ames, Canfield was known to wear a shirt discarded by a Kroger grocery store employee named Chuck. Naturally, on his office door in one of NASA's most prestigious research institutes was not Donald Canfield but a certain Chuck Kroger. Few people got the joke.

Although drawing up an inventory of the abundance of sedimentary rocks for each slice of geological time, and the average carbon and sulfur content each type contains may seem like a tall order, much information was already available from oil companies who had exhaustively scoured the world for fossil fuels over the last century. Armed with what amounted to a geological history of the relevant rocks of our planet, Berner and Canfield simulated changes in the Earth's atmospheric oxygen content over the past 540 hundred million years (Fig. 3). The curve they obtained showed a major oxygen pulse created as levels rose sharply to peak at around 35% in the middle of the Carboniferous, 300 million years ago, and then fell to around 15% 200 million years ago. Remarkably, this rise and fall in oxygen was beautifully synchronized with the evolution and extinction of Brongniart's gigantic insects, and the other animals that came to light subsequently (Fig. 3). The curve also hinted at a second minor oxygen pulse during the Cretaceous. Other than these two features, the general picture at most other times during the last half billion years was one of stability. In other words, oxygen production and consumption, inexorably linked to the life cycle of sedimentary rocks, have been finely balanced for a very long time indeed, except during the Permo-Carboniferous and perhaps part of the Cretaceous.

The deceptively simple curve is now an iconic image of our atmosphere's past representing the triumphant alliance of theory and observation. Behind it, though, lie some difficult-to-test assumptions and uncertainties, as Berner and Canfield were quick to acknowledge. What it badly needed was another independent way of unravelling Earth's oxygen history to crosscheck the first. For that, attention switched from the abundance of rocks to the abundance of individual atoms. In this scheme of things, the key to tracking ancient oxygen levels lies in the

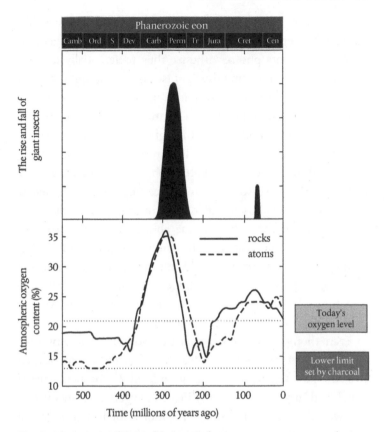

Fig. 3 Changes in the Earth's atmospheric oxygen content and giant insect abundance over the past 540 million years. The lower graph depicts two curves, one calculated using the 'rock abundance' approach, the other using the 'atomic abundance' approach. The lower horizontal line indicates the minimum level for oxygen set by the presence of fossil charcoal. Camb = Cambrian, Ord = Ordovician S = Silurian, Dev = Devonian, Carb = Carboniferous, Perm = Permian, Tr = Triassic, Jur = Jurassic, Cret = Cretaceous, Cen = Cenozoic.

abundance of heavy and light isotopes of carbon and sulfur atoms in ancient limestones and evaporites.[34] It is an ingenious idea. When plants photosynthesize they preferentially take up carbon dioxide containing the light isotope of carbon instead of the heavy one because the reactions involved need less energy and so proceed more rapidly.[35] This means that organic matter contains a high abundance of the light carbon isotope which, when buried in sediments, results in more heavy carbon being left behind in the oceans and atmosphere. In a similar fashion, and for the same reasons, microbes feasting on organic matter and producing hydrogen sulfide utilize the lighter isotope of sulfur. When pyrite is buried, the abundant heavy isotope of sulfur is left behind, altering the sulfur isotope composition of the oceans and atmosphere. We can predict, then, that the big oxygen pulse in the Carboniferous left an unmistakable signature in the isotopic composition of both limestones and gypsum deposits left over from the ancient oceans.

Even though the theory appeared sound, all attempts to exploit it and reconstruct a history of atmospheric oxygen over the past 540 million years foundered. It produced curves which had large catastrophic swings in oxygen levels incompatible with life on Earth. Some simulations, for instance, gave a negative amount of oxygen in the atmosphere—problems echoing Priestley's attempts to patch up his beloved phlogiston theory. Others implied very low oxygen levels that did not square with the known facts. Fossil charcoal is a valuable geological indicator of the minimum amount of oxygen in the atmosphere because charcoal is produced by forest fires burning in an atmosphere with a minimum of 13% oxygen (see Plate 5).[36] So the presence of charcoal in rocks and sediments dating back over the last half billion years tells us that atmospheric oxygen levels probably didn't dip below 13%.[37]

Evidently, something was wrong with the 'atomic abundance' approach as it was formulated at that time, but what? The critical problem lay with the plants, and in particular, an appreciation of how oxygen influences the business of photosynthesis. It was only uncovered after plants were grown in modern laboratory experiments in atmospheres artificially enriched in oxygen to mimic conditions 300 million years ago. During these experiments, high levels of oxygen greatly lowered plants' efficiency to absorb carbon dioxide and manufacture biomass.[38] The reason is that the process of photosynthesis is catalysed by a very ancient enzyme known as Rubisco,[39] believed to have been inherited from cyanobacteria some 2700 million years ago. But Rubisco suffers from a biochemical indecision: it has an affinity for carbon dioxide and also for oxygen. When carbon dioxide is plentiful, the issue holds no special significance, because it easily claims the enzyme's attention and feeds into the biochemical pathways that manufacture sugars. When oxygen is plentiful, Rubisco's dark side emerges, with its affinity for oxygen disrupting the smooth operation of photosynthesis. The whole business is fairly complex but the critical outcome of the experiments was to show that oxygen alters the abundance of the heavy and light carbon isotopes of plant organic matter.[40] The logical extension of this discovery was that high oxygen levels in the past should leave a distinctive carbon isotopic fingerprint in the fossilized remains of plants.[41]

When these previously hidden details of oxygen's influence on plants were fed into the 'atomic abundance' model, they allowed the isotopic fractionation of plants to change as oxygen changed. This new feedback between plants and the atmosphere brought stability to the calculations, and saw a fresh curve depicting a change in Earth's atmospheric oxygen emerge. This time it closely tracked that obtained from the 'rock abundance' method (Fig. 3): each history of oxygen supported the other.[42] Plants had picked the lock of the 'atomic abundance' method, and both methods (rocks and atoms) revealed a prominent oxygen pulse 300 million years ago.

If the conclusions of experiments with living plants are robust, they point to another means of testing the contentious Carboniferous oxygen pulse by searching for its fingerprints in the isotopic composition of fossil plants. Indeed, modern geochemical techniques revealed its signature in 300-million-year-old fossils collected from rocks around the world.[43] It may seem like a giant leap of faith to move from laboratory experiments on modern plants to predicting what we might find in fossil plants sandwiched into rocks some 300 million years old. Yet plants back then, as now, employed the same enzyme, Rubisco, to catalyse photosynthesis, and this biochemical similarity between the past and the present offers a basis for optimism in such an enterprise. Giant insects, the matching histories of atmospheric oxygen, and shifts in the isotopic composition of fossil plants—the evidence to win over the critics begins to mount.

Another, more direct, means of inferring the oxygen content of the ancient atmosphere also surfaced when scientists analysed tiny samples of 'fossil air' trapped in bubbles of fossil resin (amber).[44] These tiny time capsules reportedly contained samples of the Cretaceous atmosphere with 30% oxygen. Inevitably, in spite of a number of careful precautions and crosschecks, the results came under fire. Doubts arose over the capacity of the amber bubbles to effectively seal off a pristine sample of air for such a long period of time; oxygen is a reactive gas and amber is a porous resin.[45] Some suggested that the bubbles leaked the gases trapped inside. The effect of this leakage could be that very slowly, over the millennia, the original gaseous composition of the bubble becomes distorted, giving potentially misleading conclusions concerning composition of the ancient atmosphere. Still, all of these points have been rebuffed and it remains to be seen if the amber bubbles will one day yield up reliable secrets about the oxygen content of the past atmosphere.

Regardless of the amber resin issues, it seems quite clear that the delicate balance between oxygen production and consumption was upset during the Carboniferous. The traditional explanation is that slow tectonic movements over millions of years altered the continental configurations to create and destroy conditions suitable for waterlogged basins favouring the formation of swamplands. But another intriguing possibility is that the plants themselves were responsible.[46] The leading theory argues that, by the start of the Carboniferous, plants had evolved the ability to synthesize a tough molecule called lignin. Lignin provided plants with much needed structural support as they grew ever taller, but it presented the microbial world with a new challenge. Its complex molecular structure required a special suite of enzymes to break it down, and some scientists believe that nature's decomposers, fungi and bacteria, lacking this biochemical pathway, struggled to digest the rain of lignin-rich forest debris.[47] If correct, then global indigestion might have ensued. Swamplands bulged and the Earth's atmospheric oxygen content rose upwards. After microbes and fungi had risen to the metabolic challenge of digesting lignin, they could begin consuming the vast reservoirs of organic matter embedded in sedimentary rocks as it became exposed over millions of years by uplifting continents and falling sea levels. Only then could oxygen's status quo be slowly restored. To be sure, this elegant version of events may even be true, but it remains highly speculative and leaves several questions unanswered. Why, for example, did it take microbes, with large populations and short generation times, millions of years to evolve the trick of digesting lignin?

Whatever the reason, the Carboniferous coal measures testify to the enormous amount of carbon that was once buried and the models indicate their formation pushed oxygen levels upwards. Brongniart's giant insects that had once swarmed the air space of the ancient tropical swamps probably had enjoyed a higher atmospheric pressure and denser atmosphere due to more oxygen. In fact, in an atmosphere containing 35% oxygen air density increases by a third over today's value, permitting larger animals to fly more easily.[48] As I outlined earlier, this is in part because a wing 'pushing' against a denser atmosphere generates lift more effectively. The same reasoning explains why few modern helicopters can fly above 4 km—the air is too 'thin'. The downside to a dense atmosphere is a greatly diminished top speed as it increases air resistance on the wings and bodies of flying animals. The drag created by greater resistance effectively slows the forward motion of the animal and limits maximum speed and acceleration. Given these considerations, we should imagine our Carboniferous giants cruising majestically through the tropical swamplands; swarming now seems inappropriate.

A century on and the Harlé brothers' speculative hypothesis that an increase in atmospheric pressure aided the flight of the giant insects looks more reasonable. But better aerodynamic lift generated by a higher atmospheric pressure is only part of the story. Bathing creatures in an oxygen-rich atmosphere also benefits their respiration and physiology, because powered flight is a costly business and oxygen is the currency that counts when it comes to paying for it. Insects obtain oxygen to power their flight tissues through a branching system of rigid tapering tubes (the tracheae) penetrating into the abdomen. Ventilation of the network of tubes in modern insects generally occurs around the openings of this branching network, but is also achieved in the deeper system of tubes by body movements that cause tracheal compression and expansion in a fashion analogous to the inflation and deflation of vertebrate lungs.[49] Even so, it is not a very efficient means of 'breathing' and air trapped in the narrowest interior regions is replenished only slowly by diffusion, which can limit the supply of oxygen to the muscles. So it may be that the tiny terminal tubes limit how long insects can fly around for and how big they can grow.[50] For giant dragonflies with correspondingly longer tracheae, refreshing the air of the respiratory system might have been difficult. Living in an oxygen-rich atmosphere probably provided a means of overcoming this physiological barrier for supplying fuel to demanding flight muscles.[51] Oxygen also fuels energy production in the power plants of cells—the tiny organelles called mitochondria—and more oxygen means more energy.

Based on this line of reasoning, it is tempting to speculate that an oxygen-rich atmosphere permitted larger flying insects by relaxing the physiological constraints imposed on today's insects. So, if evolution was given another shot, would more oxygen and a denser atmosphere see insect gigantism happen again? In fact, this evolutionary experiment may have been repeated, when a second, less dramatic pulse occurred later in the Cretaceous, around 70 million years ago (Fig. 3). Remarkably, sparse fossilized Cretaceous insects, especially mayflies, suggestively hint that once again gigantism was adopted as a way of life at a time when oxygen rose to a more modest high of around 25%.

A Carboniferous atmospheric oxygen pulse could also help explain the unusual evolutionary patterns of gigantism in other animals. Amphibians benefited because they breathe through their skin and more oxygen in the air means more can diffuse into their bodies to fuel their energy requirements. Many amphibians and arthropods on land, new to the game of breathing air, used a combination of primitive gills and lungs. More oxygen improves the efficiency of primitive lungs that struggle to get rid of waste carbon dioxide produced by respiration. Air-breathing creatures benefited from the Carboniferous pulse in oxygen in other ways. It enabled muscles to recover more quickly after a sudden turn of speed used to capture prey or evade capture. Perhaps it is even possible that reptiles evolved from amphibians thanks to an oxygen-rich atmosphere. Land-conquering reptiles appeared in the Carboniferous and, unlike their amphibious cousins, reproduced with hard-shelled eggs. Could the high oxygen content of the atmosphere have facilitated diffusion of oxygen through the hard protective shell to the developing embryo inside?

Fascinating though these ideas are, they are also worryingly circumstantial. Direct evidence linking oxygen to body size comes from investigations of natural variations in the size of amphipods, small shrimp-like creatures found in the world's oceans, lakes, and river systems. In these different habitats, they experience varying amounts of oxygen because the solubility of the gas depends on temperature and salinity. Saltwater holds less oxygen than freshwater, cold water holds more oxygen than warm water. So in the cold waters of Antarctica, amphipods experience an oxygen-rich environment, but in the warm tropical seas off the coast of Madagascar they dwell in an oxygen-poor environment. And it appears amphipod size is directly related to the amount of oxygen dissolved in the water: Antarctic amphipods reach five times the size of their tropical cousins.[52] The world's largest amphipods live in the freshwaters of Lake Baikal, enjoying even more oxygen than their Antarctic relations. In fact, the identification of giant marine animals in polar regions is easily explained by oxygen, for there is a very strong relationship between the amount of dissolved oxygen in the water and maximum body size of amphipods.[53] The ability of oxygen to account for worldwide differences in amphipod size elegantly points to its capacity to raise the bar on body size.

The conclusion we reach from all this is profound. Earth's atmospheric oxygen content has had a guiding hand on the evolution of terrestrial animals in the Carboniferous, and possibly also at other times in Earth history as well. And plants played the starring role. But it is important to recognize that oxygen by itself will not necessarily cause larger animals to evolve. It simply offers the environmental opportunity for it to happen by relaxing physiological constraints and, for flying animals, aerodynamic limitations as well. The ability of nature to exploit that capacity when given the chance is admirably demonstrated in a series of ingenious experiments conducted by Robert Dudley when at the University of Texas. Dudley investigated how fruit flies responded to an oxygen-rich atmosphere typical of the Carboniferous and made a breathtaking discovery. After just five generations flies that had bred in the oxygen-rich air grew larger by 14%.[54] Oxygen, it seems, really did alleviate constraints on body size.[55] Furthermore, when the flies originally grown in a higher-pressure atmosphere were withdrawn and bred under normal atmospheric conditions, the male offspring were 14% larger than normal. In other words, oxygen relaxed the genetic control of body size, allowing the trait to be inherited and passed from one generation to the next. If this actually happened during the Carboniferous, it is easy to picture how competition between individuals could have set in train an upward spiral in body size. Rising oxygen entrained in biology an inevitable march towards gigantism.

Certainly, further imaginative experiments with modern insects will continue to yield deeper insights into the effects of the ancient oxygen pulse on animals. It is well known that oxygen is a double-edged sword: it is required to fuel energy production but is at the same time a source of toxic compounds known as reactive oxygen species. Reactive oxygen species are a major threat to the survival of cells because they damage DNA, proteins, and fats and perhaps even underlie ageing and death.[56] Organisms strive to critically maintain cellular oxygen levels sufficient for power production and yet as low as possible to minimize damage by reactive oxygen species. How did giant insects protect themselves from the seemingly inevitable onslaught of reactive oxygen species as oxygen levels ramped upwards to 30% 300 million years ago? Again experiments with living insects offer some tantalizing clues. Resting moth pupae experiencing atmospheric oxygen concentrations of up to 50% maintain levels inside their tracheae at a constant value close to 4%. As oxygen levels rise, pupae limit oxygen uptake by closing the endings of the tracheae for as long as possible, only opening them to get rid of waste carbon dioxide that has accumulated in the meantime.[57] Moth pupae have a minimum requirement of 4% oxygen, far below the modern atmospheric concentration of 21%, implying higher concentrations are noxious. Whether giant insects behaved in a similar fashion is unknown. Many modern insects are unable to close their tracheae and only avoid suffering from oxygen toxicity because the higher metabolic activity required by flying consumes it fast enough to prevent accumulation. Evolution surely coordinated breathing-system design and metabolic activity to match the demands and capabilities of each other in giant dragonflies.[58]

Pursuing the logical connection between oxygen and animal size to its extreme leads to an obvious question: what happened when oxygen levels plunged to perilously low values at the end of the Permian, 250 million years ago (Fig. 3)? The same issue— extinction by asphyxia—had in fact been raised some 30 years earlier, long before an accurate history of oxygen was established.[59] Palaeontologists noted that, coincidentally, living representatives of certain groups of animals susceptible to extinction also had the highest demand for oxygen. In the history of life on Earth available to them at the time, extinction rate and oxygen requirement were highest in squids and amphibians and declined through a series of animal groups by way of corals, starfish, and ending with anemones. Novel though it is, the oxygen-extinction hypothesis only really began to take hold much later with the realization that all forms of giant insects and animals went extinct after oxygen levels plunged from a heady 35% to a lung-sapping 15%. By the close of the Permian, the global phenomenon of gigantism was over.

We are unsure how the insect giants met their demise during the end-Permian oxygen crisis. It is often suggested that the beneficial effects of high oxygen on insect inhalation through tracheae were reversed as the atmospheric supply dwindled; oxygen delivery through the body of giant insects slowly became less and less efficient. Yet observations on modern grasshoppers question this simple notion because adults experiencing atmospheric starvation simply ventilate their breathing systems more vigorously to compensate.[60] Juvenile grasshoppers, though, seem quite unable to respond in this way, and this increases their susceptibility to asphyxia. It is interesting to speculate from this, that giant insects survived dwindling oxygen levels, only to be doomed to extinction by the failure of their young to do the same.

Prompted by the realization that oxygen starvation took hold at the close of the Permian, palaeontologists have taken another look at whether it was implicated in the extinction of animals at that time.[61] The starting point is the observation that every animal has a minimum requirement for oxygen and this sets the maximum altitude it can survive at. Humans, for instance, live and reproduce in the Peruvian Andes at their altitudinal limit of 5 km above sea level.[62] For lowland animals living during the end of the Permian, the oxygen crisis is equivalent to being transported to the thin altitude found 5 km above sea level. Falling oxygen levels through the Permian forced the mountain dwellers to migrate to the lowlands or risk asphyxia. The ensuing mass migration is proposed to have led to overcrowding, and even extinction, as burgeoning animal populations outstripped the food supply. Is extinction by overcrowding really a credible idea? Without more evidence, many scientists are understandably reluctant to embrace it just yet, even if it apparently sheds some light on a number of seemingly odd facts about the end-Permian extinction. One of the main surviving vertebrates, for example, was a pig-like creature called *Lystrosaurus*, with a barrel-chest that perhaps gave it an increased capacity for deeper breathing. Did *Lystrosaurus* owe its survival to being fortuitously well adapted to low oxygen conditions? Also, fossils of surviving vertebrates have been found 'crowded' up in the high latitudes where they experienced a cooler climate. Again, it is possible to interpret what appears to be animal migration to a cooler climate as a response to low oxygen. The cooler climate slows an organism's metabolism and reduces its oxygen demand.

We should recognize the pioneering nature of both the 'rock abundance' and 'atomic abundance' approaches for reconstructing the history of atmospheric oxygen as multicellular life blossomed. As I have described, before the advent of these methods we had little idea of what really went on. Afterwards, the models clarified the picture by using a mathematical framework to describe those processes regulating Earth's atmospheric oxygen content over millions of years. The skill is to reduce the complexity of the real world to the minimum number of fundamental processes required to open a realistic window on the history of oxygen. Of course, aspects of many natural phenomena will be lost in this kind of simplification. But we can glimpse the possible role of other feedbacks between the different components of the Earth system by taking a step back from this reductionist philosophy and relaxing some of the underlying simplifying assumptions.

Fire was one of the most controversial of these sorts of feedback.[63] It was put forward by James Lovelock and Andrew Watson when at the University of Reading, and the microbiologist Lynn Margulis at the University of Massachusetts in 1978. When suggesting a role for fire in regulating the oxygen content of the atmosphere, the trio had the Gaia hypothesis uppermost in their minds. Named after the Greek Goddess of the Earth, it proposes that 'the climate and chemical composition of Earth's surface are kept in homeostasis at an optimum by and for the biosphere'.[64] To support the notion that life maintains the Earth in a condition favourable for itself, they were looking for a way to keep oxygen levels constant over time; or put another way, to show self-regulation of the Earth's atmospheric oxygen content.

The question is: how might fire influence the oxygen content of the atmosphere over millions of years? Lovelock's idea was that as oxygen levels crept up, forests became increasingly flammable when struck by lightning. It followed from experiments showing that when paper tape is lit with a spark it ignites and burns more readily in a high-oxygen atmosphere.[65] He conceived of a negative feedback tightly constraining oxygen levels whereby too much oxygen leads to wildfires that devastate plant life. As we have already seen, less vegetation means, in the long run, less carbon burial, reducing the addition of oxygen to the atmosphere. Based on this reasoning, Lovelock and colleagues proposed that the very existence of forests throughout the Carboniferous implied that oxygen levels never rose above 25%. Lovelock himself later claimed that if oxygen levels went 'above 25 per cent very little of our present land vegetation could survive the raging conflagration which would destroy tropical rainforests and arctic tundra alike'.[66]

Obviously, an atmospheric oxygen content of around 35% in the Carboniferous is at odds with the fire feedback invoked by the Gaians. The reason for the discrepancy seems to lie in Lovelock's extrapolation from laboratory-controlled fires using paper tape to actual 'real world' forest fires. Paper tape is a poor substitute for natural bits of forest like bark, leaves, and wood, because they tend to be thicker and hold more water, giving them different combustion characteristics. Later investigations burning more realistic forest-floor materials showed that the spread of fires is unaltered in an atmosphere with 30% oxygen compared to one with 21%.[67] For drier fuels, an oxygen-rich atmosphere did seem to promote ignition. So, while the ignition of forest fires following lightning strikes appears to be more likely in an oxygen-rich atmosphere, the subsequent spread of fire will be curtailed by the overriding influence of fuel moisture. Support for the notion that, in Lovelock's memorable words, a 'raging conflagration which would destroy tropical rainforests and arctic tundra' during the Carboniferous now seems rather thin.

Other intriguing feedbacks between biology, geology, oceanography, and the oxygen content of the atmosphere are also emerging. Another team of scientists pointed out that rising levels of oxygen increase the concentration of it dissolved in seawater, thereby chemically locking up phosphorus in sediments.[68] Phosphorus is an important nutrient for marine phytoplankton but frequently limits its growth because it is always in short supply. Higher oxygen levels, it is argued, meant fewer nutrients, less growth, reduced burial of organic matter on the seafloor, and a slowing in the rate of oxygen production. This chain of events—more oxygen, less phosphorus, less phytoplankton growth, less carbon burial, less oxygen—forms a negative feedback loop that might act to slow down the rise in oxygen. In fact, phosphorus supply to the oceans and fire on land may themselves be linked through a chain of events that is postulated to end with rising oxygen levels promoting more forest fires.[69] After a fire, decimated forests fail to break down the minerals in soils to release phosphorus. So fire could retard the delivery of phosphorus into the rivers and the oceans, which in turn limits phytoplankton growth and the burial of organic carbon at sea. The net effect is a reduction in oxygen production. It all sounds a neat way of connecting events on land and in the sea, but suffers from the problem that rapidly regenerating forests recovering from fires actually develop more roots to support their expanding biomass. This effect increases, rather than decreases, the amount of phosphorus being weathered out of soils. Also, the production of charred forest remains leads to burial of microbially-resistant charcoal, creating another positive feedback pushing oxygen upwards.[70]

Interesting though feedbacks of this sort may be, their speculative nature makes it hard to incorporate them in a realistic way into the simple class of Earth system models represented by the 'rock abundance' or 'atomic abundance' approaches. However, one attempt modelled a feedback allowing more forest fires as oxygen levels began to rise, this curbs the burial of organic matter and slows oxygen production. Encouragingly, the

resulting reconstruction of Earth's oxygen history is similar to
that obtained from the earlier approach, although the inclusion
of a fire feedback damps the Carboniferous oxygen pulse to 27%.
In this more moderate form, the negative feedback of fire is
gaining credence. After all, the trunks of Carboniferous plants
were protected from fire by very thick bark, and the abundance
of fossil charcoal in Carboniferous sediments testified to out-
breaks of forest fire even in swampy environments.[71]

Speculation has long surrounded oxygen's role in the evolution
of life on Earth. In the middle of the nineteenth century the
Victorian scientist Richard Owen (1804–92) argued that the
'Creator' had chosen the Mesozoic era for the dinosaurs because
it was deficient in oxygen. This suited them because as reptiles
they had a slower metabolism than the mammals that evolved
later on. Owen then argued that oxygen levels rose throughout
the Mesozoic to make the atmosphere more 'invigorating' and
that this rise in oxygen was the ultimate cause of the extinction
of dinosaurs and the other prehistoric reptiles.[72] Owen had
no evidence for any of this and merely advanced the notion
believing a benevolent Creator matched living things to suitable
physical environments.

The present situation could hardly be more different. As the
weight of evidence accumulates, a significant rise and fall in the
oxygen content of the atmosphere during the Carboniferous
and Permian is getting harder to refute. Sceptics are finally
being won over, persuaded to accept the idea that the oxygen
content of the air we breathe is not fixed at 21%. Indeed, it is
remarkable that every time scientists tax their ingenuity to look
at something different, one more piece of supporting evidence

falls into place. In palaeontology, for example, researchers are incubating alligator eggs in atmospheres containing up to 35% oxygen. Oxygen is found to cause distinctive changes in the bone structure of the developing alligators, with those from higher oxygen growing faster and larger. If correct, it casts the growth rings in the fossil bones of long-extinct reptiles in a new light—as another source of information on the oxygen content of the ancient atmosphere.[73]

We have seen a remarkable alliance of evidence from the disparate fields of geology, physiology, palaeontology, and atmospheric chemistry. It points to the exciting possibility that the oxygen content of the atmosphere played a role in the evolution of large complex animals. Once furnished with reconstructions of oxygen's highs and lows over the last half billion years, palaeontologists have creatively generated hypotheses linking evolutionary novelty with oxygen's influence on animal physiology and ecology. Ebelmen laid the foundations for understanding oxygen over a century and a half ago. He remarked in his 1845 paper, 'many circumstances none the less tend to prove that in ancient geological epochs the atmosphere was denser and richer in carbon dioxide and oxygen, than at present.... The natural variations in the air have probably been always in keeping with the organisms that were living at each of these epochs.'[74] Only now are we finally beginning to appreciate the deeper significance his comments hold for how life in an oxygenated world evolved.

4

An ancient ozone catastrophe?

The stratospheric ozone layer shields Earth's surface-dwelling life forms from the damaging effects of harmful excess ultraviolet radiation. Its apparent fragility was emphasized all too clearly with the shock announcement in 1985 of the discovery of an ozone 'hole' above Antarctica. Could, then, natural phenomena such as massive volcanic eruptions, asteroid impacts, and exotic cosmic events like supernova explosions have damaged the biosphere's Achilles heel in the distant past? If so, are lethal bursts of ultraviolet radiation penetrating a tattered ozone layer implicated in mass extinctions? Definitive evidence of an ultraviolet radiation catastrophe is hard to come by. But contentious claims that mutated fossil plant spores point to exactly that possibility during the greatest mass extinction of all time, 251 million years ago, are now under close scrutiny.

We are all agreed that your theory is crazy. The question which divides us, is whether it is crazy enough to have a chance of being correct.

Niels Bohr (1958), *Scientific American*

THE University of Cambridge is one of the oldest seats of learning in the world and is, as befits such an august institution, steeped in tradition and history. One of the more curious traditions, which survived until 1909, was that of publicly ranking undergraduates who had taken the Mathematical Tripos, the oldest and most demanding examination of its kind.[1] Candidates concluded 10 (now 9) semesters of intensive study by sitting a gruelling series of eight lengthy papers, each more difficult than the last, undertaken over a period of nine days. In rank order, the first 30–40 were called wranglers; the man gaining the highest marks of the year held the enviable position of Senior Wrangler. By tradition the positions were published in the London *Times*, with the accompanying list carrying pictures and short biographies of the top finishers; being a wrangler conveyed a certain degree of national honour and university distinction.

Competition to become Senior Wrangler was intense. The examinations involved a test of knowledge, power of recall, concentration, and nerves, and a system of private coaching developed in response to the demands among the elite mathematicians to be Senior Wrangler. Coaches were often those who had previously placed well in the wrangler competition, with good ones able to teach essential mathematics and an ability to produce stock answers concisely so that as many problems as possible could be solved in the time available. The wrangler system evolved its own natural life cycle, ensuring its perpetuity, for a good coach could charge a tidy sum for seeing a student twice weekly over a year and usually had several candidates on his books. A certain William Hopkins was a superlative tutor who had, by 1849, coached 17 Senior Wranglers and 44 top three places. Wranglers in the top few places had the opportunity to take up a pleasant college fellowship or work as a coach fashioning a career to produce more wranglers.

Women in the days of Victorian and Edwardian Cambridge were not awarded a degree but were, from 1870 onwards, permitted to sit the Tripos examination. One of the first women to be highly placed was Charlotte Scott, whose outstanding abilities in mathematics blossomed with skilled coaching and saw her earn eighth place amongst the wranglers. As we might expect, the success of women in the Tripos examination didn't stop there, and a decade later Scott's notable achievement was overshadowed by the spectacular triumph of one Philippa Fawcett, who in 1890 was placed 'above the Senior Wrangler'.[2] Fawcett's success caused a sensation and was greeted with extraordinary scenes in Senate House.[3]

Among the Senior Wranglers of the past we find some of the greatest names in nineteenth- and twentieth-century science, including the foremost British astronomers of their time, John Herschel (1792–1871) and Arthur Eddington (1882–1944), and the mathematicians and physicists George Stokes (1819–1903) and John Strutt (the 3rd Lord Rayleigh) (1842–1919). If we add in the second and third wranglers to the list of Senior Wranglers, the names read like a roll call of some of the most distinguished British mathematical and physical scientists of the nineteenth and twentieth centuries. The list of second wranglers includes John 'J.J.' Thompson (1856–1940), discoverer of the electron, James Maxwell (1831–79), arguably the greatest physicist of his

82 THE EMERALD PLANET

era, and William Thomson (Lord Kelvin) (1824–1907). A popular anecdote once circulated concerning Kelvin's second wrangler place. The great man was apparently so confident of coming top in the exam that he asked his servant to join the gathering at Senate House and check who the second wrangler was. The servant returned and informed him, 'You, sir!' The Senior Wrangler that year (1845), William Parkinson of St John's College, had gained an edge by committing to memory a proof Kelvin had originally formulated and which came up in the exam. Parkinson was therefore able put it down without trouble whereas it had not occurred to Kelvin to learn his own proof; he then had to waste precious time deriving it from scratch.

Rayleigh was evidently a mathematical genius, and one of his examiners remarked, after Rayleigh had become Senior Wrangler in 1865, that 'Strutt's papers were so good that they could have been sent straight to press without revision.' That year he topped even this achievement by winning the Smith's Prize examination, which came after the Tripos and was believed to be a deeper, more profound, test of a gentleman's mathematical skills and scientific understanding. The first child of Rayleigh and his wife Evelyn was Robert Strutt (1875–1947) (Plate 6), who evidently inherited some of his father's prodigious gifts and went on to become an accomplished physicist. Strutt succeeded to the title, becoming the 4th Lord Rayleigh, on the death of his father in 1919, and paved his own road to scientific glory, despite the serious handicap of having a famous father, by discovering a way to date rocks by measuring the concentration of radioactive helium in uranium minerals.[4] He is less well remembered for his part in bringing about a scientific understanding of the ozone layer.

Ozone consists of three oxygen atoms bound together in a manner that transforms them into a single highly reactive molecule. The German chemist Christian Schönbein (1799–1868) noticed a pungent smell during his electrolysis experiments with water, the breaking apart of water molecules with an electric current.[5] He also noticed the same odd smell pervading the air after thunderstorms—themselves phenomena that act like large-scale electrolysis experiments. Proposing that electricity itself did not have an odour, in 1839 he named the unidentified component of air it created ozone, from the Greek *ozon* for smell (ozon).[6] Ozone soon became a subject of great scientific interest following the seminal contributions of the Frenchman Marie Alfred Cornu (1841–1902) and the Englishman Walter Hartley (1845–1913). Cornu noticed that sunlight split into its constituent wavelengths with a quartz prism was missing that part of the spectrum sitting in the ultraviolet range[7] and postulated the presence of a gas in the atmosphere blocking it by absorption.[8] A clue to the identity of the gas emerged when Hartley found ozone to be a very effective absorber of ultraviolet radiation, and this led him to propose that ozone was present somewhere in the atmosphere.[9]

Rayleigh's interest in the gas was sparked into life when his colleague Alfred Fowler (1868–1940) showed him the spectra of light from various stars, which had a detailed fingerprint pattern of ultraviolet radiation absorption that was seen as regularly spaced bands.[10] Investigating further, the duo then built a miniature 'ozonizer', producing ozone and allowing them to observe first hand in the laboratory the detailed characteristics of ozone's absorption spectrum. What they found is best described in Rayleigh's own words: 'we saw at once that the bands in the solar spectrum fitted those in the ozone absorption. It was a dramatic moment.'[11] Rayleigh's next contribution, as if this earlier advance was not enough, was to pursue an ingenious line of enquiry to find out if the ozone doing the absorbing sat in the lower atmosphere. To do this, he set up an experiment in

1918 to photograph the spectrum of a mercury-vapour light source at night using a small quartz prismatic camera across the Chelmer Valley in Essex, a distance of 'almost exactly' four miles.[12] Special permission was required for the experiment because a blackout rule was being enforced against the Zeppelin airships. Fortunately, however, he was well placed to obtain it through his uncle, Arthur Balfour (1848–1930; Prime Minister, 1902–1905), who was then Foreign Secretary in the administration of Lloyd George (1863–1945; Prime Minister, 1916–1922). The resulting photographs revealed the atmosphere to be comparatively transparent to ultraviolet radiation at ground level across a distance of four miles. This research, conducted while Rayleigh was on holiday, 'proved for the first time that there was little ozone in the lower atmosphere and that the ozone seen in the spectrum of the low Sun must be located at higher altitudes.'[13]

Where the bulk of it was actually located in the atmosphere did not get definitively sorted out until 1929, following the invention of a workhorse instrument for reliably measuring ozone by the Englishman George Dobson (1889–1976). Dobson's instrument monitored relative changes in the light from the Sun at various adjacent wavelength bands that are related to the total amount of ozone between the instrument and the Sun, that is, from the bottom of the atmosphere to the top. Accurate measurements of this sort require clear skies, something of a premium in Oxford, so at the invitation of his collaborator Paul Götz (1891–1954), Dobson decamped to Arosa in the Swiss Alps, where the crystal-clear mountain air afforded them greater opportunities. In the Alps the duo made detailed ozone measurements at different times of the day, when the path length between the Sun and the instrument varied. Their ingenious approach revealed that most of the ozone was situated at an altitude of around 25 km.[14]

Looking back, we see that within a dazzlingly productive period in the late nineteenth and early twentieth centuries, a handful of visionary scientists with considerable ingenuity had not only identified ozone in the air, but also measured its total concentration in the atmosphere and identified where most of it was placed, high up in the atmosphere. Moreover, the pace of advancement didn't stop there. The final dénouement—how ozone came to be there—crystallized in the decades that followed when the pioneering geophysicist Sidney Chapman (1888–1970) formulated his concise theory (in just five reactions) for the formation and destruction of ozone by the action of sunlight on oxygen.[15] Chapman's theory explains why the ozone layer sits between 15 and 50 km above sea level with the brilliant insight that ozone is constantly being created and destroyed by the action of ultraviolet radiation on itself and oxygen. His scheme predicted that ozone production stops below about 15 km because the presence of large quantities of it higher up in the atmosphere screen the energetic ultraviolet radiation required for its formation. It also correctly predicts that the absence of ozone above about 50 km occurs simply because insufficient oxygen exists in the thin air up there for production to exceed destruction.

Chapman's theory also explained an important finding eventually reported in 1902 by the cautious Frenchman Léon-Phillippe Teisserenc de Bort (1855–1913). Teisserenc de Bort was a pioneer of high-flying instrumented balloons and had noted what he perceived to be a rather odd trend in air temperature with increasing altitude. As he anticipated, temperature fell with altitude but then, at ~11 km and above, it stabilized and occasionally even rose. At first he thought the pattern might be an effect of solar heating on the balloon instruments and wrapped them in cork housing. He also conducted balloon flights at night in an effort to eliminate the effect. But still the same trend persisted. Eventually, after diligently accumulating data from 236 balloon ascents, he suggested in 1902 that the atmosphere could be divided into two layers—a warmer layer of relatively still air above a cooler, denser, more turbulent layer below.[16] The upper layer he dubbed the stratosphere and the lower one the troposphere. The temperature inversion, as it is known, is due to the stratosphere being heated by ozone's absorption of ultraviolet radiation—the warmth makes it comparatively 'still' because it is not subject to the vigorous convective mixing characterizing the troposphere below. In honour of his pioneering contributions to meteorology, craters on both the Moon and Mars are named after him.

Elegant though it was, Chapman's scheme was found wanting in due course—it predicted five times more ozone than the meteorologists' measurements indicated was actually up there. In other words, there had to be some as-yet unappreciated means by which ozone was being destroyed. The fundamental leap required to solve the problem was taken comparatively recently, in 1970, by a then young scientist called Paul Crutzen. Crutzen showed that, remarkably, the oxides of nitrogen, produced by soil microbes, catalysed the destruction of ozone many kilometres up in the stratosphere.[17] Few people appreciate the marvellous fact that the cycling of nitrogen by the biosphere exerts an influence on the global ozone layer: life on Earth reaches out to the chemistry of the stratosphere. Establishing this connection between the Earth and the sky earned Crutzen a share of the 1995 Nobel Prize for chemistry. Reflecting on it years later, he commented:

> I started reading the literature and I got this strange feeling something wasn't right with the theory of ozone in the atmosphere. I thought the answer involved nitrogen oxide, though I

doubted for a long time that I should publish my ideas. I had no data. Nobody knew if there was any nitrogen oxide up there. I still thought that proper scientists were Einstein-like figures, not people like me.[18]

We can see that by the late 1970s scientists had a reasonable grasp of the fundamentals of the ozone layer, with answers to questions such as where it was, why it was there, and how it was formed, all seemingly wrapped up. In short, roughly 90% of the Earth's ozone layer resides in the stratosphere where it is continually being produced by the action of sunlight on molecular oxygen and itself. What happened next shocked the world. The revelatory findings stemmed from meticulous observations by British Antarctic Survey scientists who had been patiently measuring the total amount of ozone in the atmosphere since 1956 at the Halley station. The Halley Bay research base is afloat on an Antarctic ice shelf, and the buildings themselves need to be periodically renewed as they slowly disappear beneath snow only to be occasionally disgorged years later from the edge of the ice shelf, embedded in icebergs. When the time came to analyse the Halley records, rather belatedly as it happens because of the need to clear a backlog of data, Joseph Farman, Brian Gardiner, and Jonathan Shanklin found, much to their amazement, a dramatic depletion of springtime ozone levels from 1976 onwards (Fig. 4). Convinced the trend represented a genuine phenomenon and was not due to a 'drift' in the calibration of the instruments operating in adverse weather conditions,[19] the team announced the shock discovery of an ozone 'hole' over Antarctica in a landmark 1985 paper published in the journal *Nature*.[20] Immediate concerns were raised about the reality of the phenomenon and why no other group had reported dramatic springtime losses of ozone—especially those working with data from satellite-borne ozone monitoring systems.

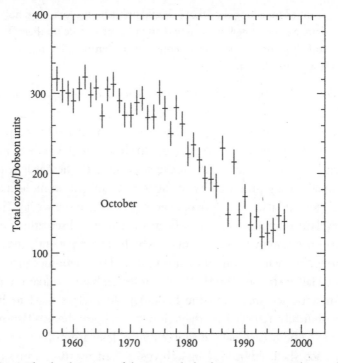

Fig. 4 The development of the ozone hole above Antarctica. The graph shows monthly mean (horizontal bar) and range (vertical bar) of total ozone measured at Halley Bay, Antarctica for October between 1957 and 1997.

It's a good point, and one that has caused consternation at NASA's Goddard Space Flight Center ever since because it ran a space programme monitoring ozone between October 1978 and February 1987—exactly the time the springtime ozone hole developed. The story widely reported at the time was that the satellite computer systems were programmed with an algorithm dismissing the anomalously low ozone readings in the datasets.[21] However, this doesn't quite ring true because the NASA team reanalysed its satellite data and published a paper in 1986 revealing that the ozone hole was a large-scale phenomenon covering much of Antarctica.[22] Still, mythical or not, a convincing explanation of why the NASA team failed to discover the Antarctic ozone hole is lacking[23] and it is tempting to reflect that the episode is simply a case of small-budget science winning out over its big-budget cousin. NASA's ozone-monitoring satellites generated 50 million ozone readings a year, and the feeble processing capabilities of computers in the late 1970s and early 1980s couldn't cope with such enormous datasets. By contrast, Farman and colleagues could readily sift through meticulously collected datasets from a single location to spot springtime depletion; no one at NASA had such a cue pointing them to the relevant portions of their vast datasets.

The surprise discovery of the ozone hole caught the world's attention and the public became fascinated and appalled by what was happening at the bottom of the world. The culprit was found to be the accumulation of chlorine in the stratosphere produced by the decomposition of the long-lived industrial chlorofluorocarbons (CFCs), used at that time in refrigerators and air conditioners, by sunlight.[24] Ozone destruction proceeds when special high-altitude clouds of ice particles form under the very cold winter and early spring conditions in the polar stratosphere; ice particles provide surfaces for rapid reactions that convert the inactive chlorine derived from the CFCs into its ozone-destroying form.[25] The reactions only take place in the presence of sunlight, which is the reason the hole appears in the spring.[26]

Warnings over the possible threat CFCs posed to the ozone layer had been issued a decade earlier, but the models, as they were then formulated, predicted rather minor ozone depletion occurring at lower latitudes than the Antarctic.[27] So, the surprise was twofold—the location of the hole and its severity. Further-more, because CFCs have a long average lifetime of 50–100 years, significant amounts remain in the stratosphere long after emissions have been halted thanks to the implementation of the Montreal Protocol which legislated for reducing the production of ozone-depleting substances like CFCs. This means the recovery of the ozone hole will take decades.[28]

The discovery of the springtime Antarctic ozone hole emphasized the apparently fragile nature of the thin ozone shield protecting life on Earth from harmful excess ultraviolet radiation. Indeed, collected at sea level and compressed by the higher atmospheric pressure there, it would only be 3 mm thick. I use the word 'shield' because the maintenance of this thin layer of gas protects life on Earth from excess high-energy short-wave ultraviolet-B radiation.[29] Indeed, without it, terrestrial animals and plants could not exist. Life shows this sensitivity because excessive amounts damage the nucleic acids RNA and DNA—the building blocks of life—leading to mutations and carcinogenesis. In humans, excess ultraviolet-B radiation causes sunburn and prolonged exposure can increase the chances of skin cancer. It is estimated that a 1% reduction in the ozone layer enhances the effective ultraviolet-B radiation dose by 2%—enough to increase some forms of skin cancer by 4%.[30] Ozone loss over Antarctica has led to a worrying increase of 12%[31] in DNA-damaging ultraviolet-B radiation since 1980 and already there are signs that the unique communities of plants and animals in the region are being adversely affected. On the coastal fringes of the Antarctic Peninsula, only two native species of vascular plants cling on and resist disturbance by seals and penguins: Antarctic pearlwort (*Colobanthus quitensis*) and Antarctic hairgrass (*Deschampsia antarctica*). Careful monitoring has revealed that the already glacial-paced growth of both species is being stunted by the extra springtime ultraviolet-B penetrating the thinner ozone shield.[32] Elsewhere, around the margins of the ice sheets in the Southern Ocean, elevated ultraviolet-B radiation disrupts the daily business of photosynthesis undertaken by tiny surface-dwelling phytoplankton positioned at the base of the food chain.[33]

The shock discovery of an ozone hole at the bottom of the world raised the important possibility that the ozone shield might not, as generally assumed, have always been present during the evolution of complex plants and animals over the last half billion years. Perhaps the asteroid strikes and epic volcanic eruptions that have repeatedly punctuated Earth's history temporarily annihilated it? If so then blistering bursts of ultraviolet radiation might be implicated as a contributing factor to the troublesome great mass extinctions that have befallen complex life.[34] In fact, the concern is a reversal of one first mooted in 1928 by the palaeontologist Harry Marshall at the University of Virginia.[35] Marshall ventured that prolonged bouts of intense volcanism produced enormous dust clouds that screened out ultraviolet radiation and caused rickets, sterility, and disturbance to the metabolic and nervous systems of the animals. 'Cold blooded animals', he claimed 'chilled by a drop in temperature [owing to the dust], were too sluggish and helpless to search for vitamin-containing foods, and their vital processes were so slowed down as to reduce the rate of evolution of new adaptations.' These ideas are now generally regarded as wide of the mark, but an alternative possibility, that life suffered the consequences of sudden ozone depletion events, should give us pause for thought. It holds profound significance for our understanding of the possible environmental forces shaping the evolutionary trajectory of life on Earth. Ultimately, it was, as we shall see, unusual fossil plant remains which offered a controversial pointer to an ancient ozone catastrophe.

What evidence is there that asteroid strikes and volcanic events periodically destroy the Earth's ozone shield? One way forward is to actually observe what happens when one or other of these phenomena occur. For asteroids, at least, this is not a seemingly likely or desirable possibility, but on the morning of 30 June 1908, inhabitants of the Stony Tunguska River Basin in central Siberia witnessed first hand the spectacular effects caused by a large extraterrestrial object entering Earth's atmosphere at supersonic speed.[36] The scale of the event is unprecedented in recorded history. Eyewitnesses reported that a massive pale blue fireball swept out of the sky and exploded high above the Tunguska River valley. When the stony Tunguska asteroid exploded, it is estimated to have released up to 20 megatons of energy.[37] The resulting blast wave flattened 2000 square kilometres of dense Siberian forest—an area about the size of Greater London—snapping tree trunks like matchsticks. Elsewhere, 70 km away, people in the town of Vanavara were knocked to the ground by the force of the shock waves, which travelled around the Earth twice over.

As the asteroid seared through the atmosphere, the shock waves surrounding it heated the air in its path to tens of thousands of degrees, breaking apart the paired atoms of the molecular gases oxygen and nitrogen. As the trail expanded and cooled, individual oxygen and nitrogen atoms split apart by the intense heat recombined with themselves and, with each other to form tens of millions of tonnes of nitric oxide.[38] The atmospheric disturbance triggered brilliant night-time crimson 'sky-glows' for two weeks after the event throughout Europe and Asia. It may also have depleted northern hemisphere ozone levels by 45%,[39] but it happened before the widespread ozone-monitoring network was established.[40] Claims for a dip in ozone levels based on datasets from the Smithsonian Astrophysical Observatory at Mount Wilson, California, are unconvincing because of the unreliability of the technique employed.[41] Still, an ecological benefit of the event was its fertilization of the taiga forests with nitrogen, as the nitric oxide was washed out of the atmosphere by rainfall.[42]

Other than Tunguska, we have witnessed no major asteroid strikes and so have been unable to actually observe how the ozone layer might be affected.[43] However, this is not the case for major volcanic eruptions. For on 15 June 1991, the world witnessed one of the most destructive volcanic eruptions of the twentieth century: Mount Pinatubo on the island of Luzon in the Philippines sprang into life after having been dormant for over 600 years. Satellites recorded the astonishingly rapid spread of enormous quantities of dust, ash, and gases around the world.[44] The eruption caused the immediate loss of more than 300 lives as homes collapsed under the heavy ash fall and torrential rainfall from the simultaneous passage of Typhoon Yunya: the disastrous conjunction of the two events forced the evacuation of more than 200 000 people.

The explosive eruption blasted 30 million tonnes of sulfurous gases, ash, and other debris into the stratosphere to an altitude of 40 km.[45] In the stratosphere, the sulfur dioxide rapidly became chemically transformed into droplets of sulfuric acid and fine particles known as aerosols. Both the droplets and the aerosols provided surfaces for chemical reactions to activate the large stratospheric chlorine reservoir, put there by the breakdown of CFCs, into its ozone-destroying form.[46] Not long after the eruption, a small tear appeared in the ozone layer above the tropics, lasting for a couple of weeks in mid-June.[47] Ozone levels the following winter and spring dipped to record lows.[48] Similar ozone depletion followed the injection of sulfur dioxide into the stratosphere by the El Chichón eruption in Mexico during March and April of 1982.[49] Yet in spite of these disturbances to the ozone layer, negligible extra ultraviolet radiation reached Earth's surface because the dusty atmosphere that followed the eruptions scattered and reflected incoming sunlight, offsetting the effects of a thinner ozone shield.[50]

Spectacular though these two recent explosive eruptions were, even they don't amount to much when measured against the volcanic yardstick of Earth history. Huge volcanic eruptions have repeatedly taken place during the last half billion years, when millions of cubic kilometres of basalt poured out onto the Earth's surface in brief periods of geological time. Many of these exceptional flood basalt eruptions, involving hundreds of repeated effusions of lava, have been provocatively linked with mass extinctions,[51] and it is tempting to think they wrought widespread destruction of the ozone layer. Yet the basis for this tempting extrapolation is far from straightforward. We have to be careful not to confuse the contrasting styles of the two different eruptions. Large flood basalt eruptions are not usually explosive affairs like those of Pinatubo fame, but comparatively sluggish efforts whereby lava is slowly haemorrhaged through fissures in the Earth's crust. Slow haemorrhaging of lava is a far less efficient means of injecting large quantities of toxic gases and other chemicals into the stratosphere. And, as with impact events, there is no direct observational evidence that they can damage the ozone layer. The only eruption of comparable style in recent history was when a paltry 15 cubic kilometres of lava poured out from a 25-km-long fissure at Laki, Iceland between 1783 and 1784, well before we even knew about ozone.[52]

Whether typical flood basalt eruptions emitted sufficient quantities of ozone-destroying chemicals like halogen gases and sulfate aerosols remains uncertain.[53] As I have already outlined, the explosive eruptions of El Chichón and Pinatubo damaged the ozone layer because they blasted sulfate aerosols into the stratosphere, activating inert chlorine reservoirs derived from CFCs. In the ancient atmosphere, obviously unaltered by our use of spray cans and refrigerators, the crucial issue is how much chlorine they released.

Fortunately for organisms living on the surface of the Earth, initial concerns over the thin ozone shield being our planet's Achilles heel have proved to be unfounded. The ozone hole created by the CFCs is an exceptional case. Current knowledge suggests that natural phenomena like asteroid impacts and huge flood basalt eruptions may be comparatively minor threats to its integrity. Imagine the surprise, then, when an international team of palaeontologists led by Henk Visscher at the University of Utrecht in the Netherlands sensationally claimed that the stratospheric ozone layer had been all but obliterated 251 million years ago, at the end of the Permian.

The disastrous end-Permian mass extinction took life to within a hair's breadth of complete annihilation.[54] The biggest mass extinction since the advent of animal life on Earth, it wiped out around 95% of all species, causing the collapse of biological communities that took tens of millions of years to recover. For many decades, researchers traditionally focused on documenting the severity of the extinctions in marine and terrestrial animals chronicled in the fossil record at the close of the Permian. Visscher and colleagues, however, decide to look at the end-Permian apocalypse in a different light, tracing instead the trials and tribulations of terrestrial floras as animal diversity plummeted.[55] The switch paid off when fossils from rocks in eastern Greenland told the story of the protracted dieback of ancient conifer forests.[56] As the forests deteriorated, the landscape opened up and the lycopsids, a pioneering group of herbaceous plants, proliferated, colonizing the decimated terrain.

Yet there was something strange about the spores left behind by the humble lycopsids, something that provided intriguing clues to the environmental stresses the parent plants had endured. For the spores in the Greenland sediments were in the form of an unusual mutated variety, one characterized by four individual spores clustered together into a single disfigured clump (see Plate 7).[57] Usually this sort of unseparated clump is only seen during normal spore development and is not released. Something appeared to have happened that disrupted the usual pattern of development and release of individual spores—but what?

The mutation story doesn't end there, because further scrutiny of the mutant fossils revealed the clumps of spores were stuck together at the very points where germination of each individual usually takes place. Paradoxically, the mutations had sterilized the proliferating lycopsids. The situation is analogous to what can happen when geographically isolated plant populations dwindle and become subject to excessive inbreeding. The Saharan cypress (*Cupressus dupreziana*), endemic to the Algerian Tassili N'Ajjer desert of the Mediterranean basin, is a classic modern example. The Saharan cypress is under serious threat of extinction, with only 230-odd native trees remaining.[58] As a result of inbreeding, normal pollen development is impaired and abnormal grains are produced, composed of different numbers of subunits stuck together in a way that inhibits germination. The future is bleak for the critically endangered cypress and so it might have been for the lycopsids except that they had a neat biological trick up their sleeves—they could employ an asexual means of reproduction,[59] ensuring their own ecological success at the expense of the ecological disaster that befell the conifer forests.

The isolated discovery of a high abundance of mutated spores from a few layers of rock in eastern Greenland hardly amounts to much, but it took on an altogether more profound significance when the investigators realized that the same mutant spores could be traced across the world in rocks of similar ages. Rocks dating to the end-Permian as far apart as North America, Europe, Asia, and Africa all contained exceptionally high frequencies of mutated spores (Fig. 5). Earlier investigators failed to appreciate the significance of these fossil mutants,

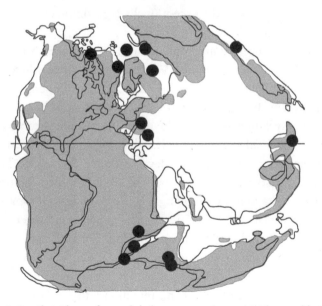

Fig. 5 Fossil evidence for a global mutagenesis event? The worldwide occurrence of mutated fossil plant spores and pollen. Each black circle represents a locality with a high abundance of mutated fossil spores/pollen in end-Permian rocks. The circles are plotted onto the continental configuration for 251 million years ago.

mistaking them instead for new species, but Visscher's team quickly realized they might offer an astonishing insight into the effects of an environmental catastrophe on the genetic make-up of plants. The following year, other researchers uncovered independent fossil evidence of genetic mutations in end-Permian rocks, only this time in the form of abnormal pollen grains released by surviving pockets of conifer forests in Russia and Northwest China.[60] Normal conifer pollen grains have two air sacs to aid wind dispersal, whereas the distinctive abnormal forms isolated from Russian and Chinese rocks comprise multiple air sacs stuck together. By itself, this is not so unusual. Abnormalities in pollen production occur all the time in nature, but not in the exceptionally high abundance found in end-Permian rocks.

The existence of numerous mutated fossil spores in rocks of a similar age worldwide led Visscher's team to conclude that they were looking at evidence for a global environmental muta-genic event. In a proposition echoing the sentiments of Otto Schindewolf (1896–1971), a doyen of German palaeontology, they boldly advanced the idea that the mutations can be explained by a near-global devastation of the ozone layer. This loss allowed increased harmful ultraviolet-B radiation to induce mutations in surviving plants. Schindewolf, famously anti-Darwinian all his life, developed his ideas on mass extinc-tions following his appointment as Professor of Palaeontology at the University of Tübingen after the Second World War. He argued for sudden bursts of cosmic rays from supernova explo-sions as a means of increasing mutation rates within different groups of organisms;[61] but with little evidence to support this theory, his critics were rightly sceptical.

Ironically, now that evidence for mutations appears to have come to light, we can probably rule out exotic bursts of cosmic rays from outer space as the cause.[62] When a star blows up, the core collapses into a black hole or a neutron star, and the titanic explosion emits galactic cosmic rays (charged atomic nuclei) and a burst of gamma radiation (bright, very energetic flashes of radiation). A couple of years ago, astronomers got a glimpse of the phenomenal energy involved when their instruments detected a bright gamma ray flash from a neutron star, which released in a fraction of a second as much energy as the Sun releases in a quarter of a million years.[63] Although the first serious investigation of what might happen if Earth experienced a nearby supernova explosion provoked a furore in the scientific press, with claims for the long-term annihilation of the ozone shield, it turned out that crucial chemical reactions responsible for regenerating ozone had been ignored. When factored into the calculations, these reactions buffer the ozone shield against the worst excesses of the cosmic storms.[64] In fact, as we might expect, the susceptibility of Earth's ozone shield to a supernova explosion depends on our proximity to it. Given the location of Earth in the Galaxy, the likelihood of us being sufficiently close to one to have it generate serious ozone depletion is vanishingly small, with an estimated probability of it happening approximately once every one and a half billion years.[65]

Undaunted, astrophysicists have continued to let their imaginations run wild,[66] and only recently creatively offered us yet another way for astrophysical events to trigger global extinctions via the destruction of ozone.[67] This one recognizes that Earth is usually shielded from cosmic rays by the magnetic properties of the solar wind and Earth's magnetic field, and that, in the past, both shields have been switched off. When both magnetic shields are down,[68] Earth is vulnerable to severe ozone destruction and surface life to a sharp increase in ultraviolet radiation. Moreover, the harsh ultraviolet environment might persist for as long as it takes the Earth's core to re-establish a magnetic field, 1000–10 000 years—more than enough time to induce ecological trauma. But the chances of both shields being down simultaneously are slim. It takes about a million years to pass through the dense interstellar clouds that suppress the magnetic field, and magnetic reversals occur, on average, about every 200 000 years, giving odds of one in five that both will happen at the same time.[69] Not exactly a long shot. The give-away might be the coincident timing of a magnetic reversal read from the minerals in rocks and a mass extinction, but the only example we have dates to a very minor extinction event late in the Eocene, 33 million years ago.[70]

It seems, then, that Schindewolf's exotic cosmic rays are an unlikely explanation for the mutated fossil spores. And I have already outlined some of the difficulties with invoking asteroid strikes and epic volcanic eruptions as effective destructive agents of Earth's ozone shield. So what weapon of mass ozone destruction did Visscher's team envisage was unleashed 251 million years ago? Their proposal centres on the environmental consequences of one of the most massive volcanic eruptions in Earth history, an epic bout of volcanism in Siberia that produced the Siberian Traps, magnificent terraces of ancient layers of solidified basalt.[71] The remnants of the eruptions cover an area of at least 5 million square kilometres and suggest that up to 4 million cubic kilometres of lava erupted onto the Earth's surface through fissures in the ground in less than half a million years. Furthermore, to the delight of Earth scientists studying the event, the peak eruption phase is contemporaneous with the mass extinctions.[72]

But Visscher and colleagues dismissed the Siberian Trap eruptions as a direct threat to the ozone shield. Instead, they pointed out that beneath Siberia's icy surface lie the remains of the largest coal[73] and salt deposits[74] on Earth, two types of sediments that might enhance the capability of the Siberian eruptions to damage the ozone layer. Prospectors drilling for oil on the Siberian plateau discovered long ago that the main eruptive flows welled up beneath thick layers of older rocks rich in coal and oil.[75] In fact, about 50% of the volume of the Traps lying buried in Siberia is sandwiched into these organic-rich sediments. As the eruptions proceeded and lava ascended from the depths, it didn't all flow directly onto the surface, but instead percolated between lenses of ancient coal and salt beds, and mixed with hydrothermal water. The widespread heating of these sediments, and the action of hot groundwater dissolving the ancient salts, was a subterranean pressure cooker synthesizing a class of halogenated compounds called organohalogens, reactive chemicals that can participate in ozone destruction.[76] And in less than half a million years, this chemical reactor is envisaged to have synthesized and churned out sufficiently large amounts of organohalogens to damage the ozone layer worldwide to create an intense increased flux of ultraviolet radiation.[77]

The question is this: does this mechanism really have the capacity for widespread ozone damage, and can we dismiss a more direct role for the Siberian Traps so easily? The Siberian eruptions are characterized by unusual and intensive phases of explosive activity. Volcanologists are resistant to the idea because, they argue, low viscosity basaltic lava degasses rather easily, defusing the potential for explosive eruptions. But the geological evidence is clear, the volcanism proceeded with unusual ferocity. The basalts are associated with a great deal of material flung out from the volcanic vent called tuff.[78] The tuff itself contains large amounts of rock originating from a kilometre beneath the vent, with some rare fragments originating from a staggering depth of ten kilometres. In fact, it is estimated that around 20% of the total volume of material extruded during the eruptions had pyroclastic origins.[79] By analogy with a better understood younger eruption that produced the Columbia River Basalts Group in North America,[80] the eruptions may have released unusually large amounts of the chlorine, much of it injected with explosive force straight into the stratosphere where it could exert maximum damage to the ozone layer.

Even though there are uncertainties involved in dealing with the masses of different gases released and the timescale of the eruptions, a collaborative team of scientists from the Universities of Cambridge and Sheffield realized sufficient information exists to investigate these issues using a mathematical model simulating the chemistry of the atmosphere.[81] They addressed two straightforward questions. First, what effect did the organohalogens released by heating the ancient coals and salt beds have on the ozone layer? Second, how did this compare with the simulated effects of the Siberian eruptions? The findings of this group offered a unique insight into the possibility of a pre-anthropogenic ozone hole opening up towards the close of the Permian.

The research team found that the original idea of heating coals to produce organohalogens simply wasn't up to the task of widespread ozone depletion. It wasn't big enough or bad enough to severely jolt the chemistry of the atmosphere out of whack, even if a proportion of the organohalgens were injected straight into the stratosphere. The eruptions on the other hand had far greater potential, for widespread and severe ozone destruction. If we accept dating evidence suggesting that the main eruptive phase happened within a few hundred thousand years,[82] the enormous quantity of chlorine launched into the stratosphere by the eruptions annihilated the ozone shield. Throughout the mid- to high-latitudes of both hemispheres, ozone loss was simulated to be a staggering 60 to 80% (Fig. 6), in effect the ozone layer had all but collapsed.[83]

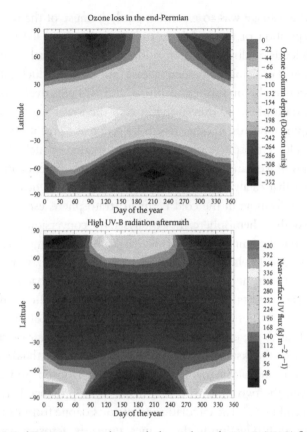

Fig. 6 End-Permian ozone loss and ultraviolet radiation-B (UV-B) fluxes. The upper panel illustrates the computer simulated depletion of the ozone layer caused by the eruption of the Siberian Traps, 251 million years ago, relative to background levels. The lower panel illustrates the ultraviolet-B (UV-B) radiation reaching the surface of the Earth. Experiments with living plants indicate that UV-B fluxes in excess of 17 kJ m^{-2} d^{-1} cause heritable DNA damage. Note that regions with severe ozone depletion and high UV-B fluxes are extensive in both the northern and southern hemispheres.

The damage was so extensive partly because of the unusual composition of the end-Permian atmosphere and led to very high levels of ultraviolet-B radiation reaching the Earth's surface (Fig. 6). It probably contained at least four times as much carbon dioxide as the present atmosphere, and this effectively cooled the stratosphere by preventing heat in the lower regions from escaping upwards, in other words by acting as a greenhouse gas. A cooler stratosphere promoted the formation of high-altitude clouds of ice particles, the same type of clouds we encountered earlier that catalysed ozone destruction over Antarctica. Except that by forming in the cold Permian stratosphere, especially in the southern hemisphere, they catalysed the conversion of chlorine from the eruptions into its active ozone destroying form.

Oxygen levels towards the end of the Permian were also thought to be at an all time low of 15%, a low that had seen off the giant insects (Chapter 3). Not only had this suffocated the giant insects, but it also limited the capacity of the ozone layer to 'self heal' as the eruptions progressed. The auto-repair mechanism arises thanks to the slow movement of air called the Brewer-Dobson circulation, named after Gordon Dobson and Alan Brewer, the two men who first deduced its presence between 1949 and 1956.[84] The circulation transports ozone from its main region of production in the tropics, upwards, across-ways and down towards the mid- and high-latitudes, with the bulk of it eventually descending in the polar regions. The Brewer-Dobson circulation explains why, paradoxically, the ozone layer is thinnest in the tropics, where most ozone is produced (in chemical reactions driven by sunlight), and thickest at the poles. When oxygen levels are low, ozone production in the tropics dips and less is transported to the poles.[85] We should note in passing that plants are the primary reason oxygen levels were so low. The long-term shift in global vegetation patterns from large forests to herbaceous vegetation throughout the Permian, as continental climates got progressively drier, was helped along by regional destruction of forests during the end-Permian. All this greatly reduced the burial of organic matter on land, the primary source of oxygen into the atmosphere on a timescale of tens of millions of years.[86]

The difference between the capacity of the two mechanisms, the Siberian eruptions and the 'heating of coals', to actually destroy the ozone layer lay in the masses of halogen, each one generated and released into the atmosphere. The fraction of chlorine emitted from the Siberian Traps reaching the stratosphere was estimated to be three times the total amount of chlorine produced by heating the organic-rich sediments beneath Siberia.

It appears, then, that worldwide depletion of the ozone layer during the end-Permian some 251 million years ago was a real possibility (Fig. 6). If it did occur, then our present-day ozone hole has an ancient precedent.[87]

We cannot leave our theme of environmental degradation and its link with the ozone layer just yet, however, because geologists attempting to explain the environmental circumstances that brought life to the brink of annihilation are not content with environmental damage wrought by a massive bout of volcanism. They have constructed what has been gruesomely termed a 'killing model' that invokes two further agents of environmental degradation: a massive release of methane from the seafloor and global stagnation of the world's oceans.[88] The suggested involvement of a third, an impact event, remains highly controversial. Every aspect of the evidence for it, from the sensationalist claim of having found the impact crater buried offshore of north-western Australia,[89] to the detection of supposed extraterrestrial gases trapped in cage-like carbon molecules,[90] has been roundly criticized. Even if it did occur, for reasons I described earlier, the impact is unlikely to be the culprit responsible for the mutant fossils.

Both methane and stagnation of the oceans might have had a detrimental effect on the ozone layer. The argument runs that initial global warming caused by the release of large amounts of carbon dioxide by the huge Siberian eruptions melted frozen reservoirs of methane gas trapped in ice beneath the oceans on continental margins.[91] Stupendous amounts of methane were suggested to have been released within a few tens of thousands of years as the ice melted, amplifying global warming. This may then have triggered more warming, by strengthening the greenhouse effect. In the context of the ozone layer, the fate of the methane is crucial. Some of the methane released would be oxidized to water vapour in the lower atmosphere (the troposphere) within a few years, but not all of this will reach the stratosphere because the air has a tendency to be 'freeze-dried' by the cold temperatures at the top of the troposphere. Methane escaping the troposphere intact and reaching the stratosphere would be oxidized and provide a source of water vapour which, in the presence of sunlight, is then broken down into small, highly reactive fragments that destroy ozone. A sufficiently massive methane release could deplete the ozone layer. How much methane it takes to trigger a collapse in the ozone layer is still under investigation.[92] If it lies towards the upper end of the estimated range, then it might be sufficient to thin the ozone layer, exacerbating ozone damage already caused by the Siberian Traps.[93]

The third element of the killing model, joining the Siberian eruptions and the methane burp, is stagnation of the world's oceans—a phenomenon deduced from careful observations of Permian rocks from before the extinctions and Triassic rocks afterwards.[94] Before the extinctions, the rocks that formed on what was then the seabed contained a huge diversity of fossil shells, skeletons, and burrows of bottom-dwelling invertebrates that fed and shuffled through the sediments. Afterwards, the deposited sediments were very different, being much darker in colour and largely devoid of animal life. Faunal changes, combined with geochemical analyses of the rocks, revealed a dramatic switch in ocean conditions, from waters that were well-oxygenated to those having virtually no oxygen at all. The switch is characteristically found in rocks from locations as far apart as China, Italy, and North America. Conditions of extreme oxygen deprivation—anoxia—are likely attributable to warmer oceans holding less dissolved oxygen and the onset of more sluggish circulation patterns.[95]

One consequence of widespread anoxic oceans might have been a build-up of hydrogen sulfide produced by bacteria, a build-up thought by some to have led to an abrupt discharge into the atmosphere of huge amounts of hydrogen sulfide. The resulting toxic atmosphere may be capable of destroying the ozone shield.[96] As far as it goes, the hypothesis is intriguing but also very preliminary. It is far from clear whether sufficient hydrogen sulfide could be produced by a catastrophic degassing from anoxic oceans to cause serious ozone depletion on a global scale.

As I hope will now be obvious, the detailed picture of catastrophic environmental degradation painted by palaeontologists to explain the great end-Permian mass extinction includes three factors that could all have contributed to damaging Earth's ozone shield. If we arrange these in a hierarchy of importance, based on what we know so far, then the Siberian Traps top the list. The other three candidates, the heating of coals, the massive release of methane and hydrogen sulfide are possibly suspect, even if they actually happened.

Accepting that there are grounds for believing in an ancient ozone hole we can next consider the important question of whether the mutations observed in the fossil record are really consistent with the effects of enhanced ultraviolet-B radiation. In this context, information from living plants informs the debate by shedding an illuminating light on the biological processes involved. We can envisage that at least two conditions are necessary, maybe even sufficient, for mutations to occur. First, there must be evidence that spore/pollen mutations are under genetic control, and second, that enhanced ultraviolet-B radiation decreases the stability of plant genomes. How do these requirements stack up against what we know?

As we saw in Chapter 2, molecular geneticists aim to elucidate and understand the genetic networks and regulators used and reused to build organisms. On the plant side of this grand endeavour, a particular favourite is the simple angiosperm thale cress (*Arabidopsis thaliana*), which belongs to the mustard family (Cruciferae or Brassicaceae). Everything about the unassuming thale cress is small, except the unusually large amount of attention lavished upon it by scientists. It has a small physical size, a few centimetres, and a small genome, containing an estimated 20 000 genes organized into five chromosomes.[97] Probing the genome of thale cress, a group of molecular geneticists at Stanford University discovered that just two genes are required for separating pollen grains during normal development.[98] When the function of these two genes is knocked out, pollen grains remain connected and are released as a characteristic clump of four grains stuck together, a pattern reminiscent of those found in Permian sediments. Further analysis hinting at the function of the two genes has also revealed why. Lesions in either gene lead to defects in the degradation of the sticky polymer, pectin, which binds the pollen mother cell to the daughter pollen grains, thereby preventing separation.[99] The gene function of the two genes seems to be in degrading or modifying construction of the cell wall of the pollen mother cell, this suggests that abnormal pollen production is under genetic control.

The second requirement, that plant genomes are unstable when exposed to harmful ultraviolet-B radiation, is well established by a wealth of experimental and observational evidence. Indeed, only a few years ago scientists studying vegetation beneath the ozone hole over Tierra del Fuego, deep within the Fuego National Park in southern South America, showed that local populations of a native herbaceous species (*Gunnera magellanica*) are already experiencing DNA damage as a consequence of increasingly severe doses of ultraviolet-B radiation.[100] This is probably the tip of the mutational iceberg compared to that expected from the intense ultraviolet-B radiation resulting from the worldwide destruction of the ozone layer.[101] In experiments with thale cress, exposure to high levels of ultraviolet-B radiation produced mutations in the DNA that worsened with each subsequent generation. This suggests that the plants underwent heritable and cumulative changes in the expression of genes involved in DNA repair. If the ozone layer was destroyed for a few hundred thousand years by the Siberian eruptions, and the results of these sorts of experiments are anything to go by, the prolonged harsh ultraviolet environment might have had devastating mutational consequences (Fig. 6).

Evidence supporting the idea that ultra-high doses of radiation induce abnormalities in pollen has also come from ecological monitoring after the Chernobyl nuclear power plant disaster. The disastrous meltdown of one of the four nuclear reactors at Chernobyl took place on 26 April 1986, a consequence of human error in temporarily shutting down the cooling system. The error led to an explosion and nuclear fire lasting for 10 days and saw the evacuation of 135 000 people from the surrounding region.[102] Large amounts of radioactive caesium, strontium, plutonium, and other radioactive isotopes were released into the atmosphere and rained out across an area of some 200 000 square kilometres. The local area, within a 30-km radius, received the highest doses. Ecologists monitoring the effects on the surviving plants and animals in the decades that followed reported a number of striking effects,[103] especially the sudden appearance of albinism in the barn swallow. Partial albinism is caused by mutations that are usually rare, but in barn swallows in the Chernobyl region the frequency increased nearly tenfold compared to elsewhere.[104] Of particular interest, though, was the finding that pine trees had begun producing abundant pollen with abnormalities.[105] The abnormalities appeared in the form of distinctive multiple air sacs—a mutation reminiscent of the fossils dating to the environmental catastrophe that unfolded at the end of the Permian.

We have seen that exposing living plants to severe doses of radiation can induce mutations analogous to those observed in fossil spores and pollen. The reasons why are quite well understood, even down to the molecular level of individual genes, and they strengthen the case for linking the fossil mutant spores to a hypothesized end-Permian ozone catastrophe. Nevertheless an even more satisfying connection might be established if empirical data directly reflecting past ozone levels could be brought to bear on the matter. This seemingly unlikely possibility may now be a step closer thanks to recent work showing that plants develop more protective pigments when exposed to increased ultraviolet-B radiation.[106] The pigments, in the forms of complex organic molecules, are generally chemically stable and found in the leaves, spores, and pollen of land plants; they have even been discovered surviving in fossil plant materials.[107] These exciting findings suggest fossil plants are an untapped archive of Earth's past ultraviolet radiation environment.[108] A small step towards that goal was taken when researchers showed that investment in protective screening pigments in the spores of clubmosses on South Georgia today has increased by 40% compared to 40 years ago—the rise corresponds to a 14% reduction in ozone levels.[109] Clubmosses are living cousins of the herbaceous lycopsid pioneers that profited at the expense of the conifer forests as the Permian drew to a close, so the finding raises the tantalizing possibility that the mutant spores themselves archive past ozone levels. Reliably retrieving this valuable information will be technically demanding, requiring the application of cutting-edge analytical techniques from across the Earth sciences.

We have travelled on a remarkable scientific journey, one that is far from over. What began with the pioneering achievements of nineteenth- and twentieth-century atmospheric chemists and physicists led to the astonishing finds and sensationalist claims of twenty-first-century palaeontologists. The journey has taken us to a scientific frontier, and like all such frontiers, it is littered with contention. The opening quote of this chapter is Niels Bohr's (1885–1962) attributed response to Wolfgang Pauli (1900–58), a titan in the golden age of physics early in the twentieth century. Bohr was commenting after Pauli had given a lecture claiming to have solved all the unsolved problems in elementary particle theory. Bohr believed Pauli's proposed solution was indeed crazy, but in his view it was not nearly 'crazy enough' to stand a chance of being correct. The revered mathematical physicist Freeman Dyson, also present at the meeting, added more diplomatically,

> when a great innovation appears, it will almost certainly be in a muddled, incomplete and confusing form. To the discoverer himself, it will only be half understood; to everyone else, it will be a mystery. For any speculation which does not at first glance look crazy, there is no hope![110]

The parallels here with the gauntlet thrown down by the palae-ontologists are obvious. Is the exciting idea that the ozone layer collapsed 251 million years ago really credible? Visscher's kindly demeanour belies a keen intellect and he has a reputation for meticulous attention to the scientific details of his arguments, so his team's ideas deserved serious consideration. According to the latest simulations with models of the chemistry of Earth's atmosphere, the eruption of the Siberian Traps appears to be the strongest contender for bringing about widespread damage to the ozone layer. If further investigations support this claim, we have to applaud the team's insights into appreciating the signifi-cance of mutated spores and pollen that suddenly appear in the end-Permian fossil record. New insights are sure to lie ahead. Ultimately, corroboration between theory and observations would see a prolonged bout of fierce ultraviolet radiation added to the other agents of environmental degradation surrounding the great end-Permian mass extinction.

5

Global warming ushers in the dinosaur era

Terrestrial plants survived the five great mass extinctions that have befallen marine life over the past half billion years, and their fossil remains hold precious clues for unlocking the hidden environmental mysteries surrounding the extinctions. Those from the rocks of Greenland's eastern coastline illuminated the debate about the causes of the mass extinction at the boundary between the Triassic and Jurassic, 200 million years ago. They revealed that the extinction had been accompanied by soaring carbon dioxide levels and a switch to 'super-greenhouse' conditions. The warming put paid to their primitive reptilian competitors, and allowed the dinosaurs' spectacular rise to ecological success. The primary cause of the greenhouse warming appears to have been the build-up of carbon dioxide in the atmosphere, propelling the Earth inexorably towards a warmer state. Humankind's activities are starting to replicate this scenario. We would be wise to recognize such events as a wakeup call for anticipatory action on global warming.

Thus life on Earth has often been disturbed by terrible events: calamities that perhaps shook the entire crust of the earth to a great depth.

Georges Cuvier (1812), as translated in *Georges Cuvier: fossil bones and geological catastrophes*

R EVEREND William Buckland (1784–1856), a British vicar and palaeontologist, was the first Professor of Geology at the University of Oxford (1813) (see Plate 8). Charming and eloquent, Buckland was also an accomplished lecturer. His biographer summed him up rather well, remarking in 1894 'it is impossible to convey to the mind of any one who had never heard Dr. Buckland speak, the inimitable effect of that union of the most playful fancy with the most profound reflections which so eminently characterized his scientific oratory'.[1] Brilliant and famously eccentric, he once offended stuffier colleagues at a British Association meeting in Bristol by strutting around the lecture theatre imitating chickens to demonstrate how prehistoric birds could have left footprints in the mud. On another occasion he:

> attracted an audience totalling several thousand for a lecture in the famous Dudley Caverns, specially illuminated for the purpose. Carried away by the general magnificence, he was tempted into rounding off with a shameless appeal to the audience's patriotism. The great mineral wealth lying around on every hand, he proclaimed, was no mere accident of nature; it showed rather, the express intention of Providence that the inhabitants of Britain should become, by this gift, the richest and most powerful nation on Earth. And with these words, the great crowd, with Buckland at its head, returned towards the light of day thundering out, with one accord, 'God save the Queen!'[2]

Buckland also claimed to have eaten his way straight through the animal kingdom as he studied it and, allegedly, part of Louis XIV's embalmed heart, pinched from the snuffbox of his friend the Archbishop of Canterbury. He was aided in the eccentric culinary consumption of animals by his son Francis Buckland (1826–80), the celebrated Victorian naturalist and one-time Inspector of Her Majesty's Salmon Fisheries, who evidently inherited his father's eccentricity. Francis Buckland lived amongst beer-swilling monkeys, rats, and hares and regarded firing benzene at cockroaches through syringes as a fine sport. Francis arranged with London Zoo to receive off-cuts from the carcasses of unfortunate animals. This arrangement saw sliced head of porpoise occasionally arrive at the table of the Buckland household along with the more mundane delicacies such as mice en croûte and roast mole.

Lyme Regis in Dorset, southwest England, renowned for its Cobb Harbour and inspiring John Fowles' *The French lieutenant's woman*, held a special place in the heart of William Buckland. It lies at the point where the Triassic rocks disappear below the sea and the younger Jurassic rocks, thick clays and thin limestone of the blue 'Lias', lie above, forming the cliffs. Deposited in a moderately deep tropical sea that teemed with marine life, the cliffs contain abundant fossils. Some, such as pencil-shaped belemnites and coiled ammonites, are easy to find but others, like ichthyosaurs (marine reptiles) and fish are far rarer. The rocks also contain evidence of life on the land in the form of fossil wood, insects, and dinosaurs washed into the sea 200 million years ago. Buckland often visited the cliffs there during his childhood, having been born six miles away in Axminster, and later wrote, 'They were my geological school...they stared me in the face, they wooed me and caressed me, saying at every turn, Pray, Pray, be a geologist!'[3] The cliffs yielded many magnificent dinosaur specimens to Mary Anning (1799–1847), the remarkable nineteenth-century fossil hunter. To Buckland, they also offered strange deposits he identified as fossil droppings (coprolites) of extinct giant saurians; his pioneering work opened up an entirely new field of study—identifying animals' diets—and speaks to his eccentricity. Buckland's connection with the locality can still be found today in the Philpot Museum, which lies in the town behind Cobb Harbour. In the museum sits his desk, inlayed with sectioned, polished coprolites (see Plate 9). The coprolite table was, Francis remembered, 'often admired by persons who had not the least idea what they were looking at!'

For all this, Buckland is perhaps most well known for inaugurating the scientific investigation of dinosaurs. In around 1818, he was asked to identify the fossil bones of a giant carnivorous reptile discovered in Jurassic rocks. After six years of deliberation on the matter, and spurred on by the possibility of younger rivals claiming priority, he named the animal *Megalosaurus* or 'great lizard' on account of its huge size. Unfortunately, *Megalosaurus* turned out to be not a lizard at all, but a large meat-eating dinosaur some 12 m long. Nevertheless, Buckland still goes down in history as the first individual to give a scientifically valid name to a species of dinosaur, because the name *Megalosaurus* is still used for the dinosaur represented by his specimens.

Although Buckland is credited as naming the first dinosaur, it was the brilliant Victorian anatomist and palaeontologist Richard Owen (1804–92) who in 1842 gave us the word dinosaur. With a flash of insight, Owen realized whilst studying fossil specimens of *Megalosaurus* and *Iguanadon*, collected largely by his rival Gideon Mantell (1790–1852), that he was looking at a distinct group of animals. These vertebrates had scaly skin, sometimes with armour plates, laid eggs, and had femur (thigh) bones that curved inwards at the top to fit into the hip socket like mammals. This latter adaptation, Owen reasoned, allowed them to walk with their legs held upright beneath them rather than being splayed out like those of crocodiles. To distinguish them from all other creatures he coined the term 'Dinosauria', from the Greek *deinos* meaning 'terrible' or 'fearfully great' and *sauros*, meaning 'lizard'. Owen, as unpleasant and fearsome as the beasts he described, consistently attempted to discredit his rivals and reap the scientific rewards of their painstaking collecting efforts. He died in 1892, still clinging to his anti-Darwinian views, his reputation in tatters, torn apart by Thomas Huxley (1825–95) and other enemies made throughout the hostilities marking much of his career.

The extraordinary discoveries of *Megalosaurus* bones, and later those of other giant prehistoric creatures, captured the imagination of Victorian England and fuelled the heroic age of British palaeontology. Charles Dickens felt compelled to work the theme into *Bleak House*: 'Implacable November weather. As much mud in the streets as if the water had but newly retired from the face of the earth, and would not it be wonderful to meet a *Megalosaurus*, forty feet long or so waddling like an elephant lizard up Holborn Hill.' The opportunity to showcase the exciting new discoveries of the pioneering fossil hunters came in 1854 after Joseph Paxton's (1803–65) Crystal Palace building—the largest iron-framed glass building ever constructed—was relocated from Hyde Park to the suburbs of South London. Relocation saw 200 acres of landscaped parkland set aside for geological exhibits. Hugely popular, the exhibition included several full-sized dinosaur reconstructions displayed on artificial islands, including *Megalosaurus* and *Iguanodon*, built by the sculptor Waterhouse Hawkins (1807–89). Guided by Owen, who used known parts of the skeletons to estimate the sizes and possible overall shapes of the animals, they became wrongly portrayed, with several being given the appearance of clumsy rhinoceros-like beasts walking around on four legs. Hawkins' prehistoric sculptures can still be found in Sydenham Park, London. Battered by the elements and stranded in time, they are a testimony to the Victorians' mistaken concept of the dinosaurs. It was a noble attempt to breathe life into the fossil bones of long extinct animals they believed had been destroyed by a biblical deluge.

Since those early days, our understanding of dinosaur natural history has been transformed by fossils recovered by ongoing palaeontological explorations worldwide. Stunning finds have revealed details about the structure of their eggs, embryos, and hatchlings, as well as clues to posture, locomotion, and herding. Indeed, palaeontologists have probably learned more in the last two decades than in the last two centuries. We now know that the dinosaurs, and indeed half of life on Earth, famously went extinct at the boundary between the Cretaceous and the Tertiary, 65 million years ago.[4] Details of the extinction come from sites in the western interior of North America that contain especially rich deposits of Cretaceous dinosaur fossils. These have yielded some of the remarkable skeletons of *Tyrannosaurus* that adorn museums worldwide. Acrimonious debate breaks out even today over the question of whether the dinosaurs were wiped out in a geological instant or suffered a drawn-out demise. The fossil evidence is notoriously capricious—dinosaur fossils are, relatively speaking, rare even in the western interior of North America—allowing one group to argue forcibly for a gradual pattern of die-off and the other for a sudden extinction.[5] The current situation divides the community into a minority of 'gradualists' and a majority of 'suddenists'.[6]

The big question of what caused the dinosaur extinction still causes sparks to fly and is fiercely contested; the issue is not easily divorced from the divisive arguments of the 'gradualist'

and 'suddenist' camps.[7] Luis Alvarez (1911–88) and colleagues at
the University of California, Berkeley sparked the controversy
into life by boldly hypothesizing that a large meteorite impact
wiped out the dinosaurs.[8] The evidence for this was their
remarkable discovery of a 60-fold jump in the concentration
of the rare element iridium. The large rise in iridium is suggestive
of an impact by an asteroid or meteorite because these objects
can deliver the element to Earth in high concentrations. It has
subsequently been discovered at over a hundred sites worldwide
and has a chemical signature confirming its extraterrestrial origins. Alvarez's original apocalyptic vision saw vast quantities of
dust and debris lofted into the atmosphere by the impact,
blocking out incoming sunlight to cause a deep freeze and
prevent photosynthesis. Under these conditions of a 'nuclear
winter', the food chain collapsed, condemning the dinosaurs
and other animals to extinction by starvation. Raging wildfires
ignited by the re-entry of ejected material, heated to incandes-
cence as it entered the atmosphere, and acid rain, further add to
the vision of mayhem and destruction caused by an asteroid
impact.[9] Although initially greeted with fierce criticism, dramatic confirmation of an impact came a decade later with the
discovery of a giant crater of the right age buried beneath rocks
off the Yucatán Peninsula in the Gulf of Mexico.[10] Measuring
180 km across, the crater is thought to have resulted from a large
extraterrestrial object some 10 km in diameter—as wide as Mt
Everest is tall—slamming into the Earth. Even though aspects of
how an asteroid impact could actually cause a mass extinction
are undergoing reconsideration,[11] the totality of the evidence
begins to stack up; contingent events happen and affect the
evolution of life on Earth.

The iconic mass extinction of the dinosaurs at the Cretaceous–
Tertiary boundary constitutes only the most recent of several

great extinctions experienced in the history of complex life on Earth over the last half billion years.[12] The British geologist John Phillips (1800–74) can be said to have launched the study of mass extinctions in 1860 with the publication of a landmark book entitled *Life on earth: its origin and extinction*. In the book, Phillips traces the ups and downs in the diversity of marine life from fossil sediments in Great Britain. An innovative and careful aspect of Phillips' pioneering contribution was his attempt to correct for the number of fossils in a given thickness of sediment (per hundred feet) that represented a geological stratum. This volumetric correction essentially gave him a crude expression of diversity per unit time. It is an endeavour greatly occupying palaeontologists today, who recognize that because the fossil record doesn't give us the real history of life but a distorted version of it, we need to account for the known biases.[13] The best way to take account of these, as Phillips realized, is to correct for the varying amount of exposed fossil-bearing rocks found in different geological time intervals. The reward for his attention to detail was the identification of the great mass extinctions at the end of the Permian and the end of the Cretaceous.[14] Although, in fact, it is thought that Phillips was more concerned with demonstrating to fellow geologists the need to divide geological time up into three major portions—eras—by showing each had a distinctive assemblage of fossils that applied worldwide. Nevertheless, whether he realized it or not, his research revealed two very real mass extinction events.

By the twentieth century, palaeontologists had assembled modern databases cataloguing the estimated times of the first and last appearances of different organisms in the fossil record to reveal the detailed history of evolution and extinction of life on Earth. Marine shelly organisms, readily preserved in the fossil record, are traditionally the focus of greatest attention, and surveys of changes in their diversity identified five mass extinction events (the so-called 'big five') that stood out from the normal background level of extinction.[15] Even today, the information incorporated into such databases is known to be imperfect, not least because only fossils of a small fraction of all animals that once lived are ever unearthed. Still, 10 years' worth of corrections and additions to the leading database of marine animal diversity failed to change the original patterns displayed, charting the ups and downs of marine biodiversity.[16] It prompted the creator of the database Jack Sepkoski (1948–99), at the University of Chicago, to sagely observe that 'real events in the history of life, involving major radiations and mass extinctions, shine through even faulty data'. Encouraged by the robust nature of the patterns, in spite of the imperfections, scientists embarked on the grand challenge of cataloguing changes in the diversity of the fossil record of all life, including microbes, algae, fungi, plants, and animals. The catalogue revealed that the persistent ongoing rise in global biodiversity over the last half billion years had been interrupted by the same big five mass extinctions originally identified in marine animals.[17]

Of the big five mass extinctions, perhaps the most enigmatic and least studied is the one that dates to the boundary between the Triassic and Jurassic periods, 200 million years ago.[18] According to some, the end-Triassic mass extinction saw off a fifth of all families of marine animals.[19] Clams, bivalves, and gastropods that once swarmed the primitive oceans were decimated, and coral reef-building animals disappeared worldwide. Giant marine reptiles that had dominated the seas for millions of years were eliminated and with them went the conodonts, a ghostly group of primitive eel-like fishes already in decline well before the end-Triassic. On land, animals simultaneously suffered a crisis comparable to that of their marine cousins, with a quarter of all families going extinct. Indeed, the mysterious end-Triassic event killed off a greater proportion of creatures than its Cretaceous–Tertiary boundary counterpart.[20] If we rank the big five mass extinctions by looking at the severity of the ecological impacts—such as, for instance, estimating how quickly ecosystems recover—then the end-Triassic event weighs in as the third worst event of all time.[21]

There is compelling evidence, then, of a major crisis for Earth's biodiversity at the transition from the Triassic to the Jurassic. But not everyone agrees: controversy hangs over the timing and magnitude of the extinctions. Is it possible the extinction event identified in the global compilations is simply a statistical artefact? Dissenters point out that the databases are unable to detect gradual declines in animal diversity leading up to the end of the Triassic that are revealed by a more detailed sift through the evidence from the handful of well-recognized Triassic–Jurassic boundary sites around the world.[22] If this is the case, then claims for a 'sudden' mass extinction at the close of the Triassic may not be warranted. Other palaeontologists, conducting new analyses of Sepkoski's listing of fossil marine genera, raised doubts about the legitimacy of including the end-Triassic extinction among the big five, and propose that it should be downgraded to a 'mass depletion' event.[23] The debate rumbles on, and lest this seem like a fuss about semantics, we should recognize that present losses in biodiversity, and our guide to likely rates of recovery, can only be gauged against past extinctions. These natural 'experiments' open a window onto the vulnerability of life to global environmental change and classifying the true extent is central to the endeavour. Future discoveries will yield an increasingly precise view of what actually happened 200 million years ago. For now, most parties agree that as the Triassic drew to a close, biodiversity declined in a stepwise fashion, culminating in a mass depletion at the boundary.

Whatever the magnitude of the end-Triassic extinction, the event holds special significance for the dinosaurs recognized by Victorian dinosaur hunters like Buckland, Mantell, and others. For not only did the dinosaurs survive, to emerge unscathed as many marine animals and primitive reptilian-competitors died away, but afterwards their fortunes also took a decided turn for the better. At the start of the Jurassic, dinosaur populations and diversity rose as a panoply of predators and prey filled the landscape, establishing their supremacy by invading freshwater marshes, forests, and estuarine creeks. Big predatory meat-eating dinosaurs also began to take centre stage as they came into their own. It seems that the elimination of their competitors fuelled the diversification of the dinosaurs by freeing up resources to give them ecological elbow-room to fill out the terrestrial landscape. Early in their evolutionary history, fate favoured the dinosaurs by events as accidental and opportunistic as those that saw their ultimate demise and replacement by mammals.[24]

The enigmatic nature of the end-Triassic extinctions stems, in part, from uncertainties over the causes. Impacts by meteorites, devastating volcanic eruptions, and global climate change are all postulated as possible agents of mass destruction. Frustratingly fragmentary, the fossil evidence by itself limits our capacity to distinguish between the different ideas and better identify the likely perpetrators. But by marshalling evidence from a wide variety of fields, we can begin to make progress in clarifying what went on and at least bring the picture into sharper focus. Arguably the most plausible scenario may also be the most worrying. It takes us on a journey across the vast supercontinent of Pangaea and down through the Earth's crust, forcing us along the way to recognize a dormant agent of catastrophic climate change that current trends in global warming once again threaten to disturb.

🌿

Pangaea, from the Greek for 'universal world', was the largest landmass in Earth's history; an undivided giant continent for the evolution of gigantic dinosaurs. Assembled during the Permian, 250 million years ago, Pangaea stretched from pole to pole with a vast sprawling inhospitable desert bisected by the equator, while being surrounded by a great global ocean known as Panthalassa (Fig. 7). Lacking the thermal inertia of the oceans,

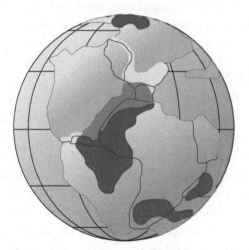

Fig. 7 The ancient supercontinent Pangaea, which stretched from pole to pole and was made up of all the major continents sutured together. The shaded areas in the north and south represent lava extruded during the eruptions of the Siberian Traps (251 million years ago) and the Deccan Traps (65 million years ago), respectively. The shaded area in the centre shows the greater extent of the lava forming the Central Atlantic Magmatic Province, which dates to the Triassic–Jurassic boundary, 200 million years ago.

the unusual landmass experienced intense continental heating and cooling which generated an extremely seasonal climate and monsoonal circulation weather patterns. Average summer temperatures[25] over large regions in the interior exceeded 40 °C, with daily highs over 45 °C. Deep wind-blown sand dunes buried by the annual arrival of extreme summer monsoonal rains accumulated on the low-latitude desert of Pangaea's western margin. Over the lifetime of Pangaea, sands reached a depth of two and a half kilometres, the thickest dune deposits in the world. The remains of these ancient dunes form the spectacular Navajo sandstones in the canyons and cliffs of Utah, southwestern North America.[26] There is little doubt that life on Pangaea endured extremes, finely balanced on the margin between life and death.

Quite how the hot arid climate changed at the end of the Triassic, when global biodiversity plummeted, is unclear. Tectonic unrest in the intervening 200 million years has destroyed forever much of the critical portion of Earth history laid down on the seafloor. That part of the rock record which survives is found at only a handful of far-flung localities in western North America, Peru, Australia, and Western Europe. Whether it retains a faithful archive of ancient climatic change after being eroded over millions of years is open to further investigation. With the marine archive of this critical sliver of Earth history mostly 'missing in action', we are forced instead to recover it from the fossil record of the plants because the plants escaped extinction while around them animals on land and in the sea perished. The survival of land plants is not unexpected. Plant life is generally more resistant to mass extinctions than animals because the continuity of future generations in plants is generally assured by them shedding seeds or else producing underground corms and rhizomes from which the shoots of recovery can sprout when the time is right.[27] This is not to say they are immune to extinctions, far from it, but it was the tenacity of plant life to cling on in the face of environmental adversity 200 million years ago that bequeathed palaeontologists a rich legacy of fossil remains during the end-Triassic. The challenge is to decode the climatic and atmospheric information embedded within the matrix of the fossils and piece together the ancient puzzle of the changes in Earth's global climate.

Greenland offers us one of the richest legacies of fossil plants straddling the Triassic and Jurassic. The largest 'island' in the world, it is broader in the north than the south,[28] and the whole interior of the mountainous plateau is permanently buried under a shield-shaped ice-sheet over 3 km thick in its central portion that forms an unbroken expanse for hundreds of kilo-metres. The ice-sheet finds outlets to the sea through deep fjords, discharging flotillas of icebergs that, when released on the west-ern side, can find their way into the mouth of the St Lawrence River. In the warm summer months, it is possible to gain access to coastal exposures on the eastern margins where rocks and fossils can be excavated. Indeed, the layers of fossil plants are so spectacularly thick that they form coal seams which once, in 1823, misled the English arctic explorer William Scoresby (1789–1857) and the Scottish mineralogist Robert Jameson (1774–1854) into mistaking them as being from the Carboniferous Period, a time some hundred million years earlier. The determined efforts of the distinguished palaeobotanist Thomas Harris (1903–83) corrected these problems of age in 1926–27 and went on to establish them as one the best-studied Arctic fossil floras in the world.[29]

Harris's investigation of fossil plants from Greenland rocks later proved pivotal to unravelling environmental changes at the end of the Triassic, and yet his interests in them came about quite by chance while a student at Cambridge University in 1925. Twenty-five years earlier, the Geological Survey in Copenhagen accidentally sent Triassic–Jurassic rocks from Greenland, instead of Cretaceous materials, to the Cambridge palaeobotanist Albert Seward (1863–1941). On learning of this, it was suggested that Seward hold on to them until he found the 'right man' to work on them. When Harris came along he was in Seward's eyes evidently the right man and assigned the task. The haul of fossils contained in 15 packing cases had been collected from an earlier expedition to Scoresby Sound, East Greenland and provided a mere glimpse of the treasures to be had. Shortly afterwards, Harris left for Greenland at the surprise invitation of the director of the Greenland Geological Survey. He later commented 'he [Koch] asked me—could I come with him on an expedition to East Greenland for a year, starting next month. It appeared to me instantly that it was one of those situations where thought doesn't lead to a wiser decision, so I said "yes"!' The summer of 1925 saw Harris leave for the north-east coast of Greenland to collect extensively from fossil-bearing beds in Scoresby Sound.[30] The expedition yielded tons of rock and fossils. Under his close attentions, the Scoresby Sound fossils soon offered up their secrets and he published the results between 1931 and 1937 in his monumental five-volume work *The fossil flora of Scoresby Sound, East Greenland*. Thore Halle (1884–1964), one of the foremost Arctic palaeobotanists of the day, later wrote of it, 'I have no hesitation in stating...the whole of the palaeobotanical literature contains no account of a Mesozoic flora which can compare with the excellent treatment of the Rhaeto-Liassic flora of East Greenland.'[31]

Some 70 years later, the surprise announcement of the discovery that trees produced leaves with fewer pores (known as stomata) as levels of carbon dioxide increase (see Chapter 2)

attached special significance to the collections of Harris's Greenland fossils.[32] The relationship provided palaeontologists with a
simple, elegant means of breathalyzing the carbon dioxide content of Earth's ancient atmosphere, and this crucial information
was codified into the fabric of Harris's fossil leaves. Physiologists
have shown that plants respond in this manner to conserve
precious water supplies in their constant battle to avoid desic-
cation. For as stomatal pores open to take up carbon dioxide,
water from the humid interior of the leaf streams out into the
drier atmosphere. Reducing the number of pores as carbon dioxide levels increased allows photosynthesis to proceed unhindered
whilst restricting accompanying water losses. Stomatal pores
govern the economics of photosynthesis and carbon dioxide
is the currency.

When re-examined in the light of this new discovery, the
forgotten haul of fossil leaves from Greenland's eastern coastline
exposures, which had languished for decades in the basement
of the Danish Geological Museum, Copenhagen, contributed
dramatic evidence of climate change. For the fossils indicated
for the first time that atmospheric carbon dioxide levels had
soared across the Triassic–Jurassic boundary.[33] Fossil leaves
from Swedish rocks backed up the remarkable conclusion. The
carbon dioxide content of the ancient atmosphere tripled within
a few hundred thousand years before dropping back down again
(Fig. 8). Because of the greenhouse effect, a burst of global
warming would be expected to accompany the rising carbon
dioxide levels, raising global temperatures by up to 8 °C (Fig. 8),
some 10 times the amount of warming we have experienced in
the past century.[34] But a legitimate concern with this claim is
that in the absence of suitable marine sediments that normally
chronicle the vicissitudes of ancient climates, it hasn't been
verified against independent evidence.

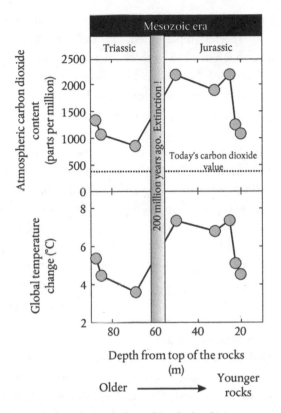

Fig. 8 Dramatic changes in carbon dioxide levels (upper panel) and global temperatures (lower panel) across the Triassic–Jurassic boundary as recorded by fossil leaves from Greenland. Mean changes in global temperature are calculated from the carbon dioxide-induced 'greenhouse effect'. Note that overlying rocks get progressively younger.

In the search for evidence for greenhouse warming, Greenland fossils once again remained centre stage, only this time it was not the pores in the surface of the leaves that held the climatic clues but the shape of the leaves themselves. In his original investigations, Harris noted a striking and distinctive pattern of species replacement in different beds moving upwards through the 90-m-thick rock sequence.[35] Fossils in the lower 30 m of Triassic rocks indicated forests dominated by species with large entire leaves, including relatives of the maidenhair tree (*Ginkgo*). Those in the top 60 m of Jurassic rock revealed forests composed of species with small or more finely dissected leaves. Harris later commented 'when one put together the floras of a lot of beds it became obvious that there were two floras, one in the top 60 m and the other in the lower 30 m. There was real mixing of the two floras over about 5 m at the boundary but no more'.[36] The rapidity of species replacement is probably due to a slowing in the deposition of material, telescoping a great interval of time into a thin series of rocks.

Still, although the 'sudden' switch in the composition of Greenland floras is startling and quite real, does it square with what we might expect from the sudden arrival of a hot, carbon dioxide-rich climate? A team of researchers from the University of Sheffield pursued the logical consequences of their own observations on fossil-leaf anatomy to decide if such an interpretation was warranted. Piecing the evidence together, they reasoned that large-leaved tropical trees stayed cool late in the Triassic as high numbers of pores on the surface of the leaves allowed plants to release more water and cool off, in much the same manner as sweating cools mammals. The transition to a carbon dioxide-rich atmosphere changed all this. It caused a decline in the number of stomatal pores that left leaves unable to shed heat absorbed from sunlight by 'sweating', and vulnerable to lethal overheating. As the climate got even hotter, it exacerbated the problem until in the grip of an extreme end-Triassic heatwave, only trees with small or dissected leaves escaped the heat to stay cool.[37] Not only did they intercept less sunlight, making them less prone to overheating, but their narrow leaves also created more turbulent air, allowing them to shed heat more effectively. Viewing the switch from large- to small-leaved forests seen in the Greenland rocks from this perspective reveals it to be an entirely reasonable ecological response to extreme global warming. Evidently, with more carbon dioxide, the hot Triassic climate got hotter still. Large-leaved trees went into decline as things simply got too hot for photosynthesis, and those with small leaves, able to shrug off the heat, took over—natural successors in the super-hot world.

As is so often the case, the unfolding narrative painstakingly pieced together from the fossil record inevitably raised as many questions as it answered. What could credibly deliver the enormous volume of carbon dioxide required to drive up the atmospheric content in this way? Coincidentally, in the same year that evidence for rising carbon dioxide levels came to light, a team of geoscientists at the University of California reported the exciting discovery that the remnants of a huge volcanic eruption dated to the mass extinction event, 200 million year ago.[38] This particular one was the remains of a large igneous province (LIP) called the Central Atlantic Magmatic Province that turned out to be bigger and badder than anyone had previously suspected.[39] Dismembered lava features found in several countries formed a pattern that radiated outwards from the central Atlantic, given the position of the continents 200 million years ago. Much of the lava is deeply buried or eroded, making it difficult to estimate the original volume of basalt extruded. It is likely, though, that some 2–4 million cubic kilometres poured out of the Earth's interior within a few hundred thousand years.[40] The lavas are preserved today in North and South America, and East Africa, and cover an area of 7 million square kilometres—over three times the size of Greenland. Remnants of it can still be seen along the shores of the Hudson River when you cross the George Washington Bridge on the way out of Manhattan.

The eruption and formation of the Central Atlantic Magmatic Province in a brief moment of geological time requires further explanation. Lava flows out of the rifts that dissect the seafloor worldwide continually, but this 'background' volcanism is quiet and routine and cannot account for the swift outbursts of eruptive activity that created the extensive lava fields of the Central Atlantic Magmatic Province.[41] A new explanation was needed. In the late 1960s, several scientists, notably the Canadian geophysicist Tuzo Wilson (1908–93) at the University of Toronto, followed by Jason Morgan at Princeton University, proposed a radical hypothesis that fitted the bill.[42] Wilson and Morgan suggested that the Earth's mantle circulated in two modes. The mantle is a zone of ductile material 3000 km thick which sits above the hotter inner core and beneath the thin outer crust. In the more usual mode, large-scale convection generated within the mantle due to heating from the hot core nudges the plates floating on the surface, causing them to move. But in the other mode, narrow, deep-rooted plumes of hot material rise through the mantle. The plume heads act like blow-torches, burning their way through the overlying plates to form long-lasting regions of volcanic activity, so-called 'hot spots'. As the plate moves, so the hot spot produces a trail of volcanic islands. A hot spot located beneath the Pacific Ocean is responsible for creating the Hawaiian archipelago. The north-west orientation of the island chain reveals the passage of the ocean plate over a deep-seated hot spot that has persisted for some 70 million years.

The Central Atlantic Magmatic Province was likely created by volcanic eruptions produced by an abnormally hot mantle plume rising up beneath the continental plates of Pangaea. The plume would have produced a large mushroom-shaped head beneath the overlying continental plate hundreds of kilometres across, sustained by its slender tail delivering hotter molten materials from the Earth's mantle. It formed beneath continental crust already stretched and thinned as North America started to rift apart from South America and Africa to give birth to the Atlantic Ocean.[43] The coincident formation of a hot mantle plume during the slow break-up of Pangaea pushed the crust upwards beneath the plume head, creating a huge, bulging convex dome hundreds of kilometres across. As the hot mantle material rose up from deep within the Earth it underwent decompression, much like old-fashioned diving bells on being brought up to the surface from depth. But in the mantle, decompression generated heat that eventually 'burned' through the overlying continental crust. Massive outpourings of lava poured through fissures in the mantle, initially streaming down the slopes of the uplifted region to spread out over parts of the interlocking continents that were to form North America, West Africa, and South America. Later, as the region collapsed and subsided, lava would have flowed in more spectacular explosive style.

These eruptions heralded the disintegration of Pangaea and simultaneously belched out lava and huge amounts of carbon dioxide, sulfur dioxide, and other gases, altering the chemistry of the atmosphere and the oceans. We can make some attempt to estimate the volume of carbon dioxide released by analogy with emissions from the Kilauea volcano, Hawaii, which releases about 10 million tonnes of carbon dioxide for every cubic kilometre of basalt produced. This suggests that as roughly 4 million cubic kilometres of basalt flowed onto the Earth's surface from the rift created as Pangaea foundered, it discharged some 40 billion tonnes of carbon dioxide, 10 times the amount contained in our fossil fuel reserves. An equivalent amount of sulfur dioxide could also have been emitted, which would have rapidly been converted to sulfuric acid and removed from the atmosphere as acid rain.[44] Yet more carbon dioxide could also have been disgorged into the atmosphere and oceans as the hot plume of magma moved up through the crust, melting the carbonate-rich layers as it went.[45]

The revelatory discovery of the true extent and age of the Central Atlantic Magmatic Province suggests it is the 'smoking gun' for the end-Triassic mass extinction, equivalent to the giant Chicxulub impact crater dating to the dinosaur extinctions at the Cretaceous–Tertiary boundary. Indeed, this tempting connection between massive volcanic eruptions and extinctions has been made before.[46] Of the big five mass extinctions, four apparently coincide with the eruption and formation of LIPs, although whether the link is one of cause and effect remains highly controversial.[47] Indeed, 7 of the last 10 supposedly 'major' mass extinctions are controversially associated with episodes of massive volcanism, apparently strengthening the link. Yet establishing a plausible mechanism for wiping out large numbers of species, especially in the oceans, by massive volcanic events is problematic.[48]

Nevertheless, we can now conceive of an apparently reasonable scenario for a climatic catastrophe at the end of the Triassic that connects the ecological stability of life on Earth with much of the geological evidence. Imagine a mantle plume rising up from deep within the Earth's interior to create a hot spot beneath a crust thinned by the slow disintegration of Pangaea. The volcanic eruptions pour forth huge volumes of molten rock onto the land and spew out enormous quantities of gases into the oceans and atmosphere. The release of sulfur dioxide perhaps created acid rain and a sulfate aerosol haze, briefly cooling the planet before the rise in carbon dioxide strengthened the greenhouse effect and produced global warming. Indeed, some favour the beguiling idea of poisonous acidic rains as the prime candidate for dealing a major blow to life, leading to the collapse of ecosystems and extinction.[49]

But despite its elegant simplicity, the seemingly attractive hypothesis linking the end-Triassic extinctions solely to a massive bout of volcanism has to be rejected. It simply cannot be the whole story because it fails to explain away a stubborn detail that turned up in the fossil leaves from Greenland. Those leaves right at the boundary between the Triassic and Jurassic showed a curious abrupt increase in the abundance of the light, stable isotope of carbon, denoted ^{12}C. The same spike has subsequently been identified in sediments of the same age traced across the world from tidal rock pools in the Queen Charlotte Islands, Canada via the wooded hills of Hungary, to St Audrie's Bay on the Somerset coastline of south-west England.[50] Increasingly, it looks like a global phenomenon. Every time investigations are undertaken searching for it at some new locality with rocks spanning the Triassic–Jurassic boundary, the spike turns up. A universal isotope anomaly of this kind tells us that the physical processes cycling carbon between the atmosphere, ocean, and rocks underwent a dramatic upheaval during the extinctions.

Only a massive change in the isotopic composition of atmospheric carbon dioxide absorbed by terrestrial and marine plants to synthesize biomass could have caused the simultaneous appearance of an isotopic spike in rocks around the world. What could actually accomplish such a change? Those aligned to the 'massive volcanism' point of view believe that the huge amounts of carbon dioxide delivered to the atmosphere during the eruptions added sufficient quantities of ^{12}C to do the trick. Others think differently, arguing that an additional carbon dioxide source had to be involved in some way.[51] One way to put the volcanism idea to the test is by formulating a mathematical model describing how carbon dioxide injected by eruptions is cycled between the atmosphere, oceans, and rocks in the greenhouse world at the end of the Triassic.[52] The upshot of this mathematical modelling is that we find volcanoes have a minor influence on the isotopic composition of atmospheric carbon dioxide and could not have been responsible for the global isotopic spike. The reason is straightforward: volcanoes emit carbon dioxide with an isotopic abundance similar to that already in the atmosphere; even injecting unrealistically large quantities fails to exert much of an effect. Volcanoes, then, account for less than a third of the observed spike, at best.

But the models also reveal another equally serious snag with the idea of volcanoes as agents of dramatic global warming during the close of the Triassic. They release carbon dioxide relatively slowly over hundreds of thousands of years. This means Earth's normal feedback processes can cope. So, as the carbon dioxide content of the atmosphere rises, the greenhouse effect strengthens and warms the climate, which in turn accelerates carbon dioxide consumption by enhancing the chemical weathering of continental silicate rocks. Atmospheric carbon dioxide is effectively dumped into ocean sediments as carbonate. This is the familiar feedback loop regulating the Earth's climate we encountered in Chapter 2. It acts as a thermostat, helping to cool the Earth when the climate system warms and it means that the volcanic activity that created the Central Atlantic Magmatic Province by itself cannot push atmospheric carbon dioxide content up to the same high levels indicated by the fossil leaves. Obviously, if the fossil leaves are reliably recording past atmospheric conditions,[53] then the mismatch and the isotopic anomaly imply the involvement of a supplementary source of carbon dioxide. And one possible source produced by methanogens lies entombed in ice on the ocean floor.

Methanogens occupy one of the deepest branches in the tree of life—the Archaea—and by implication are some of the earliest life forms. They get their energy by making methane in the absence of oxygen, a very basic sort of metabolism. Hydrogen gas or the breakdown products of dead plants and animals provide the fuel. Methanogens can thrive at great depths below the seafloor, living hundreds of metres down in the sediments where they produce large volumes of methane gas, all of it rich in ^{12}C. Under the immense pressures and cold temperatures of the seafloor, methane is trapped and concentrated in cages of water molecules to form a solid crystalline compound called methane hydrate. Brought to the surface, methane hydrate looks like ice and evaporates as a white mist that burns if ignited (see Plate 10). Discovered primarily in waters around the margins of most continents, methane hydrates are especially abundant in the cooler polar regions and lock up anywhere from 5000 to 9000 billion tonnes of carbon—several times the amount of carbon stored in the world's forests and soils.[54]

Methane frozen into ocean-floor sediments as gas hydrates is a prime source of ^{12}C. When released it is quickly converted to carbon dioxide and a rapid outburst of massive amounts can easily alter the isotopic composition of the carbon dioxide in the atmosphere. Biomass manufactured from it by marine and terrestrial plants is also affected. This scenario still requires an initial trigger, and for that volcanoes can be brought back into the fold.[55] For it is possible that the eruption of the Central Atlantic Magmatic Province led to a gradual build-up of carbon dioxide in the atmosphere, which started to warm the climate. Over thousands of years, oceanic circulation patterns could have brought the warmed surface waters into contact with the seafloor, destabilizing the frozen gas hydrates and triggering the release of methane gas. In the numerical models, the sudden release of massive amounts of methane from its frozen cage in the seafloor sediments produces soaring carbon dioxide levels consistent with those indicated by the fossil leaves. It also produces a spike of ^{12}C in the atmosphere and oceans that explains its presence in the fossil leaves from Greenland and marine organisms of sediments around the world.

In this revised version of events, we have finally arrived at a unifying theory for the mass extinction of 200 million years ago satisfying all of the available evidence. It is a speculative scenario, no question, but each link in the chain of logic fits the facts. Moreover, the sequence of events on which it rests is not unprecedented and finds a parallel with a better understood sudden global warming episode that took place at the end of the Palaeocene, 55 million years ago.[56] The end of the Palaeocene is marked by a sharp rise in sea-surface temperatures, up to 6 °C within a few thousand years, and is linked to the sudden addition of methane by a ^{12}C spike in ocean sediments worldwide,[57] comparable to that found in Triassic–Jurassic boundary rocks. Oxidation of the methane to carbon dioxide rapidly acidified the oceans, which burnt through thick layers of calcium carbonate on the seafloor in chemical reactions that neutralized the acidity.[58] Indeed, this is the likely consequence of carbon dioxide being released into the atmosphere as we continue to burn fossil fuels and could spell trouble in decades to come for coral reefs, calcareous plankton, and other marine organisms whose skeletons or shells contain calcium carbonate.[59]

Pinning down the environmental trigger that released stupendous amounts of methane from the seafloor 55 million years ago is a difficult task. One possibility gaining ground makes the connection with enhanced carbon dioxide emissions from the eruption of the North Atlantic Volcanic Province (NAVP). Chemical analyses of the shells of tiny marine organisms—foraminifera—revealed that the climate and oceans gradually warmed, perhaps in response to carbon dioxide released from NAVP, until a threshold was breached that triggered a switch in the circulation of the Pacific.[60] The reorganization of ocean circulation patterns then transferred warmer waters to the deep ocean, thawing the frozen masses of ice and gas compressed into the sediments, to release vast quantities of methane gas and promote further warming. So the sequence of events leading to the pulse of 6 °C warming at the end of the Palaeocene mirrors that suggested for the end of the Triassic.

Of course, the Triassic–Jurassic boundary extinction is four times older than the warming event at the end of the Palaeocene, and correspondingly less of the evidence remains; what does survive is harder to interpret reliably. We do not yet know if gradual warming by carbon dioxide 200 million years ago upset the great ocean currents of Panthalassa, but computer simulations guided by the evidence from the fossil leaves have revealed a glimpse of Pangaea's climatic response to soaring carbon dioxide levels. In the model simulations, millions of square kilometres of Pangaea's landmass underwent unrelenting heat and extreme drought, subjecting organisms to severe water and heat stress.[61]

We can put everything together to paint a portrait of progressive environmental degradation that pushed species over the edge at the end of the Triassic. The key event seems to have been the eruption of the Central Atlantic Magmatic Province, when more than half of the 4 million cubic kilometres of basalt poured out onto the Earth's surface in a few hundred thousand years.[62] Brief intense episodic eruptions pumped a mixture of gases, including sulfur dioxide and carbon dioxide, into the atmosphere. Initially, the sulfur dioxide may have caused acid rain but carbon dioxide had the longer-term effects leading to prolonged warming that propagated into the deep ocean and triggered the catastrophic release of methane from frozen gas hydrates. The release of huge volumes of methane and carbon dioxide appears to have rapidly acidified the oceans, causing an ecological crisis among coral reefs and calcareous plankton dependent upon calcium carbonate to construct shells and skeletons.[63] The further input of greenhouse gases into the atmosphere may then have led to more warming, further melting the gas hydrate reservoirs, and locking Earth into a positive feedback loop that further amplified greenhouse warming. As the oceans warmed, their oxygen content may have declined, subjecting marine faunas to anoxic conditions.[64] These extreme climatic conditions may have pushed species beyond the limits of their survival, causing populations to crash, the collapse of the ecological communities, and species extinction, first on land and then in the oceans.[65] Animals facing the stark situation as the Triassic drew to a close had to 'adapt or perish',[66] and as the extinctions swept aside dominant groups of animals, they freed up resources and fuelled the diversification and successful rise to ecological dominance of the dinosaurs.

All of this begs the question of how the climate system rallied from such a dramatic upset to allow the ecological recovery of the surviving animals and plants. The answer lies once again in the capacity of Earth's thermostat to re-establish order from climatic chaos given a sufficient length of time. As the climate got hotter, it accelerated the weathering of continental rocks, helping to scour carbon dioxide out of the atmosphere and establish a new, cooler, climatic equilibrium. When the release of carbon dioxide is, in a geological context, slow and steady, like that from the eruptions of the Central Atlantic Province, the system copes. When the release is massive and sudden, like methane from the seafloor, it overwhelms these feedback processes. The recovery of the climate system probably took hundreds of thousands of years and we can follow the sequence of dramatic heating and recovery in abrupt changes in the isotopic composition of elements washed out of the eroding rocks to enrich the seawater, which have become preserved in the Triassic–Jurassic boundary sediments of south-west England.[67]

It remains to be seen whether or not this scenario for the end-Triassic extinctions stands the test of time. All the pieces of the puzzle have only come to light from research in the last decade. Not all scientists are prepared to accept the proposed connection between greenhouse warming and the Triassic–Jurassic boundary mass extinctions. The publication of evidence for a dramatic rise in the carbon dioxide content of the atmosphere understandably prompted a search for the same signal with an independent means of breathalyzing the atmosphere to evaluate the claim. For this, a team led by scientists from Bloomsburg University turned to the record preserved in fossil soils. Announcing their findings in *Nature* in 2001, the team claimed fossil soils showed no change in atmospheric carbon dioxide levels across the Triassic–Jurassic boundary.[68] But does this deal a fatal blow to the super-greenhouse warming-extinction theory?

At issue here, and what separated the studies of fossil soils and stomata, is a huge vacuum of geological time. Rigorously evaluating claims for a cause (carbon dioxide) and effect (global warming) requires knowledge of the timing and duration of each factor. Soaring carbon dioxide levels lasted probably less than a hundred thousand years, peaking with maximum volcanic activity and methane release. After this, as we have seen, Earth's climate naturally rebounded towards a new climatic equilibrium. The fossil soils analysed for a high carbon dioxide signature in the early Jurassic appeared to have formed millions of years after levels returned to 'normal'.[69] However, suppose for a moment that the study of fossil soils is correct and carbon dioxide levels remained unchanged across the Triassic–Jurassic boundary. We are then left with the difficult task of explaining the terrestrial and marine extinctions and the carbon isotope anomaly. The Bloomsburg team address this issue by linking the marine extinctions to changing sea levels, driven by a flexing of the Earth's crust related to the activity of the mantle plume head. As the sea level changed it altered the availability of habitats suitable for marine organisms to live in. This may well be true up to a point, but sea-level fall hardly explains the disappearance of land animals or the change in the forest composition on Greenland. Acid rain may seem like a good solution to this problem, but its rapid removal from the atmosphere makes it an unlikely candidate for driving ecological change in Greenland's forests over tens of thousands of years.[70] Moreover, neither sea-level changes nor acid rain can account for the worldwide appearance of a negative carbon isotope spike.

Some scientists champion alternative theories of what might have happened. Paul Olsen at the Lamont–Doherty Earth Observatory, Columbia University is one the most eloquent of these and believes that a large asteroid or meteorite hit the Earth with enough force to annihilate life at the end of the Triassic.[71] Olsen and his team investigated footprints, dinosaur bones, and plant remains in rocks that are the remnants of the Central Atlantic Magmatic Province of eastern North America. Sandwiched between the Triassic and Jurassic rocks, in the sediments where the reptiles and dinosaurs swapped places, they discovered a modest six-fold increase in the concentration of iridium.[72] But a small increase is simply suggestive of an impact event, and the level could easily have been concentrated by natural processes or by the outpouring of lava from the massive eruptions. As I mentioned earlier, those levels following an undoubted asteroid strike at the end of the Cretaceous rose by a factor of 60, a rise that has been discovered at over a hundred sites worldwide. Only further chemical analyses of the Triassic–Jurassic boundary rocks will pin down whether the rise in level has an extraterrestrial origin.

Olsen's team also turned up two other telltale characteristics of an impact in North American Triassic–Jurassic boundary rocks: ferns and shocked quartz (quartz whose microscopic crystalline structure has undergone deformation characteristically caused by intense pressure). Rocks in Pennsylvania contain a Triassic–Jurassic boundary layer rich in fern spores.[73] Ferns are opportunistic colonizers of recently devastated landscapes. Regenerating from the wind-blown spores produced by distant parent colonies, they are often among the first plants to appear on areas cleared by volcanic eruptions—nature's clue that ecological recovery is under way. Flourishing ferns coinciding with an iridium peak might, then, be taken as evidence for an impact-disturbed landscape, as seen at Cretaceous–Tertiary boundary localities near the impact crater. The discovery of shocked grains of the mineral quartz forms the second strand of evidence supporting the suggestion of an impact; these are usually only found near nuclear test sites or known impact craters. But shocked quartz has only turned up in the end-Triassic rocks of the Northern Apennines in Italy and not, as yet, anywhere else.[74] Again the situation contrasts with that at the Cretaceous–Tertiary boundary: rocks near the relevant crater contain shocked quartz in abundance.

Other aspects of the impact theory also leave room for doubt. No suitable impact crater or 'smoking gun' has yet been discovered. Five craters caused by extraterrestrial impacts date to the

Late Triassic. The largest is the impressive Manicouagan structure in Quebec, in north-eastern Canada. At 100 km across, in its partly eroded state, it is one of the largest known impact craters on the surface of the Earth.[75] But its age, and that of the other four, pre-dates the Triassic–Jurassic boundary mass extinctions by 10 million years or more.[76] There is, of course, ample scope for concealment or erosion of an impact crater, so perhaps it lays hidden, awaiting discovery. The latest tantalizing piece of evidence is the discovery of an unusual sediment layer spread out over a quarter of a million square kilometres in central England that bears all the hallmarks of being disturbed by shock waves similar to those found after major earthquakes. Intriguingly, the disrupted sediment is tied more or less to the end-Triassic extinctions, but we lack evidence of an impact crater of equivalent age that might have caused the shock waves.[77]

The fern spike and quartz grains found in Italy are not yet a fully convincing story documenting an impact at the Triassic–Jurassic boundary. Only time will tell if it is all too tentative to take seriously or adds up to something more substantial. One attempt at reconciling the clues to an impact by a meteorite or comet with evidence for a mantle plume is the so-called 'Verneshot hypothesis'.[78] Named after the famous French science fiction writer Jules Verne (1828–1905), who originally discussed them, 'Verneshot' events are postulated explosive releases of pressurized gases that build up beneath the crust and propel materials into suborbital trajectories. Like many intriguing ideas, though, this one is hard to evaluate properly, and remains highly speculative.

The Scottish geologist and natural philosopher James Hutton (1726–97) laid the foundations of modern geology in two papers read before the Royal Society of Edinburgh in 1785, which presented his case for the principle of uniformitarianism. This concept, Hutton's great idea, was later published in his seminal two-volume text, *Theory of the Earth*, in 1795. It states that the present-day appearance of the Earth is explainable in terms of the same processes that can be observed today. Marshalling his arguments with care and supporting them with a wealth of observational facts, Hutton argued that, given enough time, modern-day processes could erode mountains, and deposit sediments on the sea floor that eventually become uplifted to produce mountains again. This endless cycling is implied in his famous phrase 'we find no vestige of a beginning—no prospect of an end'. The upshot of this revolutionary idea was, of course, that the length of time required to create Earth's modern appearance far exceeded the 6000 years allowed by the standard interpretation of the Bible. No great acts of catastrophism, like the biblical flood, were required to explain the geological record.

All this flew in the face of popular opinion at the time and Hutton's proposals came under vigorous unfounded attack from others championing the notion that explaining the geological record required invoking short-lived catastrophes of the Earth. John Playfair (1748–1819), a Scottish mathematician and geologist and a friend of Hutton's, did him the great service of writing a masterly and engaging synthesis of his impenetrable work. Playfair's book, published in 1802 as the *Illustration of the Huttonian theory of the Earth*, helped increase the popularity of Hutton's original ideas. Regardless of what those in denial believed, the idea of uniformitarianism had arrived, and Charles Lyell (1797–1875) skilfully picked up the baton, fleshing out and extending Hutton's concept in his influential book *The principles of geology*, a text based on evidence gathered from his own field studies while travelling around Europe. Lyell, like Hutton before him, realized that the same processes at work today shaping the land surface have operated over the immense span of geological time. These ideas were neatly summed up by Archibald Geike (1835–1924), who coined the memorable phrase 'the present is the key to the past'.

We can see that in the context of escalating concerns over humankind's alteration of the global environment, Hutton's and Lyell's profound contributions to geology can be (and often are) readily extended to the current debate: 'the past is the key to the future'. The phrase is not meant to suggest that some past warm greenhouse climate state is an analogue for our future warmer Earth,[79] but rather that by identifying and understanding the processes that created past warm climates, we can better understand our climatic future.

The global carbon dioxide content of the atmosphere continues to increase with our continued combustion of fossil fuels since the mid-nineteenth century. Indeed, the present atmospheric content exceeds anything Earth has experienced in the last million years, and possibly the last 20 million years. Leading climate scientists agree that the global climate system is responding to the accumulation of carbon dioxide and other greenhouse gases in the atmosphere.[80] Earth's mean surface temperature has increased by an average of 0.8 °C since the beginning of the twentieth century, a rise exceeding the natural range of climate variability and sitting outside anything experienced in at least the last thousand years.[81] The world's oceans are also beginning to warm.[82] In the last 40 years, the oceans have soaked up 85% of the extra heat absorbed by the Earth due to greenhouse gases, with the remainder melting glaciers and warming the atmosphere and continents. Most of the warming has taken place in the upper layers of the ocean, with the Atlantic Ocean taking up more than the Pacific and Indian

Oceans combined. Our oceans are probably warmer now than at any time in the last few million years.

As the surface oceans continue to warm, and the heat propagates into the deep ocean to reach the seafloor, there is increasing concern over the stability of the frozen gas hydrates, the icy reservoir of greenhouse gas containing an amount roughly equivalent to that stored in all our fossil fuel reserves put together. The lesson from the past for our climatic future is that the reservoir may be very sensitive to ocean warming. Current models suggest the timescale of ocean warming and hydrate meltdown is crucial. If gradual, then the consequences are unlikely to be dramatic. If rapid, the situation could be more serious, locking us in to a positive feedback cycle of more warming driving more melting and further warming.[83] Opinion over the potential for this nightmare scenario to unfold is divided, but we must be wary of courting disaster. The gas capacitor on the seafloor has been recharging for thousands, probably millions, of years. The slow burning fuse to our own global warming time bomb is now lit, paced by the unhurried motion of the deep ocean circulation. As the warming caused by human activities shows signs of penetrating deep into the oceans,[84] we would be wise to recognize it as a wake-up call for anticipatory action on global warming.

6

The flourishing forests
of Antarctica

Heroic polar explorers recovered fossils indicating that forests once occupied ice-free continental landmasses extending to within 1000 km of the poles. These finds were soon followed by the revelation that the warmth of the ancient polar climates more closely resembled today's northern Mediterranean coastline with mild winters and warm summers. As the story of the polar forests unfolded, a strange ecological puzzle materialized—why were those in the northern hemisphere mainly deciduous? The issue sparked a transatlantic polemic for nearly a century and acceptance of the mistaken notion that it was a survival strategy for coping with the long, dark, polar winters. New research is now making this received wisdom obsolete and offers promising pointers to alternative explanations for the spectacular vestiges of ancient plant life at the poles.

The great tragedy of science—the slaying of a beautiful hypothesis with ugly data.

Thomas Huxley (1870), *Collected essays*

The past seizes upon us with its shadowy hand and holds us to listen to its tale.

Albert Seward (1926), The Cretaceous plant-bearing rocks of western Greenland. *Philosophical Transactions of the Royal Society* **B215**, 57–173.

B Y arriving at the South Pole on 14 December 1911, the Norwegian explorer Roald Amundsen (1872–1928) reached his destination over a month ahead of the British effort led by Captain Robert Falcon Scott (1868–1912).[1] As Scott's party approached the South Pole on 17 January 1912, they were devastated to see from afar the Norwegian's black flag. On arrival, they discovered the remains of his camp with ski and sledge tracks, and numerous dog footprints. Amundsen, it turned out, had used dogs and diversionary tactics to secure victory while the British team had man-hauled their sledges. These differences were not lost on *The Times* in London, which marked the achievement with muted praise, declaring it 'not quite in accordance with the spirit of fair and open competition which hitherto marked Antarctic exploration'. Exhausted, Scott and his men spent time the following day making scientific observations around the Pole, erected 'our poor slighted Union Jack', and photographed themselves in front of it (Plate 11). Lieutenant Bowers took the picture by pulling a string to activate the shutter. It is perhaps the most well known, and at the same time the saddest picture, of the entire expedition—a poignant image of the doomed party, all of whom look utterly fed up as if somehow sensing the fate awaiting them. The cold weather, icy wind, and dismal circumstances led Scott to acerbically remark in his diary: 'Great god! This is an awful place and terrible enough to have laboured to it without the reward of priority.'[2]

By this time, the party had been hauling their sledges for weeks, and all the men were suffering from dehydration, owing to fatigue and altitude sickness from being on the Antarctic plateau that sits nearly 3000 m above sea level.[3] Three of them, Captain Oates, Seaman Evans, and Bowers, were badly afflicted with frostbitten noses and cheeks. Ahead lay the return leg, made all the more unbearable by the crippling psychological blow of knowing they had been second to the Pole.

After a gruelling 21-day trek in bitterly cold summit winds, the team reached their first cache of food and fuel, covering the distance six days faster than it had taken them to do the leg in the other direction.

On 8 February 1912, with the physical well-being of the men in serious decline, the Pole team decided to spend the day searching for geological specimens around the Beardmore Glacier at 82 °S before proceeding further. The pause to hunt for precious fossil specimens served to give the men a much needed rest in the warmer weather off the dreaded plateau, and help them recover from the brink of exhaustion. Scott wrote in his diary of that day 'the moraine was obviously so interesting that when we had advanced some miles and got out the wind, I decided to camp and spend the rest of the day geologizing'. It was to be the only day on the arduous homeward march that they devoted to geology.[4] Thrilling fossil finds temporarily reinvigorated the party. Wilson, the group's geologist, enthusi-astically wrote in his diary of the 'magnificent Beacon Sandstone cliffs... masses of limestone in the moraine... had a regular field day and got some splendid things in the short time'.[5] The geological specimens included 'beautifully traced leaves in lay-ers'[6] and added 16 kg (nearly 35 lbs) to their already-overloaded sledges. Although this may seem like a bad idea for the fatigued group, it is probably not as foolish as it sounds. The weight of the rocks was but a small fraction of the total the polar party dragged behind them. What counts when man-hauling sledges is not the deadweight but the friction between the runners of the sledges and the snow.

Just over a week later, 'Evans collapsed, sick and giddy, unable to walk even by the sledge on ski'.[7] He died quietly on 17 February 1912. Over the past three and a half months he had marched a staggering 1200 miles (1930 km). The sad end was in sight for the rest of the party. Steadily dropping temperatures, and Oates' deteriorating condition, slowed the progress of the other men. By 16 or 17 March (the date is uncertain) Oates could continue no longer, and realizing he was becoming a terrible hindrance to his comrades, crawled through the snow and out into the blizzard to meet his certain death, with the famous parting words to his companions: 'I am just going outside and may be some time.' Scott recorded the noble gesture of his companion in his diary entry: 'We knew that poor Oates was walking to his death, but though we tried to dissuade him, we knew it was the act of a brave man and an English gentleman. We all hope to meet the end with a similar spirit, and assuredly the end is not far.'[8]

The remaining three members of the party jettisoned non-essential equipment, but at the urging of Wilson continued to drag their burden of geological specimens. By 19 March, 11 miles from the food and fuel depot that could have saved their lives, a blizzard blew up. The plan for Wilson and Bowers to leave a frostbitten Scott, find the depot, and return, was abandoned. In the end, Scott and his two companions died frozen and starving in a frail tent soon covered by drifting snow. It was not until eight months later, on 12 November 1912 that the remains of the polar party and their equipment were discovered a metre or so down in the snow by the relief expedition. Notebooks, letters, diaries, chronometers, and two rolls of film containing Bowers' photographs were recovered from the tent. At the back of Scott's last notebook they found his 'message to the public' and his last entry, simply 'for God's sake look after our people', a plea capturing the awful truth of their situation. On the sledge outside, buried beneath the snow, the search party recovered the fossil specimens, dragged to the very last, that were later to make a valuable contribution to our understanding of Earth history.

Much has been written about the travails of Scott's ill-fated expedition. The fact that it ended so tragically is usually attributed to several well-known problems: Scott relied on motor sledges, used weak ponies, and allowed a fifth man to go forward on the final assault for the Pole when supplies for the return journey had been based on a party of four. When collected together, these facts contributed to the legend that the expedition failed as a result of Scott's inept leadership. But that view is largely due to Roland Huntford's character assassination, as set out in his best-selling book *Scott and Amundsen* in 1979. This was published in paperback in 1986 under the better known title *The last place on Earth*. Fortunately, staunch allies are beginning to redress the balance, exposing Huntford's mistaken claims and misguided damaging attacks, arguing instead for us to recognize the inspirational achievements of the man and his expedition.[9] One of Amundsen's successful team later said:

> It is no disparagement to Amundsen and the rest of us when I say that Scott's achievement far exceeded ours.... Just imagine what it meant for Scott and the others to drag their polar sleds themselves, with all their equipment and provisions to the Pole and back again. We started with 52 dogs and came back with eleven.... What shall we say of Scott and his comrades who were their own dogs? Anyone with any experience will take off

his hat to Scott's achievement. I do not believe men ever have
shown such endurance at any time, nor do I believe there ever
will be men to equal it.[10]

New scientific analyses are also helping to debunk the myth
of Scott as merely competent. Susan Soloman at the National
Oceanic and Atmospheric Administration in Boulder, Colorado,
shone an illuminating new light on the expedition's fortunes:
she undertook a detailed examination of the meteorological
conditions Scott's party experienced and compared them with
subsequent datasets from scientific stations located on Antarctica. Her telling analyses recast Scott as simply unlucky.[11] In her
book *The coldest march*, Soloman reveals that Scott's party had
encountered an unusually prolonged cold spell of ferocious
weather on the Barrier towards the end of February 1912.
Temperatures there dropped to 10 °C below the long-term average, and stayed abnormally cold for three weeks. This sort of
freak weather phenomenon has been repeated only once in the
last 38 years. The problem is that the cold temperatures cause
the snow surface to freeze solid like sandpaper and, with no
melting of ice crystals to lubricate the movement of the runners,
man-hauling sledges becomes exceptionally hard work; the new
ice crystals act as brakes on the runners. Soloman claims that
slow progress back from the Pole was inevitable given the
extreme conditions endured by the party; if it had been warmer,
she argues, Scott could have made it to the critical food depot.
Scott wrote of his predicament, 'on this surface we know we
cannot equal half our old marches, and for that effort we expend
nearly double the energy'.[12] It was doubly unfortunate that Scott
suffered from frostbite and had to rest up during the blizzard.
Blizzards on Antarctica usually cease after a couple of days,
circumstances that would normally have allowed Bowers and

Wilson to press on to reach food, but they dared not leave Scott alone.

The extreme courage shown by the men of the British expedition, and the extensive scientific observations they made, was far from wasted. On recovery, their precious cargo of fossils turned out to be of the highest geological significance. Following the arrival of the specimens in London in May 1912, Albert Seward of Cambridge University (1863–1941) (Plate 12), the pre-eminent palaeobotanist of the day, conducted a scientific examination of them at the invitation of the trustees of the British Museum of Natural History.[13] A few were found to be the distinctive fossilized leaf and stem remains of *Glossopteris*, an extinct short-stature early gymnosperm tree whose name derives from the Greek for tongue (glossa), after the shape of its leaves. *Glossopteris*, a signature plant for the Permian, dated Scott's fossils at around 270 million years old. These were the first fossil specimens of *Glossopteris* found so far south, and provided tangible evidence that vegetation once existed within 300 miles (482 km) of the Pole. Evidently, Antarctica in the Permian enjoyed a far warmer climate than it does today.

The discovery of *Glossopteris* on Antarctica held great significance for the part it played in establishing the idea of 'continental drift'. Continental drift, what we now call the theory of plate tectonics, recognizes that the positions of continents are not fixed in time. Eduard Suess (1831–1914) was the London-born son of a German wool merchant, and he made an important attempt at synthesizing early ideas on the topic in the final decades of the nineteenth century. Suess, like others before him, noticed the obvious similarities between some continental coastal geometry, especially eastern South America and the western coast of Africa, which appeared to fit together like a jigsaw puzzle. Realizing they shared Permian sediments that hosted abundant *Glossopteris* fossils, he postulated that the present-day landmasses of Africa, Australia, and India once formed part of the supercontinent Gondwana. Suess was right about the existence of Gondwana, but wrong about the mechanism of its formation. He articulated the idea that mountain ranges, and other features of the landscape, were formed by the contraction of the Earth's crust as it cooled, much like the wrinkling of the skin of an apple as it dries out. Remnants of Gondwana had, in this erroneous picture, foundered into the sea.

It was not until 1915 that the German meteorologist Alfred Wegener (1880–1930) formulated a largely correct model of the 'continental drift' hypothesis that has stood the test of time. The scientific community was not ready for Wegener's brilliant thinking, rejecting the idea in its German form in 1915, again on publication of the English translation in 1922, and yet again in 1936 when the French edition was published as *La genèse des continents et des océans*. It was to take a further 40 years of heated argument before the dispute was finally laid to rest, converting the continental drift hypothesis into the theory of plate tectonics. What finally clinched it was the demonstration that thin oceanic crust is being produced at depth beneath the oceans, effectively pushing the continents apart. It was also realized that elsewhere, notably today at the western margins of the Pacific Ocean, oceanic crust is being destroyed as it dips below the edges of the thicker continental crust into the mantle. The birth and death of the Earth's outer crust through these processes elegantly solves the problem of allowing continents to move about without the need to invoke Suess's 'shrinking' of the planet.

In the early 1920s and 1930s, as fierce controversy over Wegener's theory raged, Seward boldly brought Scott's fossil finds into play at a time when intellectual heavyweights of the geophysical community still regarded continental drift as the 'impossible hypothesis'. Before Scott's finds, as Suess had pointed out, *Glossopteris* fossils occurred in countries now widely separated from Antarctica, mainly in southern South America, South Africa, India, and Australia. Afterwards, the missing piece of the puzzle fitted into place. Reassembling the continents into their positions during Permian times revealed that *Glossopteris* once occupied what appeared to be a continuous geographical distribution on a single southern hemisphere landmass. Here, argued Seward, was strong evidence for the existence of Gondwana. Scott's fossils proved to be the crucial missing piece of evidence that firmed up its existence.

Historical tendrils of fate and fortune reach out in all directions from Scott of the Antarctic, and a notable example reached Marie Stopes (1880–1958). Stopes is regarded as 'one of the most remarkable women of the twentieth century' for her pioneering contributions, proclaiming and defending women's rights to control and enjoy their sex and family lives.[14] She later became a distinguished poet and playwright, but palaeobotany was her first love,[15] and Stopes had perceptively realized the importance of finding fossil plants on Antarctica. Consequently, in around 1908 she spoke to Scott during a fund-raising dinner in Manchester for his 'Terra Nova' expedition to the South Pole, but he was unable to accommodate her wish to accompany them to Antarctica.[16] Scott did, however, return to Manchester to familiarize himself with a variety of fossil specimens already held in the University's collections, and Stopes provided expert guidance. Whether she had actually shown Scott fossils of *Glossopteris* is unclear. If she had, it seems doubtful that the information was transmitted to Wilson, for he recorded in his diary that the leaves found on their single day's fossil hunting 'were like beech leaves in shape and venation'. Later scrutiny of the same fossils by Seward showed that Wilson's 'beech' leaves were those of *Glossopteris*. Sadly, it seems that Scott and his companions died without realizing the great significance of the fossils packed onto their sledges, fossils which would later contribute to revolutionizing our view of Earth history.

Nearly 30 years before Scott's adventures on Antarctica, at the opposite end of the world, polar explorers exemplified by the Norwegian Fridtjof Nansen (1861–1930) had already discovered fossil plants within striking distance of the North Pole.[17] Nothing like it had been unearthed before on the barren Arctic landscape. Amongst the earliest finds were those uncovered in 1883 by the Lady Franklin Bay Expedition, led by the Arctic explorer Lieutenant Aldophus Greely (1844–1935) of the US Cavalry.[18] Three members of the expedition reached a latitude of 83°24′ N, the northernmost tip of Greenland, and in doing so broke the 300-year-old British record for the 'farthest North'. It was a feat not repeated for another 13 years. One of the members of the Lady Franklin Bay Expedition was Sergeant David L. Brainard (1856–1946), a square-jawed soldier who had served his time in the Second Cavalry, and had seen action against the Sioux and Nez Percé Indians. Brainard went on to explore Ellesmere Island in the Canadian High Arctic and found fossil tree stumps *in situ*—these became known as 'Brainard's forest'. Here was startling evidence that trees, perhaps even forests, once existed close to the Arctic Circle. Sadly, the expedition, though successful in its collection of meteorological information, ended in tragedy and later controversy. Supply ships failed to reach the explorers during the summer of 1882 and again in 1883. As a result 18 men perished from starvation; only Brainard, Greely, and four others survived. Sensationalist reporting of the survivors' brush with death led them to give testimonies at Congressional hearings against charges of cannibalism.[19] Greely later died a true American hero, despite allegations of mutiny and cannibalism, the execution of one soldier, and the loss of most of the expedition.

Discoveries made during the earliest days of polar exploration were only the beginning. There followed a succession of fossil finds from Canada, Greenland, Spitsbergen, and Antarctica, including surprising evidence that polar dinosaurs once populated the high latitudes.[20] In some instances, discoveries of the remains of fossil forests, rather than simply fragments of woods and leaves, were made. In 1992 researchers unearthed a remarkable Permian fossil forest in the central Antarctic Mountains, near the Beardmore Glacier where Scott's party had collected their specimens.[21] The 'forest' consisted of 15 trunks, preserved by mineral-rich groundwater infiltrating and replacing the tissues of the dead trees as they decayed with minute crystals of quartz or calcite, a process known as permineralization. Numerous fossil leaves of *Glossopteris* were also found in the substrate into which the trees were rooted. This exciting find is thought to be a rare example of a young, rapidly growing *Glossopteris* forest that had enjoyed warm summers at about 85 °S, 270 million years ago, a far cry from the climate of the region today.

Of all the polar forests discovered to date, one of the most spectacular was that found at the other end of the world in the Canadian High Arctic by the sharp-eyed helicopter pilot Paul Tudge at the Geological Survey of Canada. In 1985, Tudge spotted stumps on a ridge of Axel Heiberg Island that had been exposed by sands winnowed away in the fierce Arctic winds. A quickly marshalled team of expert geologists was dispatched to investigate. To their amazement, the team delighted upon a palaeontologist's treasure trove—the best-preserved example of a polar forest ever found.[22] Some 45 million years old, the trees originally grew between 75° and 80 °N, yet despite their proximity to the Pole, towered to heights of up to 40 m. Fossil wood on Ellesmere Island is preserved by permineralization, like that of the Antarctic forests, but that on the eastern edge of Axel Heiberg Island is preserved by mummification, dried out by the harsh climate and cold winds. Deposited in the fossil leaf litter amongst the trunks are beautifully preserved flattened layers of fossil leaves. The commonest are easily recognizable as the feathery leaves of the conifer dawn redwood (*Metasequoia glyptostroboides*), a valued ornamental species renowned in parks and gardens around the world for the fiery coppery orange colour of its foliage in autumn.

We now know that the seemingly unusual phenomenon of polar forests is in fact the more normal state of affairs for planet Earth. Our familiarity with a world of polar ice-sheets and glaciers that wax and wane with seasons makes it extraordinary to discover that, for nearly 80% of the past half billion years, forests once extended up into the polar circles and beyond.[23] With the new theory of plate tectonics in place, scientists initially pondered whether the forests had really grown in the high latitudes or whether the fossils represented plants that had once grown in more temperate climates and then been carried to their present inhospitable conditions on the drifting continents, passengers riding the stately tectonic waves of Earth's ductile mantle. However, the former positions of the polar landmasses later became established by reading off the orientation of Earth's magnetic field locked into the iron minerals of rocks. It then became clear that the forests had indeed grown inside what were then, and are now, the polar circles.

The story of polar forests is intimately linked with that of polar climates, for the climate of the past controls the northern and southern high-latitude distributions of forests, just as it does today. Expansive tracts of contemporary North American boreal forests, consisting mainly of conifers like pine, larch, and firs and a few deciduous angiosperms of birch and poplar, reach 69 °N. Beyond 69 °N the permafrost rises close to the soil surface, freezing water supplies to the trees. The lack of liquid water causes tree populations to thin out and give way to low treeless tussocks of the Arctic tundra, consisting mostly of angiosperms, mosses, and lichens. At the very highest latitudes, even the tundra succumbs to the formidable cold, surrendering to polar desert, where large areas of ground are devoid of vegetation except for a very sparse scattering of the hardiest lichens, mosses, and, occasionally, angiosperms. A similar pattern is found on the other side of the Arctic in Siberia and Russia, where sparse forests of deciduous larches extend a little further northward to 72 °N, but then they, too, dwindle and give way to the endless expanse of Arctic wilderness. These northern forest boundaries—the tree line—are held in check, prevented from advancing polewards by cold temperatures that cut down the survival, growth, and successful reproduction of trees.[24] The Arctic tree line of the northernmost boreal forest, in fact, tracks more or less the geographical line demarcated by July temperatures of 10 °C, which is also the southern limit of the permafrost. The existence of polar forests in areas that are now tundra or polar desert unambiguously tells us that conditions in the past were once considerably warmer than they are in these regions today.

An indication of how much warmer it was during the heyday of the polar forests in the Mesozoic and early Cenozoic can be obtained from fossil organisms whose modern representatives are today restricted to warm climates. An early spectacular clue was unearthed in 1890 by Alfred Nathorst (1850–1921), a Swedish palaeobotanist. Nathorst had found Cretaceous fossils of leaves, flowers, and fruits of the breadfruit tree (*Artocarpus dicksoni*) in sediments off the coast of western Greenland.[25] Today, this handsome, fast-growing tree is native to a vast area, extending from New Guinea through the Indo-Malayan Archipelago to Western Micronesia, which experiences a tropical climate of mild winters and hot summers. The breadfruit tree fossils point to a similar subtropical Arctic climate in the Cretaceous. A century later, finds of the fossil bones of champsosaurs in Cretaceous sediments along the western margins of Axel Heiberg Island in the Canadian Archipelago painted a similar climatic picture.[26] Champsosaurs, an extinct group of fish-eating, crocodile-like reptiles, were similar in size to crocodiles and similarly adapted to life in freshwater. As with plants, the same reasoning for inferring past climates applies; crocodiles are strictly confined to subtropical and tropical climates, so it is presumed that champsosaurs were limited to comparable balmy climates.[27] Their presence in the Arctic in Cretaceous times suggests mild winters, usually above freezing and possibly up to 5 °C, with warm summers reaching 25–30 °C.

Complementary to evidence for warm climates inland is that for warm oceans around the shores of the Arctic and Antarctic when pockets of polar forests flourished. The Cretaceous oceans surrounding the coastlines of Antarctica[28] and the Arctic continental landmasses[29] probably reached astonishingly mild temperatures of 10–25 °C. The picture for later on, in the early Cenozoic, 40–50 million years ago, only became clear with the recovery and analysis of sediment from the bottom of the Arctic Ocean by the Arctic Coring Expedition of the Integrated Ocean Drilling Program in August 2004. Recovering the sediments from the floor of the Arctic Ocean was remarkable enough. It required two large ice-breaker ships to keep the ice at bay and allow a third drill ship to maintain its position in heavy seas.

Geochemical analyses of these extremely hard-to-obtain sediments revealed that, as in the Cretaceous, the Arctic had experienced greenhouse conditions in the Cenozoic. Summer ocean temperatures reached 18 °C, comparable to those found in the sea today off the French coast of Brittany.[30] Imagine the Arctic Ocean being warm enough to swim in! Only geology can offer the necessary sobering perspective on climate change. Elsewhere, at the other end of the world, chemical analyses of mollusc shells recovered from sediments on Seymour Island, Antarctica, revealed that Antarctic waters had also reached similarly warm temperatures.[31]

Polar forests of the Mesozoic and early Cenozoic clearly enjoyed an exceptionally warm climate, characterized by hot summers and mild but not necessarily frost-free winters. In fact, the warmth (but not the rainfall) of the ancient high-latitude climate most closely resembled that of our modern northern Mediterranean coastline. No one knows for sure why the polar regions were so astonishingly warm at this time, but part of the explanation seems to lie with high levels of greenhouse gases in the atmosphere. The Cretaceous, for example, is renowned among geologists for being a time of greatly enhanced oceanic crust production,[32] a volcanic process creating a carbon dioxide-rich atmosphere, a strong greenhouse effect, and a warm planet.[33]

The extreme seasonality of sunlight in the high latitudes also fashioned the life of trees at the poles. Seventy per cent of the human population inhabits land between 30 °N and 30 °S. Given our mid-latitude chauvinism it is worth emphasizing for a moment the striking difference in seasonal pattern of sunlight in the polar regions. The extreme seasonality arises because Earth's rotation about its axis is tilted at an angle of 23.5 ° from the vertical. This means that as the Earth rotates, the Arctic Circle points towards the Sun in summer, giving uninterrupted illumination for several months of the year. An Earth-bound observer in the Arctic sees the Sun make a complete circuit of the sky without it ever properly dipping below the horizon. At the same time, someone inside the Antarctic Circle experiences darkness as Earth rotates with the southern hemisphere pointing away from the Sun. At 70 °N, the uninterrupted sunlight of the polar summer lasts nine weeks from mid-May until early July, with corresponding shadowy winters lasting the same duration. Move polewards by just 10 ° of latitude, and the situation rapidly becomes more extreme, with the length of austral summers and winters doubling. Astonishingly, the Antarctic forests at 85 °S lived in a region blanketed by darkness for half the year, when making a living by photosynthesis would have been impossible.

How the ancient polar forests coped with the unusual combination of climatic warmth and an extreme seasonality in sunlight is revealed by the annual growth rings preserved in fossil woods.[34] Growth rings are the visual manifestation of the environmental influence exerted each year on a tree's productivity. Large cells are laid down in the wood during the warm spring and are followed by progressively smaller cells as the rains dwindle through the summer months, until eventually winter dormancy sets in. The marked growth-ring boundaries reflect the stark contrast between the small cells produced in the previous autumn/winter and the large cells produced the following spring. In this way, nature has contrived that the detailed cellular character of individual rings records a tree's growth. Each year new cells are produced and laid down on top of existing wood, to progressively thicken the trunk. The sequence of rings from one year to the next is a signature pattern for climate;[35] wide rings indicating better climatic conditions for growth, narrow rings poorer conditions. Trees are, in the memorable phrase of William Chaloner and Geoff Creber, at the University of London, 'compulsive diarists' of past climates.[36] Their fossil woods afford us a glimpse of how forests grew millions of years ago. Much like Scott's diary, the wood chronicles events taking place unseen by others, and a skilled parsing of fossil wood offers a guide to an ancient tree's life story.

The diaries of the fossil woods revealed that the private life of plants held surprises. Both the Arctic and Antarctic forest trees had produced impressively wide growth rings, typically several millimetres across, with many a centimetre or more wide.[37] Growth rings in the Permian Antarctic forest described earlier averaged widths greater than 4 mm, with the widest reaching 11.5 mm. The fossils put the efforts of our northern high-latitude trees, whose growth is restricted by the cold, dry climate, in the metaphorical shade by comparison. Deciduous Siberian larches at 72 ° N today produce average annual growth rings up to 2 mm wide, while dwarf willows in the Canadian Archipelago at 75–79 ° N manage less than 0.5 mm each year.[38] The obvious implication of the wide fossil growth rings is that the trees were very productive. The trick to revealing how productive lies in calculating the volume of wood produced each year as the girth of the trunk thickens. Assuming a tree trunk is approximately conical in shape, the volume of wood can be calculated from straightforward geometrical relations that require growth ring width and an estimate of tree height, a feature that fortunately scales in direct proportion to the width of the stump.[39] In exceptional circumstances, where the density of trees is preserved in a particular area, it is possible to go one better and infer the productivity of the entire forest.

This approach to bringing the past to life has been taken to new heights with an extraordinary study of the fossil forest remains on Axel Heiberg Island at Napartulik, Inuk for 'the place of the trees'.[40] Researchers at the University of Pennsylvania studied fossils of scattered logs, half-buried treetops, upright stumps, and mats of foliage mostly of dawn redwood. Bringing the fossil forest to life was achieved by felling and measuring dawn redwood trees at a plantation within the Tokyo University experimental station to help fill in the missing gaps relating the geometry of tree growth to fossil growth ring analyses. Guided by the modern forests, the researchers found that the productivity of the polar forests at Napartulik that grew inside the Arctic Circle, 45 million years ago, was similar to that of the temperate rainforests of southern Chile. The conclusion is astonishing. It offers an unparalleled insight into the life of trees at the poles. Evidently, forests flourished in the warm climate near the poles aided, perhaps, by carbon dioxide fertilization.[41]

Innovative investigations like these help bring the phenomenon of polar forests into sharper focus, but the most contentious ecological puzzle of them all surrounds the issue of leaf habit. For it turns out that the majority of fossils discovered in the high northern latitudes originate from taxa that today we recognize as being deciduous, like, for example, dawn redwood.[42] Why should the northern polar forests be seemingly composed mainly of trees with the deciduous leaf habit? The puzzle defied easy explanation for nearly a century. In 1984, Leo Hickey, the then-director of the famous Peabody Museum of Natural History at Yale University, voiced an opinion on high Arctic fossil forests typical of scientists at the time. He commented,

> what is even more surprising [than the warm climate] is that if we look at the kinds of plants that were present... when the climate was supposedly subtropical, we look in vain for palms, teas, or for any other types of evergreen plant that one would normally expect to find associated with the alligators, tortoises, and other vertebrates of subtropical affinities.

A curious feature of these extraordinary forests, in Hickey's eyes, was 'the strange disparity between the deciduous flora and the subtropical aspects of the animals'.[43]

The puzzling deciduous character of northern polar forests ignited a transatlantic polemic lasting for nearly a century. It started with Seward, whose views lay on one side of the debate, and were shaped following his examination of Scott's fossils. Thinking about the significance of the expedition's *Glossopteris* fossils, he reasoned, 'there is no serious difficulty in the way of Arctic or Antarctic plants surviving the winter months either in a condition similar to that of deciduous trees in a European winter, or even as evergreens living...on the surplus food manufactured in the course of prolonged sunshine'.[44] His rationale for this statement was that we see the evergreen juniper and several species of deciduous dwarf willows existing side by side in the Arctic today. To be sure they are not tens of metres tall like those of the ancient forests, but this, Seward suggested, was because the modern climate is 'abnormally' cold compared to that in the past. Seward reiterated this opinion a decade later and it went unchallenged for the remainder of his distinguished career.[45]

Thirty years on, and with Seward's views either forgotten or cast aside, a profound shift of opinion took place. The shift can be traced back to a seemingly run-of-the-mill symposium in Boston, Massachusetts in 1946 concerned with the evolution of North American floras. The symposium was attended by the influential North American palaeobotanist Ralph Chaney (1890–1971) from the University of California, Berkeley. Chaney was one of a new generation of fossil hunters, less concerned with the 'what' and 'when' of fossil plants and more with the 'why'.[46] At the conference, he expressed an alternative argument to Seward's, commenting 'For without leaves, *Sequoia* and *Taxodium* might both be expected to survive a dark winter which was not cold.'[47] In other words, the deciduous habit of these trees was an adaptation allowing survival in the warm, dark polar winter. Chaney's former graduate student Herbert Mason (1896–1994), also at Berkeley, went further, arguing that the warm winters presented substantial problems for evergreen trees because they promoted high rates of respiration during the winter. He believed 'we must conclude that no tropical, warm temperate or even temperate forest flora, could possibly live and develop in high arctic latitudes'.[48] According to Chaney and Mason, polar forests were composed of deciduous trees because this was an adaptation allowing them to avoid an 'impossible' respiratory burden. This idea lay on the other side of the debate from Seward's viewpoint that evergreens and deciduous trees might cope equally well with the polar environment.

Chaney and Mason's 'deciduous view' is founded on the problem, as the North American scientists perceived it, of evergreen trees in the polar regions consuming their food reserves during the dark winter by respiration. Respiration is the powerhouse process by which cells burn oxygen and sugars for energy, and, as in humans, it releases carbon dioxide. But the carbon dioxide lost by respiration was originally acquired by photosynthesis, and for trees carbon is the currency for growth and the fuel for life. Imagine, then, an evergreen canopy of leaves respiring for months during the enforced dormancy of the warm, dark winters, unable to replenish stocks by photosynthesizing new carbohydrate. In Mason's opinion, a tree would rapidly consume all of its fuel reserves in this way, a difficulty exacerbated further by mild winters accelerating the metabolic processes of cells and increasing the demand for energy needed to sustain them.[49] Adopting the deciduous leaf habit seems a far better strategy. Shedding leaves and entering dormancy until the

Plate 1 A fossil of *Cooksonia*, the earliest vascular land plant. Fragile and leafless, *Cooksonia* heralded the start of greater things to come.

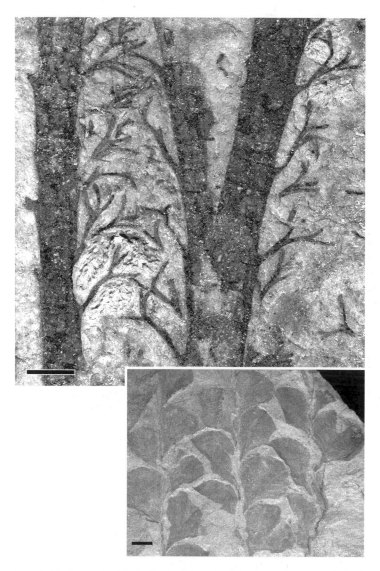

Plate 2 The leafless and the leafy. *Top*: a branch of the 380-million-year-old fossil species *Calamophyton primaevum* from Goé in Beglium. Its forking side shoots are considered to be the ancestors to leaves. Scale bar = 10 mm. *Above*: a typical example of large leaves that developed after carbon dioxide levels plummeted. This image illustrates those of *Archaeopteris obtusa*, dating to late in the Devonian, around 370 million years ago. Scale bar = 10 mm.

Plate 3 The brilliant French chemist Antoine Lavoisier.

Plate 4 Robert Berner of Yale University, who played a pivotal role in uncovering Earth's atmospheric oxygen history over the last half billion years. Berner is sporting a tee-shirt displaying the atomic details of his favourite element, carbon (C), which is number six in the periodic table.

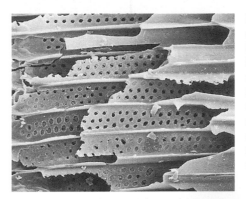

Plate 5 Fossil charcoal of gymnosperm woods from wildfire in Nova Scotia, occurring 300 million years ago. Note that the outstanding cellular preservation detail of the water-conducting cells (tracheids) is still visible. Scale bar = 100 μm.

Plate 6 Robert Strutt, 4th Baron Rayleigh, a pivotal figure in the discovery of the ozone layer.

Plate 7 Mutated fossil plant spores dating to 250 million years ago. Notice the cluster of individual spores stuck together in unusual unseparated clumps. Scale bar (top left) = 50 μm (1 mm = 10−6 m, that is, one millionth of a metre).

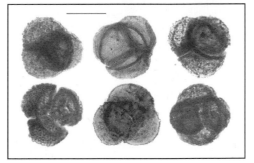

Plate 8 William Buckland in 1833. Buckland was described in the early nineteenth century as 'cheery, humorous, bustling, full of eloquence, with which he too blended much true wit; seldom without his famous blue bag, whence, even at fashionable evening parties, he would bring out with infinite drollery, amid the surprise and laughter of his audience, the last "find" from a bone cave.'

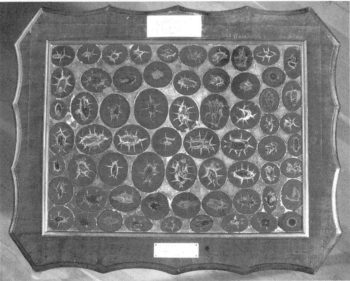

Plate 9 Buckland's table, with inlaid polished coprolites, displayed at the Philpot Museum, Lyme Regis.

Plate 10 *Left*: solid methane hydrate brought up from the depths of the ocean. *Below*: small fragments of icy hydrate burning in air.

Plate 11 *Below*: Scott's party at the South Pole. From the left: Lieutenant Bowers, Captain Scott, Dr Wilson, Seaman Evans, and Captain Oates. They stand beside the Union Jack presented by Queen Alexandra, given with the instructions to erect it if and when the South Pole was reached. The photograph was produced from the negative found inside the tent in which Scott and his companions perished.

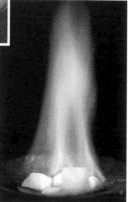

Plate 12 Albert Seward. Seward was the palaeobotanist who examined the fossils collected by Scott's party. He conjectured that trees could have survived perfectly well growing in an ancient polar environment, regardless of whether they were evergreen or deciduous.

Plate 13 Eocene-aged fossil remains of polar forests discovered in Axel Heiberg Island in the Canadian High Arctic and on the Antarctic Peninsula.

The tree stump (*above*) is thought to be of dawn redwood (*Metasequoia*), a deciduous species with feathery leaflets still widely planted today (*right*).

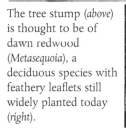

The substantial cretaceous fossil tree trunk discovered on Antarctica (*above*) belongs to the southern beech (*Nothofagus*), the relatives of which form extensive natural forests in New Zealand (*right*).

Plate 14
Renowned Irish
physicist John
Tyndall in
around 1884.

Plate 15 Martin Kamen
in 1947 (*left*) and Samuel
Ruben (*above*) in the late
1930s or early 1940s.
Kamen and Ruben were
co-discovers of the
long-lived radioactive
isotope of carbon, 14C.

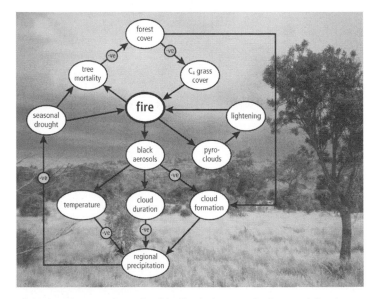

Plate 16 The complex web of feedbacks between biology and the climate
system, linked by fire, which might have accelerated the global expansion of
C4 savannas some 8 million years ago. The background photograph shows a
C4 savanna near Mt Winter, Northern Territory, Australia, in June 2000.

polar spring arrives neatly sidesteps the cost of keeping leaves alive, and conserves valuable stocks of carbon. We can see that in this context, the arguments of Chaney and Mason make sense. Certainly, deciduous dormancy for polar winter survival is an intuitively appealing notion.

Based on intuition, the deciduous view continued to receive support until it became firmly established as received wisdom. Here is Hickey again summing up the prevailing viewpoint:

> the [deciduous nature of polar forests] is simply the effect of the months-long cycle of polar darkness which would have made the evergreen habit disadvantageous...Keeping this [evergreen] tissue alive would have required the expenditure of significant portions of the food reserves of the plant, whereas a deciduous plant would avoid this problem by shedding its leaves during the polar night.[50]

Throughout the 1980s and 1990s a succession of workers on both sides of the Atlantic jumped on the bandwagon, championing the idea to explain the fossils of deciduous trees every time they turned up in sediments from high-latitude localities. Today it is established orthodoxy, with authors confidently asserting that the 'warm poles favour deciduousness as a survival strategy during long dark winters, whereas cold permits evergreeness'.[51]

The trouble with the ideas of thinkers on both sides of the Atlantic is that they rest on intuition and 'common sense', not exactly the rigorous criteria demanded of scientific hypotheses. No one had ever taken a critical look at the implicit assumptions of the 'deciduous view', or indeed those of Seward's that evergreen or deciduous trees could cope equally well with a warm polar climate. Yet the situation for the deciduous view is worrying because it has become so widely embraced by the scientific community and has limited scientific evidence to back it up. How could a 'theory' long on speculation and short on hard facts, hinging on nothing more than intuition, be embraced by so many? Passed down through the literature from one generation to the next, the idea has an allegiance closer to a religious belief than a scientific theory.

Nevertheless, following Chaney and Mason's proposal, evidence began to turn up suggesting its arguments might not be as clear-cut as originally envisaged. Some right-minded scientists urged caution at what seemed to be over-simplistic assumptions about the rates of carbon loss by respiration in a warm winter climate.[52] They drew on evidence from experiments transplanting bristlecone pines in the White Mountains of California down to sea level. The sudden switch to the warmer climate of a lower altitude decreased, rather than increased, the respiration rates of the pine needles.[53] The shift reflected alterations to the catalytic activity of the respiratory enzymes and the availability of the substrate required for energy generation.[54] Later, an inquisitive botanist, Jennifer Read, and a geologist, Jane Francis, tested the survival of southern hemisphere evergreen tree species in a simulated warm (5 ° C) polar winter of 10 weeks' darkness.[55] By far the majority of trees survived intact, none apparently the worse for the experience, a result Seward predicted in 1914. Observations from such experiments serve to remind us that evergreens might not have incurred excessively high respiration rates in the mild polar winters millions of years ago.

Other issues, too, aroused suspicions over the validity of the argument underpinning the deciduous view. These stemmed from common observations made in the glasshouses of botanical gardens. Visitors to the humid conservatories or palm houses of municipal botanical gardens are frequently rewarded with fantastic displays of lush green foliage and spectacularly colourful flowers of tropical trees and shrubs. We are, of course, witnessing for ourselves the fact that tropical plants can be raised in glasshouses at latitudes well beyond their natural range. In the Royal Botanic Gardens in Kew, near London, situated at 51 ° N, over 130 taxa of tropical and subtropical woody plants flourish in glasshouses, all of them naturally flowering and setting seed without the aid of extra illumination. Further north, at 59 ° N, in the Bergianska Botanic Garden in Frescati, along the western shore of lake Brunnsviken, north of Stockholm, it is the same story. The evidence seems clear-cut. Tropical evergreen trees and shrubs can survive in situations hundreds of kilometres north of their natural ranges, with less light in the wintertime and in the absence of sharp frosts.[56] Can we similarly expect evergreen trees in the polar regions to have displayed the same capacity to adapt to the long, warm, dark polar nights?

These different strands of evidence obviously challenge the deciduous view. Yet scientists still refused to abandon their chosen ecological 'religion', even in the face of fossil evidence pointing to evergreen trees in the high southern latitudes.[57] Instead of abandonment, they wrote 'evergreeness may be abandoned at times of high-latitude warmth, because winter respiration rates may be high enough to deplete stored metabolites'.[58] To others, evidence of this sort called attention to the urgent need for a proper experimental evaluation of the assumptions underpinning the winter dormancy as an ecological explanation allowing deciduous tree survival in the polar regions. Eventually a transatlantic collaborative team of scientists seized the baton, setting out on the ambitious path to challenge the orthodox opinion of a generation.[59]

Researchers launched the experimental challenge by cultivating tree species that have a fossil record in the high latitudes. The highly distinctive maidenhair tree (*Ginkgo*), with its fan-shaped leaves, and the needle-leaved conifers swamp cypress and dawn redwood, lined up on the deciduous side, while the broad-leaved southern beech and the needle-leaved coastal redwood lined up on the evergreen side (see Plate 13). Many of these species are what Charles Darwin dubbed 'living fossils', so-called 'resilient witnesses to the passage of time', whose fossil foliage looks similar to those of their modern representatives. Indeed, one or two living fossil species became known first as fossils and were believed to be extinct, only to be discovered alive elsewhere. Dawn redwood is a classic example, presumed to be extinct until an amazing announcement in 1946 reported that living specimens had been found in the Szechwan Province, a remote corner of central China thought to be its last natural refuge.[60] The fossils at Axel Heiberg Island and elsewhere signified its former biogeographical glory and ecological success, when it was widespread in North America, Europe, and Asia.[61] The maidenhair tree also once enjoyed biogeographical glory and ecological success in the Mesozoic, only to undergo a decline and fall that took it to within a whisker of extinction. It is thought to have been saved only by the action of Buddhist monks cultivating it in Japanese temple gardens and monasteries.[62] The maidenhair trees that adorn ornamental parks and gardens derive from seed produced by these far-flung survivors.

In the polar forest experiments, saplings of all five tree species (three deciduous and two evergreen) were raised in replicate high-tech-controlled environments recreating an approximation of the warm Cretaceous polar climate, with its mild winters and hot summers, an artificial polar light regime equivalent to a latitude 69 °N, and carbon dioxide fertilization. The researchers then measured tree growth over the following three years to evaluate the arguments of the deciduous view. If we recast this view as a formal hypothesis to make it falsifiable by observations, it really states that evergreen trees lose more carbon through respiration during the dark polar winter than deciduous trees lose by shedding their leaves. In reality, the situation is a little more complicated because the roots and bare branches of deciduous trees respire and evergreen trees shed leaves over time. If the 'deciduous view' is correct, it predicts the outcome of the experiments shown in Fig. 9. The experimental programme set the stage for pinning down which of these predictions is closest to reality.

What the researchers discovered was surprising and, in its own way, sensational. For it turned out that the profligate disposal of hard-won carbon was the preserve not of the evergreens but of the deciduous trees. Putting numbers to the argument, we find deciduous trees lost 20 times more carbon by shedding their leaves than the evergreens did by respiration in the wintertime. Representing the outcome of these experiments on our carbon balance reveals them to give completely the opposite outcome to that predicted by the deciduous view (Fig. 9). Taken at face value, the experimental results indicate the deciduous leaf habit is a false economy in a Cretaceous polar climate and that being evergreen is advantageous.[63]

Promising though these initial findings were, we cannot yet claim to have successfully overturned the argument of the deciduous camp. We have to recognize and address a number of shortcomings that leave the experimental approach wanting. An obvious issue is that the experiment only represented conditions at a single latitude, 69 ° N. Could the balance be tipped in favour of deciduous trees at higher latitudes where the longer dark polar winters cause more carbon to ebb way from evergreen trees through respiration? Another problem is that the

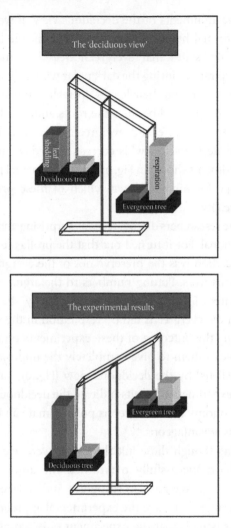

Fig. 9 Carbon balance of evergreen and deciduous trees in the polar winter. The upper set of scales shows the situation implied by the orthodox 'deciduous view', whereby wintertime carbon losses of evergreen trees outweigh those of deciduous trees because of their excessive respiratory losses. The lower set of scales represents the experimental findings. They indicate the opposite pattern, with carbon losses from deciduous trees being larger than those from evergreen trees.

experiment only lasted for three years, so the trees were still young and small. Older, larger evergreen trees accumulate a greater mass of respiring foliage, and larger deciduous trees shed more leaves each year. For this reason, experiments with saplings rather than mature trees could be misleading.

Genetic drift is yet another reason to query the experiments. It arises because no organism is an exact clone of its parents. Small random differences in the genetic code are introduced into the genetic make-up of the offspring of each generation. Over millions of years, accumulated differences result in 'genetic drift', which means that even though we suspect that the species of 'living fossil' being cultivated is the same as that preserved in the fossil record, its genetic make-up is undoubtedly different. Obviously, the similarity of fossil foliage with that of its living descendants implies slow rates of speciation among 'living fossils' for millions of years, but could genetic drift cause the physiology of the antecedent species to differ? This is quite possible, but whether it is sufficient to mislead the research effort is open to further debate.

In short, these issues mean the experiments provide no more than a snapshot of tree performance in space and time under a simulated Cretaceous polar environment. To build a more complete picture taking into account the business of tree size and latitude requires mathematical models of virtual forests.[64] These sophisticated computer models are now one of the new experimental tools of palaeobotany, complementing rather than replacing the trusty geological hammer. It's a development that should hold no surprises in the twenty-first century. Just like real forests, virtual forests have leaves, wood, and roots and require sunlight and rainfall to grow. Governed by mathematical equations describing carbon gain by photosynthesis, carbon loss by respiration, and the synthesis of new biomass, they replicate the behaviour of trees. The models attempt to describe as real-istically as current knowledge allows forest responses to climate change and fertilization by a carbon dioxide-rich Cretaceous atmosphere. Given the appropriate climatic conditions, virtual forests can be grown at any locality on Earth while at the same time computers allow us to track the detailed flow of carbon through the trees.

According to the simulations, shedding leaves is a more expensive strategy for virtual Cretaceous forests in the high northern latitudes than keeping hold of them and paying the respiratory price. The model simulations revealed that the car-bon 'costs' of mature deciduous forests are double those of evergreen forests in a polar environment, irrespective of latitude. To be sure, the respiratory burden of evergreen trees increases with latitude, as we would expect, but the increased burden is nowhere near the cost of shedding leaves. A simple cross-check on these findings can be made with a back-of-the envelope calculation to see how the information gained from experiments might translate into bigger trees. When this is done, the two approaches are found to tally with each other.[65] All of which leads us to the general conclusion that the lower wintertime respiration costs of deciduous trees are far from sufficient to compensate for the higher price they incur by annually shedding leaves.[66]

We have seen powerful evidence drawn from experiments and computer models of virtual forests which collectively spells the end of the long-held deciduous view. But the story doesn't end there because new evidence from Cretaceous fossils has come to light that hammers the final nail in its coffin. Recent excavations and re-examination of fossil floras from Antarctica and New Zealand indicate that they contain an unexpectedly high proportion of evergreen taxa.[67] Unexpected, that is, if the dormancy of deciduous trees is an adaptation to survive polar winters.

The predominant worldview of polar forests must now undergo revision as the evidence forces the rejection of a belief mistakenly fostered by scientists for over 70 years. Hickey and colleagues were right to be puzzled by the predominance of deciduous trees in forests that lived during a polar heatwave. What they did not foresee was the big difference in the small amount of carbon expended for energy production by a respiring evergreen tree relative to the large amount lost by deciduous trees shedding leaves in the autumn: the relative cost of the second strategy is at least double that of the first. With hindsight, the deciduous view was perhaps ideologically driven, a reactionary explanation to the realization that the deciduous leaf habit seems to offer an obvious explanation for how trees coped with life in long, dark polar winters. As we rethink the significance leaf habit holds for polar forest ecology in the light of this new research, textbooks may need updating.

We are left with an intriguing apparent paradox. If deciduous trees in a polar environment pay a higher price for survival in the winter, why did they prosper? Surely, this would see them quickly replaced on the ecological stage by evergreen trees. The answer is that deciduous trees have one last card to play in the tricky game of polar survival.[68] Living fast and dying young, their leaves capitalized on the short, hot summers by photosynthesizing fast and outpacing those of the evergreen trees. Over the entire year, the race between deciduous hare and evergreen tortoise is an even one. Deciduous trees put all their growth on quickly in the space of a short summer. Evergreens, altogether more sanguine in their capture of carbon dioxide, achieve slow, steady, year-round growth when the daylight and temperature permit it. In the end, trees with both leaf habits are left on an equal footing, achieving the same productivity by different patterns of carbon capture over the year.[69]

The mystery surrounding the deciduous nature of polar forests endures. But at least one erroneous explanation can be crossed off the list. New directions must be sought if we are to progress and achieve a deeper understanding of leaf habit and polar forest ecology. Nature provides a trail of varied and confusing clues.[70] Why do evergreen yews and deciduous beeches, which enjoy similar climates and soils in Britain and both cast dense shade, adopt such opposing strategies for holding onto their leaves? Why does deciduous swamp cypress hang on in the frost-free deep south of the USA while deciduous larches go further north than evergreen conifers? Although resolving these questions may seem relevant only to academics cloistered in their ivory towers, this is not the case. There are important climatic consequences to be reckoned with. In the wintertime, a landscape covered with deciduous forest allows snow to settle, which helps reflect the Sun's energy back into space, keeping the climate near the ground surface cool. But a landscape of dark, conical-shaped evergreen trees prevents snow from settling and absorbs the Sun's energy, heating up the atmosphere. Because of these differences, the possible northwards migration of evergreen and deciduous species in our boreal forests with future climatic warming will exert contrasting feedbacks on climate.[71] If evergreen forests extend into areas currently occupied by tundra vegetation, then anthropogenic warming will be amplified. If deciduous forests make the move, anthropogenic warming will be less affected. Successful predictions of climate change in the coming decades necessitate some answers to the ecological questions posed by the conifers.

One promising avenue for better understanding both polar forests and their modern-day counterparts in cold climate zones is signposted by the work of Russian botanists, who have a long tradition of studying the conifer forests of northern Siberia. For over 70 years, Russian botanists have known that patches of evergreen pines coexist alongside patches of deciduous larches.[72] The reason is believed to lie in the soils. Siberian evergreen pines occur only on well-drained sandy soils and cannot tolerate the heavier soils, preferred by the deciduous larches, that freeze more easily in the winter and thaw more slowly in the spring. The larches, then, often experience warm spring days when their roots are frozen. For evergreens, these circumstances create the disastrous situation of a frost-drought, the loss of water by transpiration that is irreplaceable through roots frozen into the ground. Inevitably, dehydration of the delicate machinery involved in photosynthesis ensues, causing irreversible damage and a terminal decline in the health of the trees.[73] Deciduous larches avoid the danger of a frost-drought by delaying their canopy development until after the soils have thawed in early summer.

Another related possibility is that the nutrient content of the soils plays an important role in determining the distribution of evergreen and deciduous conifers.[74] Having leaves that live fast and die young makes deciduous trees more nutrient demanding: the nutrients are required for enzymes to better absorb carbon dioxide and manufacture biomass over the summer. Evergreen trees, though, dwell contentedly on nutrient-poor soils that support their slower pace of life. It turns out that better drained, finely textured soils contain fewer nutrients than the heavy soils, a feature favouring pines, whereas the heavier soils with more nutrients favour the larches. There is also a subtle feedback at work here, further reinforcing the division between the 'have' and 'have not' soils. On being shed, long-lived pine needles decompose and release the nutrients stored in them very slowly, helping to starve the soils of nutrients and keeping the nutrient-demanding larches at bay.[75] Conversely, the shorter-lived foliage of the larches decomposes more readily, helping to maintain higher nutrient levels. It is quite possible that ecosystem feedbacks like this, where leaf lifespan and soil nutrient cycling mediate connections between the overlying vegetation and soils, can help explain the biogeography of polar forests. Progress in investigating this particular avenue of research is now a step nearer, thanks to the recent development of an approach for inferring leaf lifespan from the detailed cellular properties of the annual growth rings in fossil woods.[76]

The two different strategies—fast-growing deciduous trees and slow-growing evergreen trees—offer a further ecological clue to explaining why the former dominated the polar forests. Fire frequently swept through polar forests, as the abundant fossil charcoal present in the sediments testifies, just as it does in North American boreal forests today, which experience a burn on average every 50–60 years. Fast growing trees that can rapidly recolonize the devastated area are favoured in such fire-prone environments, as they quickly tower above their ponderous evergreen neighbours, forcing them to tolerate cooler, shadier conditions. If the fires are sufficiently frequent, evergreens never quite get a sufficient grip on the situation to assert their dominance, leaving deciduous trees to take over.[77]

An intriguing spin-off from thinking about controls on the distribution of conifers is that it enables predictions about what might happen with future climatic warming. Some climate modelling forecasts suggest winter snowfall in the high latitudes will decrease as the climate warms up, causing soils to get colder as the thermal insulation from the cold winter air above is reduced.[78] According to our first hypothesis, colder soils in wintertime should favour the deciduous trees, because of increased risk of frost-drought. If a warmer future climate also leads to warmer summers, as expected, then soils will warm up, stimulating the activities of the microbes involved in unlocking nutrients from the organic matter. According to our second hypothesis, soil nutrient release should also favour deciduous larches. From what we know so far, both hypotheses point towards deciduous trees advancing into the high latitudes in a future warmer world at the expense of the evergreens.

Transferring these ideas to polar forests could also offer additional insights into the question of why deciduous trees out-populated the evergreens in the high northern latitudes, alongside the fire history considerations I have already outlined. If the mild winter climates of the Mesozoic and early Cenozoic were punctuated with occasional hard frosts, frost-droughts favouring deciduous trees might be important. But with limited evidence for hard winter freezes, soil nutrient availability is perhaps more likely to hold the key. Information on the nature of ancient Arctic soils during the Eocene and Cretaceous is sparse, but what we do know suggests the soils were heavy and prone to waterlogging rather than being finely textured free-draining affairs. Heavy soils retain more of the nutrients released by decomposition in the mild winters; this situation could hand the advantage to faster-growing, nutrient-demanding deciduous trees. Past and present forest ecology may yet be united on a firmer footing, but only when intuitive guesses are combined and evaluated against observations in the real world.

There is now an urgent need to translate speculation of this sort into hard science. The polar regions are warming. In the last three decades, the northern high latitude Alaskan winters have warmed by 2–3 °C, over three times the global average.[79] Already vegetation is on the move. Comparisons of modern and historical aerial photographs of the landscape around the Ayikak River in Alaska at 68 ° N reveal that woody shrubs and spruce trees are advancing northwards, encroaching into the tundra.[80] The study area is sufficiently remote for the authors to discount human disturbance, and the finger of suspicion points towards climatic warming. What irony if the industrial activities of humans alter the chemistry of our atmosphere to warm the climate system and offer the ultimate in conservation— the resurrection of a long-extinct forest biome.

But there is more than irony at stake here. The rate of warming on the continents surrounding the Arctic Ocean is accelerating; in the 1990s it was double what it had been at any time since measurements began in the 1960s. The accelerated warming appears to be primarily the result of seasonal snow melt.[81] As the seasonal snow-cover period is shortened, it reveals progressively more of the darker vegetation and soils beneath, which in turn absorb more solar radiation and warm the atmosphere. Meanwhile, polar vegetation is changing as shrubs and trees advance northwards in tune to a warming climate. If we cross a threshold, and the replacement of low-lying tundra ecosystems with taller shrubs and trees becomes sufficiently widespread, climate will dance to the beat of vegetation. If that situation happens, the pace of climate change would be set to accelerate further.

Nearly a century has passed since Scott of the Antarctic's fossils sparked the debate about polar forests into life. That time has witnessed a tremendous growth in our knowledge of this magnificent phenomenon, which has clarified many early questions and seen many other challenging ecological questions remain tantalizingly out of reach. As the modern analogues slowly begin to reassemble in the high latitudes, with major implications for our future climate, finding some answers to those questions becomes a pressing concern. For the moment, it seems fitting that Seward should have the last word. His thoughts in 1914 after examining Scott's fossils still ring true today:

> The heroic efforts of the Polar Party were not in vain. They have laid a solid foundation; their success raises hope for the future, and will stimulate their successors to provide material for the superstructure.

The debate is far from over.

7

Paradise lost

Exotic fossil floras and faunas from the Arctic spoke clearly of balmy conditions in the Eocene, 50 million years ago, when there was little difference between equatorial and arctic temperatures. Yet these conditions are inexplicable by climate models using solely a higher carbon dioxide content of the ancient atmosphere. Remarkable climate archives written in deep ice-cores provided vital clues to the reasons why. They revealed that, as the climate had warmed since the last ice age, the atmospheric content of other important greenhouse gases, like methane, also rose. When the cascade of missing connections describing these effects is included in Eocene climate models, carbon dioxide's greenhouse warmth is amplified. Yet the models still fall short of reproducing the Eocene's balmy polar conditions. Until they can, the cautionary lesson from the past is that we may be underestimating the climatic consequences of loading our atmosphere with greenhouse gases.

One minute it darted off like a kingfisher, and the next it entirely disappeared. At times it grew as big as an ox's head, and then straightaway shrank to a cat's eye . . . finally . . . it returned to frisk in the reeds.

George Sand (1848), *La petite fadette*

THE Isle of Sheppey lies in the mouth of the Thames tucked up along the northern coastline of Kent, south-eastern England. Known to the Romans as *insula orivum*, and accessible for centuries only by ferry, the small Isle waited until 1860 for the construction of its first permanent bridge, over the River Swale to the mainland. It contains an uneasy mixture of lowland agricultural farmland, tourism, and commercial shipping activities, all divided by a diagonal east-to-west line of low hills. Elmley Marshes, situated on the southern side of the Isle, attract thousands of ducks, geese, and wading birds in the winter. Further to the east lies the Swale National Nature Reserve, a mosaic of grazing land and salt marshes that is home to short-eared owls and hen harriers. Fine beaches dotted along the northern coastline near to the traditional seaside town of Leysdown-on-Sea draw tourists whose spending boosts the local economy. Discovery of a deep-water channel off the north-west coast saw the construction of a Royal Navy dockyard at Sheerness in 1669. The new dockyard was replaced 290 years later by the commercially successful Port of Sheerness, which benefits from the capacity to accommodate large modern ships regardless of the tides.

Geology and the sea have combined to shape the cultural and economic aspects of the Isle from its earliest days.[1] In the early part of the nineteenth century, pyrite—iron sulfide—collected from the beaches and foreshore provided a source of green vitriol dye for the tanning and textiles industries. At around the same time, a small industry flourished excavating cement stones (septaria) for the manufacture of Parker's (or Roman) cement. But the supply of septarian nodules on the beaches was soon exhausted and, with the emergence of more economic means of producing cement, the industry collapsed. The fleeting septaria industry mirrors the fleeting existence of Sheppey, for the Isle is shrinking fast as wave action erodes metres of its cliffs each year. Ultimately, in no more than a geological instant, the Isle of Sheppey and its inhabitants will be gone.

The vigorous attentions of the sea that will eventually seal the fate of Sheppey also wash out of the cliffs onto the beaches a rich record of Earth's history previously locked away out of sight. The accessibility of the northern coastline between Minster and Leysdown-on-Sea, together with the steady supply of fossils released from the cliffs by erosion, has drawn fossil hunters to the region for over three centuries.[2] Indeed many of those involved in excavating septaria and collecting pyrite gathered fossils from this palaeontological treasure-trove to sell as an extra source of revenue. Edward Jacob (1710–88), the celebrated Kent antiquary, naturalist, and plant collector, published an early attempt at describing the rich variety of fossil plants and animals washed from Sheppey's cliffs in his 1777 book *Plantae Favershamiensis*. Most of the book is a catalogue of the contemporary flora of the area arranged alphabetically, but tucked away in an appendix entitled 'Establishing a short view of the fossil bodies of the adjacent island of Sheppey', are the fascinating descriptions of fossil remains from Sheppey. The appendix shows that Jacob was obviously very struck by the fact that his finds of exotic fossil *Nautilus* shells, palms, and others represented organisms that are usually found in tropical or subtropical regions. For he comments, from the contemplation of so great a variety of extraneous fossils discovered in these cliffs, which were evidently the produce of a very different climate, nothing short of universal deluge could be the cause to such an effect.

Of course, in those days it was natural to interpret fossil evidence of past life in terms of the 'universal deluge'; it had to fit with the 'known facts' as set out in the Bible. However, Jacob's great insight was to realize ahead of his contemporaries that the fossils were speaking of an ancient climatic regime, one evidently much warmer than that of south-eastern England in the latter half of the eighteenth century.

Less than a century after Jacob's efforts, the esteemed palaeobotanist James Bowerbank (1797–1877) delighted upon the rich variety of fossils to be had in the same sediments at Sheppey, writing in 1840, 'although one of the most accessible, it is probably one of the least known of any of the rich geological fields that are within a short distance of the metropolis'. He advised would-be geologists, inclined to seek out fossils on the Isle, 'having established yourself at the Inn, request Boots to desire the attendance of Mr. Hays from whom you purchase at a reasonable rate, some good fossils such as crabs, lobsters, heads and portions of fishes and numerous fossil fruits'.[3] Evidently, it was advice well heeded because interest in the fossils grew and soon led to the birth of the Palaeontographical Society in England. In the same year, Bowerbank published his classic work *A history of the fossil fruits and seeds of the London Clay*, documenting his remarkable fossil finds from Sheppey, noting particularly the numerous remains of plants whose climatic preferences were overwhelmingly tropical or subtropical.[4]

Bowerbank's classic was eventually replaced a century later when the English palaeobotanists Eleanor Reid (1860–1953) and

Marjorie Chandler (1897–1983), one the great partnerships in the history of botany, brought out their great monograph systematically describing and naming the numerous species of fossil plants.[5] In doing so, the two ladies assiduously provided a framework for interpreting the ancient environment. They showed that 90% of the genera had nearest-living relatives that are found today with a tropical distribution. Other genera were related to living plants with broader distributions taking in both the tropics and temperate regions. Seeds of the mangrove palm *Nypa* proved to be some of the most revealing fossils from this palaeontological treasure-trove; they represented a decisive indicator of subtropical climate.

We now know that Jacob and the others were collecting fossils on Sheppey from sediments comprising part of the London Clay formation, a deep sedimentary deposit stretching out in a south-westerly direction from the Essex marshes in north-east London to the westerly edge of the North Downs. The formation has long been famous for its fossils that record a remarkable slice of Earth history called the Eocene, between 55 and 34 million years ago. Fossils excavated and catalogued by fossil hunters from the London Clay formation over the past three centuries paint a picture of a southern England in the Eocene very different from that of today. Forty-five million years ago, the region was surrounded by warm tropical seas teeming with sharks, rays, swordfish, and sturgeon. Elsewhere, closer to shore, alligators, crocodiles, and turtles basked on mudbanks and lurked in *Nypa*-dominated mangrove swamps that clad the mouths of river estuaries and coastlines, dwarfed by a spectacular backdrop of humid subtropical rainforest. The closest modern analogue for this type of vegetation is that of the coastal lowlands of south-eastern Asia. The ancient shoreline ecosystems shed fragments of ferns, leafy shoots, conifer cones, and even the trunks and branches of trees, that sank and became fossilized in the sediments on the shallow seafloor, leaving behind an astonishingly rich record of paradise lost.

By the end of the twentieth century, it was becoming clear that the fossils from Sheppey were simply a foretaste of the wealth of trustworthy evidence excavated from rocks and sediments around the world by later generations.[6] In continental North America, for instance, fossil palms dating to the Eocene were found as far north as Wyoming (~45 ° N), when today they are restricted to the (usually) frost-free climates of the coastal regions of southern California and Florida.[7] Farther north, fossils of warmth-loving creatures like alligators, lizards, turtles, and tortoises turned up in sediments from the Canadian Arctic.[8] At the opposite end of the world, fossil fronds, fruits, and seeds of *Nypa* palms were discovered in Tasmania at a latitude of 65 ° S, a southern hemisphere counterpart to its English cousin.[9] Venturing further south still, to Seymour Island on the Antarctic Peninsula, we find fossil bones of extinct South American mammals and marsupials alongside strange giant penguins standing over a metre tall.[10] The mammals and marsupials are characteristic of the humid forests of Patagonia today and so testify to the unusual warmth Antarctica enjoyed 50 million years ago.

We can get a sense of perspective on just how warm the early Eocene climate was by fitting it into the bigger picture of climate change over the entire Cenozoic era (the 'Age of Mammals') that covers the past 65 million years. That Cenozoic vision is retrieved in remarkable detail from shifts in the carbonate chemistry of fossil shells of foraminifera, abundant single-celled marine organisms relentlessly archiving deep and surface ocean conditions in sediments that accumulate on the seafloor.[11] The long-term history of Earth's climate revealed by the tiny foraminifera shows us that the early Eocene was warmer than anything

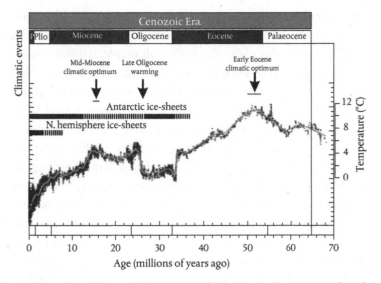

Fig. 10 Trends in global climate over the last 65 million years inferred from variations in the stable oxygen isotope composition of fossil foraminifera shells. Older than 34 million years, the line depicts the evolution of deep-ocean temperatures; younger, and the signal is a mixture of temperature and continental ice volume.

Earth experienced in the entire Cenozoic era (Fig. 10). Ever since the Eocene, global climate has lurched in a rather ungainly fashion towards cooler conditions, as it made the transition to our modern 'icehouse' world with a frozen Arctic and a deeply frozen Antarctica. So Jacob, Bowerbank, and the other fossil hunters had uncovered evidence of Earth in the grip of a prolonged heatwave lasting some 5 million years. The bigger picture afforded by 300 years of fossil discoveries from terrestrial rocks and marine sediments presented scientists with an intriguing climatic puzzle.[12] What caused this exceptionally warm episode—a 'hothouse' world—in Earth history? Why had the

Arctic and Antarctic enjoyed a balmy climate instead of being capped with ice?

Heating up the planet in such a dramatic fashion can be achieved in several ways, but there are two main contenders: altering the circulation patterns of the oceans to more efficiently redistribute heat from the tropics to the higher latitudes, or increasing the concentration of greenhouse gases in the atmosphere. Originally, the oceans were favoured as the key to the enigmatic Eocene climate.[13] Often regarded as the 'heat engine' of the Earth, the oceans have an enormous capacity for storing and redistributing tropical warmth in surface currents that migrate polewards, driven by prevailing wind patterns and changes in salinity that happen en route as water evaporates. The Gulf Stream, which flows from Florida across the Atlantic and into northern Europe, is a well-known example of this heat transport. It ensures that Europe is several degrees warmer than it would otherwise be. One idea, then, for explaining the Eocene paradise was to somehow increase the transport of heat from the equator up to the high latitudes. But there turned out to be some problems with this explanation, because later calculations showed that to get the polar regions warm enough required increasing the poleward transport of heat by a whopping 30%. And in spite of diverting the attention of the world's climate scientists for several decades, no one has yet found a satisfactory means of achieving such a massive shift in either the sprightly circulation of surface ocean or the more stately paced circulation of the deep oceans.[14]

Another oceanic idea auditioned the tropical Pacific as a means of warming the climate.[15] Its behaviour displays marked swings between a warm state—El Niño—and a relatively cool one—La Niña.[16] In the more normal La Niña state, trade winds pile up warm water in the western Pacific and drag cool, nutrient-rich water up from the depths off the coast of South America. In fact, the sea surface is about half a metre higher at Indonesia than at Ecuador. During an El Niño event, which occurs every few years, the prevailing westerly trade winds relax, allowing warm waters from the western Pacific to flow eastward. The interest in El Niño patterns for the Eocene is in how it influences climate. In La Niña mode, the eastern Pacific absorbs heat from the tropical atmosphere and dissipates it into deeper layers below. But in an El Niño year the shallow and deep oceans become so warm that the atmospheric heat has nowhere to go, causing gentle global warming and other strange climatic effects.[17] These are typified by increased heat transport away from the tropics[18] and additional continental warmth over North America,[19] both of which may help to explain the Eocene 'hothouse' climate puzzle.

Is it possible that the Earth locked into a permanent El Niño-like climate state, in which the tropical Pacific, unable to cool the atmosphere, warmed the world? It is an exciting idea because it could solve at a stroke many of the puzzles about the warm climate. Yet a series of state-of-the-art computer model simulations revealed that the tropical Pacific of the Eocene world was remarkably resistant to the idea of flipping into the permanent El Niño state. Rather than El Niño becoming a permanent fixture of the Eocene, it occurred at about the same frequency as it does now, but with a difference—the oceans underwent thermal stratification, a process that tends to slow rather than increase the export of heat to the tropics.[20]

As researchers struggled to find some answers in the oceans, they switched their attention from the sea to the sky. Carbon dioxide is, after water vapour, the most important greenhouse gas because it is very effective at providing a thermal blanket around the Earth, trapping and returning heat that would otherwise escape back to space. So it is a major contributor to global warming and this raises the question of whether higher levels of carbon dioxide might be responsible for the global warmth of the Eocene? It is an obvious question to ask, but a difficult one to answer. The Earth is reluctant to yield the secrets of her atmosphere and the business of gauging its carbon dioxide content 50 million years ago is a formidable task.

No consensus has been reached between the different approaches employed to achieve this goal, with researchers seeking clues to it in different ways obtaining different answers. One team attempted to breathalyze the Eocene atmosphere by using fossil leaves of the maidenhair tree (*Ginkgo*) from North American Eocene rocks, and calibrated their observations by growing *Ginkgo* trees under a range of atmospheric carbon dioxide levels.[21] Much to everyone's surprise, the leaves revealed carbon dioxide levels similar to those in today's atmosphere. Another group of researchers reconstructed Earth's ancient carbon dioxide content in an entirely different manner and obtained a totally different answer. They focused on the boron isotope composition of fossil carbonate shells of foraminifera.[22] The approach relies on the fact that ocean acidity is sensitive to the carbon dioxide content of the atmosphere, and that it changes the boron isotope composition of the carbonate shells of foraminifera living in the oceans. According to this rather indirect method, the Eocene atmosphere contained 10 times as much carbon dioxide as it does today. Perhaps these two wildly different estimates should most charitably be viewed as setting the lower and upper bounds on the 'actual' carbon dioxide level in the atmosphere.

Where the real value might lie between these extremes is suggested by yet another means of breathalyzing Earth's ancient atmosphere. This one relies on analysing the abundance of carbon isotopes left behind in the fossilized remains of marine algae in deep-ocean sediments. During photosynthesis, algae preferentially incorporate the lighter carbon isotope, ^{12}C, over the heavier one, ^{13}C, and the effect becomes more marked when more carbon dioxide is dissolved in the seawater from the atmosphere above. By analysing the carbon isotopic compos-ition of fossil algal remains down through successive layers of sediment, the approach generates persuasive graphs of the rise and fall in carbon dioxide levels over millions of years. However, like the other methods I have touched on, it is far from bullet-proof because other factors, like how quickly the algae grow and how big they are, can bias the results. Nevertheless, bearing these issues in mind, we find that the fossil algae suggest that the atmosphere contained up to four times as much carbon dioxide 45 million years ago as it does today.[23]

So, a carbon dioxide-rich atmosphere seems to offer a good solution to the Eocene climate puzzle. And yet...when four times the present-day carbon dioxide content is fed into the climate models, they spectacularly fail to reproduce the tropical paradise revealed by the exotic fossil floras and faunas.[24] Instead of the poles in winter being as balmy as a summer day, the models blanket the polar regions, and much of the northern hemisphere, with snow. In fact, to warm up the Earth, melt the snow, and create tropical conditions for palms and alligators to flourish in the high latitudes, it takes eight times the present-day level of carbon dioxide. But the problem with this carbon dioxide-rich solution is that it overheats the tropical oceans.[25] All of which suggests that the ancient climate of the Eocene is telling us something important—our current climate models, which rest upon a century of observations and are the best tools we have for forecasting future climate change—are still incomplete. They seem to be missing those mysterious feedbacks hidden within the Earth system that have the capacity to amplify the warming caused by carbon dioxide. Identifying these feedbacks to build a more complete picture of how planet Earth functions, remains a major scientific challenge.

In the present, politically charged debate over global warming, carbon dioxide gets all the press. Stories of movie stars and rock stars, renowned for their extravagant lifestyles, easing their consciences by planting forests or driving eco-friendly hybrid-electric cars to reduce carbon emissions make eye-catching copy, but do nothing to draw attention to the contribution of other greenhouse gases to global warming.[26] One reason these 'other gases' are often brushed aside in debates over climate change is their presence in the atmosphere in seemingly trivial amounts compared to carbon dioxide.[27] The contemporary atmosphere, for example, contains 385 parts per million of carbon dioxide, although the figure is rising at a rate of roughly 2 parts per million annually. Think of this concentration as being equivalent to diluting 385 cartons of milk, each holding 2.5 litres, into an Olympic swimming pool.[28] Methane, on the other hand, is the third most abundant greenhouse gas, after carbon dioxide and water vapour, and is present in the atmosphere in an amount equivalent to just short of two milk cartons diluted in our swimming pool. Ozone and nitrous oxide, the two other 'big three' greenhouse gases besides carbon dioxide, are present in still lower concentrations and are represented by the addition and dilution of 150 ml, and 775 ml, of milk in the pool, respectively. By themselves, these other gases don't amount to much, but the important point to grasp, as we shall see, is that collectively methane, ozone, and nitrous oxide can exert a powerful influence on the climate system.

The crucial link in the chain connecting these 'other' gases with the warm Eocene climate was forged in the crucible of Victorian science over a century and a half ago by the discoveries of the distinguished Irish scientist John Tyndall (1820–93) (Plate 14). One of the leading lights of Victorian science, Tyndall was an accomplished lecturer, experimentalist, and theorist, who succeeded Michael Faraday (1791–1867) as superintendent of the Royal Institution of Great Britain. Before Tyndall, scientists were only interested in the absorption and transmission of heat by solids and liquids because of the exceptional difficulty involved in making equivalent measurements with gases. Tyndall, though, cracked the problem for gases by constructing a new scientific instrument, the ratio spectrophotometer (Fig. 11), exclaiming in

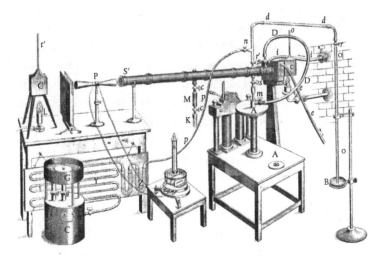

Fig. 11 John Tyndall's ratio spectrophotometer.

his journal: 'experimented all day; the subject is completely in my hands!'[29] The instrument consisted of two radiation sources, one shone through a long tube filled with the gas of interest and one directly opposite behind a simple screen. Small cones focused the radiation from each source onto a differential thermopile connected in series to a galvanometer. To determine if a particular gas absorbed infrared radiation, he would fill the tube with it and the difference in intensity of the radiation striking the thermopile caused differential heating and a deflection on the galvanometer—a large deflection meant a strong greenhouse gas filled the tube. This ingenious apparatus still survives today, housed in the Royal Institution in Piccadilly, London, where Tyndall was elected 'Professor of Natural Philosophy' and became a persuasive evangelist for the cause of science.

In his research on gases, Tyndall convincingly demonstrated that 'perfectly colourless and invisible gases' differed enormously in their abilities to absorb radiant heat. He correctly identified carbon dioxide, methane, nitrous oxide, and ozone as effective greenhouse gases and noted that neither oxygen, nitrogen, nor hydrogen had any significant greenhouse properties.[30] The energy in infrared radiation (Tyndall's radiant heat) was insufficient to break the chemical bonds within the molecules of the gases and split them apart into their constituent atoms. Consequently, it excited the molecules in those gases with an asymmetrical arrangement of three atoms (carbon dioxide, nitrous oxide, and ozone), increasing molecular vibrations that transfer heat to the air when the molecules collide. Methane, nitrous oxide, and ozone, are all potent greenhouse gases despite being present in the atmosphere at relatively low concentrations because, molecule for molecule, they are more effective 'heat trappers' by operating in an uncluttered part of the electromagnetic spectrum. In fact, methane, nitrous oxide, and ozone are, in that order, about 25, 200, and 2000 times more effective at causing planetary warming than carbon dioxide.[31]

All this must have been exciting enough, but Tyndall's most startling discovery was that of the important greenhouse gas water vapour. He was stunned to find it absorbs 80 times more heat (infrared radiation) than pure air and was quick to realize its importance for climate. 'Water vapour', he said, 'acts more energetically upon the terrestrial rays than the solar rays; hence, its tendency is to preserve to the earth a portion of heat that would otherwise be radiated into space'.[32] Alert to its profound climatic significance, he commented,

> this aqueous vapour is a blanket more necessary to the vegetable life of England than clothing is to man. Remove for a single summer-night the aqueous vapour from the air which overspreads this country, and you would assuredly destroy every plant capable of being destroyed by a freezing temperature. The warmth of our fields and gardens would pour itself unrequited into space, and the sun would rise upon an island held fast in the iron grip of frost.[33]

Understandably taken by his discovery of this suite of greenhouse gases, Tyndall speculated, presciently as it turned out, that they might explain 'all the mutations of climate which the researches of the geologists reveal'.[34] Only with the recovery and analysis of the world's most extraordinary archives of past climates, a century and a half later, did the nature of this connection begin to crystallize. The deep ice-cores extracted from the frozen caps of Antarctica and Greenland preserve not only historical records of climate but also of the atmospheric content of a handful of Tyndall's greenhouse gases. Some of the most remarkable ice-cores are those drilled from the East Antarctic ice-sheet,[35] with the deepest reaching back an astonishing 740 000 years and taking in the last eight climate cycles.[36] The Antarctic ice-cores provide compelling evidence for the link between greenhouse gases and climate change. They demonstrate that the atmospheric content of carbon dioxide, methane, and nitrous oxide has marched up and down with the repeated swings in climate for hundreds of thousands of years caused by wobbles in the Earth's orbit that perturb the geographical and seasonal distribution of solar energy striking its surface.

One of the many significant contributions these exquisite records of past global change have made to our understanding of ancient climates has been to show that, as the climate warms up, the methane content of the atmosphere rises. Looking in detail at the record in an ice-core drilled from the summit of the Greenland ice-sheet, scientists from the Scripps Institute of Oceanography at the University of California, San Diego showed that as the last ice age came to a close, the atmospheric methane content rapidly increased several decades after the climate started to warm.[37] In other words, rising atmospheric methane concentrations are a consequence of a warmer climate. Methane is, like carbon dioxide,[38] a slave to the climate system and this points to the existence of an important positive feedback loop: a warmer climate causing a higher methane concentration, which in turn drives more warming via the methane greenhouse. What's more, this mechanism for amplifying global warmth is missing from climate models. It operates because methane is produced by primitive, swamp-dwelling, anaerobic microbial organisms, dubbed methanogens; they obtain energy by consuming organic material and produce methane as a by-product of their metabolism. After the methane bubbles out from the swamps, it floats above marshes, shrouding them in an eerie pale haze that occasionally ignites and burns—the will-o'-the-wisp that intrigued the French nineteenth-century writer George Sand (1804–76).

Climate determines the amount of methane bubbling out of swamps because it exerts a strong control on the metabolism of methanogens and the decomposition of the organic matter they consume. On a grander scale, climatic change can also influence the productivity and extent of the northern wetlands. All of these factors can act to boost or slow down methane production in marshlands and other wetlands when the climate warms or cools.[39] It is also even possible that if the carbon dioxide content of the atmosphere rises, it further stimulates methane emissions from wetlands—positive feedback loops abound within system Earth. Recent research has shown that fumigating wetlands in Chesapeake Bay with carbon dioxide stimulated emissions by 80% as the methanogens benefited from the extra supply of organic matter pumped below ground by the plants.[40]

Moreover, it doesn't end there because the nature of the gas means that Tyndall's favourite greenhouse gas, water vapour, also has an important part to play. For unlike carbon dioxide, which does not significantly influence the chemistry of the atmosphere, methane is chemically reactive. Each molecule of methane survives for about nine years before being destroyed by its conversion into water vapour and carbon dioxide by highly reactive chemical scavengers known as hydroxyl radicals. Hydroxyl scavengers are produced when ozone is cleaved high up in the atmosphere by ultraviolet radiation from the Sun, in the presence of water vapour. These short-lived molecular fragments act as the detergent of our atmosphere, scrubbing out contaminants from the air. The point is that when a small fraction of water vapour derived from methane passes into the very dry stratosphere, it acts as a powerful greenhouse gas in its own right. In all, water vapour derived from methane is estimated to enhance the climatic effect of methane by about 15%.[41]

Although the picture is less clear and rather more complex for nitrous oxide, there is also evidence from ice-cores that it behaves in a similar fashion to methane, with the atmospheric content of this greenhouse gas rising as the climate warms.[42] Again, as with methane, microbes are the culprits. Those dwelling in the soils of tropical regions produce two-thirds of the nitrous oxide emitted by terrestrial ecosystems and their emissions of the gas are closely linked to precipitation and temperature. This means emissions from soils respond to climate change. Microbes in the oceans, too, are a significant source of the gas, contributing a third of the total amount released into the atmosphere.[43] As with methane, biology reaches out to climate through microbially mediated reactions in soils and water.

We can obviously see from all this that pioneering research in the laboratory of a Victorian physicist is linked with the high-tech instrument-laden laboratories of modern physicists and atmospheric chemists working on ice-cores drilled from frozen ice caps. They have uncovered possible positive feedbacks between climate and methane, and climate and nitrous oxide, that could have amplified global warmth in the Eocene. Imagine the situation. Microbial powerhouses lying within stagnant marshlands, endlessly pumping methane into the ancient atmosphere, while their counterparts in tropical soils churn out nitrous oxide. The methanogens are in turn 'fed' by a plentiful supply of organic material produced by the forests fertilized in a carbon dioxide-rich atmosphere.

What's more, all of these effects were magnified in the Eocene because the primeval wetland methane factories and the tropical forests were vast, and would have dwarfed their contemporary counterparts. Without continental ice-sheets to lock up huge reservoirs of water, high sea levels flooded coastal areas, creating extensive river deltas and tidal low-lying salt marshes of mangroves, like those uncovered by Edward Jacob on the Isle of Sheppey. Inland, the warm humid climate saw the formation of vast freshwater wetlands in the tropics and tundra alike. The widespread Eocene-aged coal deposits left behind in the rocks are the telling signature of the former glory of Eocene wetlands. Located mainly on land, but also around the margins of the ancient coastlines, the give away 'coaly' sediments covered perhaps six million square kilometres, an area over three times that of modern wetlands.[44] In fact, the true figure could be considerably higher because estimating their original extent, before erosion, is uncertain. The ancient coalfields also suggest that the low latitudes supported widespread tropical swamps, which today coincide with unexpectedly high methane concentrations in the tropical atmosphere.[45]

All of which points to a methane greenhouse as a tantalizing part of the explanation for the Eocene warmth, with the involvement of other greenhouse gases also very much on the cards. Indeed, a decade earlier a team of geoscientists, when based at the University of Michigan, were quick to recognize the signs of methane written in the rocks, and have gone down in history as early champions of methane as a powerful agent of Eocene warmth.[46] Recognizing this possibility is an important step forward, but dealing with it is quite another. The catch is that, unlike the situation that exists for carbon dioxide, we have no means of analysing geological materials to determine the content of methane, or indeed any of Tyndall's other greenhouse gases, in the ancient atmosphere beyond a million years, the age of the world's oldest ice-sheets. As a result, climate modellers investigating the causes of the Eocene warmth choose to use levels that are the same as those found in the atmosphere during pre-industrial times—an assumption that is, we have seen, almost certainly incorrect.

The issue at stake here is the role of these enigmatic gases in the Eocene climate puzzle. It is perhaps a reflection of the times that help in building a better picture came not from old-fashioned detective work in fossil-rich quarries but with the dawning of a new era in the accuracy of climate predictions—one heralded by the arrival of so-called Earth system models. In Chapter 1, I pointed out that the development of these models, aiming to describe the Earth as an integrated system, has been likened in significance to the Copernican revolution. They represent a hugely significant addition to the repertoire of those studying the behaviour of our planet. At the top end, state-of-the-art Earth system models are extremely complex and command the attentions of supercomputers. Their ambitious aim is to describe nothing less than the operation of system Earth with the mathematical equations governing the physics and chemistry of the oceans and the atmosphere, and the essential biological processes of the living world. What makes them powerful scientific tools is that the different components—the oceans, atmosphere, cryosphere, and biosphere—interact with one another as realistically as current knowledge allows to better reproduce the real world behaviour of our planet.

All such models have as their starting point solar radiation from the Sun, which drives global patterns of weather and climate in combination with the Earth's rotation. The models extend over the entire surface of the planet, from ground level up into the stratosphere; they divide space up into packets or grid cells of a few hundred square kilometres, bounded by latitude, longitude, and altitude, which exchange air parcels containing gases, particles, energy, and momentum with their neighbours. In spite of being state of the art, we should recognize that these models contain many uncertainties in determining the behaviour of critical physical and biological processes, uncertainties that reflect gaps in our knowledge. Clouds, for instance, are one of the least understood features of the climate system yet can exert a huge influence on it, often by forming and acting on scales smaller than the few hundred square kilometres of the grid cells in the models. Nevertheless, these uncertainties notwithstanding, Earth system models remain the best tools we have for conducting 'experiments' with our planet, and their capacity to make accurate simulations continues to evolve hand-in-hand with new discoveries.

The advent of Earth system models provided scientists with a superb opportunity to harness powerful computer technology to provide a more solid theoretical footing for the possible role of methane and other greenhouse gases in amplifying early Eocene carbon dioxide warmth. The first suite of Earth system model simulations investigating the lost paradise of the Eocene was reported by a UK-based team of researchers.[47] The UK scientists performed a range of Eocene climate simulations with up to six times the present-day carbon dioxide concentration to bracket the uncertainty I described earlier. In each case, the potential for feedbacks was incorporated by allowing the climate to influence the emissions and atmospheric chemistry of the gases released by the terrestrial and marine biospheres. In turn, the newly computed levels of greenhouse gases were allowed to influence climate. In other words, for the first time, microbial and vegetation processes interacted with climate and the chemistry of the atmosphere. Only when a stable equilibrium climate was obtained were the simulations halted to examine how the levels of the greenhouse gases and climate had changed.

What the Earth system simulations showed was dramatic. They revealed a high atmospheric content of a range of different greenhouse gases (methane, nitrous oxide, and ozone), higher than anything assumed in most previous investigations, and a feedback on climate strongly amplifying the warming caused by carbon dioxide.[48] Inside the virtual world of the computer simulations, the methanogens, stimulated by the warm climate, dominated the chemistry of the atmosphere, pushing the methane content up to over 3500 parts per billion—twice the present-day concentration. Its continual replenishment from the wetlands maintained the high methane content, and at the same time its continual destruction by oxidation added water vapour, a powerful greenhouse gas, into the drier stratosphere. Meanwhile, as the swamp-dwelling microbes produced methane, their microbial counterparts in tropical forest soils flourished in the warm wet climatic regime, manufacturing prodigious amounts of nitrous oxide.

Perhaps all this was, in some ways, no surprise. The Earth system simulations had revealed some of the secrets of the ancient sediments, but they uncovered an altogether more surprising one by showing that levels of the potent greenhouse gas ozone had increased by some 50% in the troposphere, the lower part of the atmosphere. The reasons for this were first exposed in a classic piece of research undertaken by the Dutchman Arie Haagen-Smit (1900–77), who had been investigating the severe damage to crops in the Los Angeles area in the mid-1940s that was traced to high ground-level ozone levels.[49] At that time, it was widely believed that ozone formed only in the stratosphere by the cleaving of molecular oxygen by sunlight, and scientists were puzzled about the origins of the unexpectedly high ozone levels.[50] Haagen-Smit showed that it was produced when hydrocarbons and nitrogen oxides released by vehicles and industry reacted together in the presence of sunlight. Ozone is a major component of urban smog and the key role played by nitrogen oxides in Haagen-Smit's 'smog reactions' was dramatically illustrated during the electrical blackout in North America in August 2003. Over a hundred power stations were affected, causing misery to millions of people in the north-eastern United States and south-eastern Canada.[51] Still, alert to the possibility of conducting a unique experiment, a research team from the University of Maryland analysed air samples taken over Pennsylvania during the blackout and compared them with those from the previous summer when the power plants were fully operational. Within a day of the blackout, air quality improved and tropospheric ozone levels fell by 50%.

Of course, we find no industrial activity emitting hydrocarbons in the Eocene. Instead, the ancient forest took the place of industry by releasing a class of hydrocarbons known as volatile organic compounds—the chemicals responsible for the fragrant smell of pine and eucalyptus forests. Forests cloaking the ancient landmasses from the equator to the poles acted as chemical factories whose emissions, stimulated by a warm climate, fuelled ozone synthesis in the troposphere.[52] Lightning flashes, and the forest fires they ignited in the stormy tropical atmosphere added nitrogen oxides to the cocktail of gases, which reacted in the presence of sunlight to produce tropospheric ozone.

Under certain circumstances, forests generate ozone in the same way today. In the foothills of the Sierra Nevada Mountains, California, oak trees naturally emit a reactive hydrocarbon called isoprene, which rapidly reacts with wind-borne nitrogen oxides from industrial pollution to form ozone.[53] In fact, on occasions, isoprene emitted by the oak forests is responsible for 70% of the ozone in the region through the reactions uncovered by Haagen-Smit's pioneering studies. It is because ozone can also contribute to smog formation that presidential candidate Ronald Reagan famously commented in 1980, 'approximately 80% of our air pollution stems from hydrocarbons released by vegetation, so let's not go overboard in setting and enforcing tough emission standards from man-made sources'.[54] The media hung Reagan out to dry for this, but he was at least partly right. Ozone's participation in the smog reaction can be delightfully demonstrated by placing squeezed orange peel into a jar of ozone. As fine jets of volatile oils spray out of the peel, they react with ozone to form white streamers in the jar.[55] The forested areas of the Smoky Mountains in the United States and the Blue Mountains in Australia are often shrouded in 'blue hazes' due to this effect.[56] Perhaps the ancient forests of the Eocene world were cloaked in a similar eerie haze?

The high ozone levels in the Eocene atmosphere pose something of a paradox because it is well known that tropospheric ozone is a pernicious pollutant that can have detrimental effects on the health of forest ecosystems. Indeed, its toxicity to plant communities was first observed downwind of Los Angeles.[57] Could Eocene ozone levels spell trouble for the health of the forests? The answer is, probably not. Borrowing from the findings of research conducted to determine how future increases in background ozone levels might affect forest ecosystems, we find that trees are afforded a degree of protection from ozone injury when carbon dioxide levels are high.[58] Protection is conferred by the partial closure of the microscopic stomatal pores on the surfaces of leaves that happens in response to a carbon dioxide-rich atmosphere. This has the effect of limiting the entry of damaging ozone into the leaf. Impressive large-scale field trials conducted with forests in northern Wisconsin have reported carbon dioxide fertilization can not only ameliorate ozone damage to forests, but actually reverse it.[59]

The question is: how did an atmosphere loaded with the greenhouse gases methane, water vapour, ozone, and nitrous oxide amplify the warming caused by carbon dioxide? A gauge for the possible effects can be obtained by estimating how they perturb the energy balance of the planet by preventing heat from the Sun escaping the Earth's surface and atmosphere back into space. The capacity of greenhouse gases to warm the planet by acting in this manner is termed the 'forcing' of climate. James Hansen, Director of the NASA Goddard Institute for Space Studies, explains the concept well by using an analogy with Christmas tree lights.[60] He points out that a miniature Christmas tree bulb emits about one watt of heat and human-made greenhouse gases are heating the Earth's surface at the rate of about two watts per square metre. Humans have effectively placed two Christmas tree light bulbs over every square metre of the Earth. This would seem to create a paradox: how can such feeble heating alter global climate? The answer is, by persistence. The extra heating is maintained day and night over many years until it eventually warms the surface oceans and thus the entire climate system.

The concept of forcing helpfully allows us to standardize the capacity of different greenhouse gases to warm the planet—it is analogous to the stock-market practice of expressing foreign currencies relative to the pound sterling or the dollar. It offers a means of divorcing the straightforward perturbation of the energy balance from the more complex response of the climate system. In the Eocene, the additional radiative forcing caused by the high levels of methane, nitrous oxide, tropospheric ozone, and, indirectly, by the production of water vapour from methane, was approximately equivalent to two and a half Christmas tree lights per square metre. If this appears rather small, remember that this represents extra climate forcing that was being missed in earlier climate modelling studies of the Eocene world. If the carbon dioxide concentration in the Eocene atmosphere was six times the present-day value, rather than double, the warmer climate produces correspondingly higher levels of all these greenhouse gases. In effect, the strength of this positive feedback loop increases with rising carbon dioxide levels.

These sorts of simulations signify a positive feedback loop between Eocene climate and greenhouse gases. Collectively, the levels of these gases that developed with an Eocene climate warmed by double the present-day carbon dioxide content contributed, on average, an extra 4.5 °C to the warming of the land surface. The higher concentrations of greenhouse gases resulting from an Eocene climate heated by four times the present-day carbon dioxide level added even more extra warming, an average of around 5.5 °C. These numbers are themselves noteworthy, but became even more so when striking regional patterns of climatic warming emerged in the high latitudes and over continental interiors; in localized hot spots, wintertime temperatures rose by over 10 °C (Fig. 12).

The degree of regional warming was far more than the radiative forcing numbers would have us believe—why the discrepancy? The answer lies in the initial conditions of the model. As I mentioned earlier, even with up to six times the present-day carbon dioxide concentration, the simulated climates, though warm, were still cooler than implied by the exotic fossil floras and faunas. Consequently, this led to a reflective blanket of snow in the continental interiors at the high latitudes during wintertime, which efficiently returned the Sun's gaze, much as a white tee-shirt keeps us cool in summer. The strong amplification of the warming caused by the higher levels of the other greenhouse gases was caused by it melting snow cover to reveal the dark vegetation beneath, which efficiently embraced the Sun's gaze to

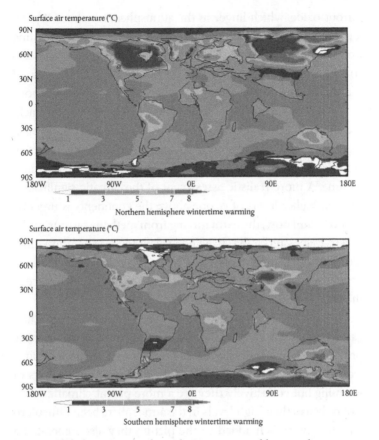

Fig. 12 Global warming in the early Eocene caused by greenhouse gases other than carbon dioxide. The upper map shows the warming during the northern hemisphere winter (December–February), and the lower map warming during the southern hemisphere winter (June–August).

warm the atmosphere. This cycle of warming causing snow melt leading to more warming forms another positive feedback loop.

Exciting though these developments are, they are really only a first step. They do not yet do full climatic justice to the chemical composition of the Eocene atmosphere. Unlike methane and nitrous oxide, which linger in the atmosphere for about 8 years and 120 years, respectively, ozone has a lifetime typically lasting only a few weeks. This means that large geographical and seasonal gradients in its concentration can develop. In the case of the Eocene, extensive forested regions emitting hydrocarbons raised the ozone concentration considerably above the background level (in the presence of sufficient amounts of nitrogen oxides), especially in the polar regions where the long hours of summer sunlight enhanced ozone's formation and chemical lifetime. A more realistic assessment of the climatic significance of these higher levels of ozone above the continents is urgently needed. Until now, the extra forcing from methane nitrous oxide, except water vapour, has been equivalent to raising carbon dioxide levels. But the extra forcing from the ozone would redistribute heating towards the continents and may be a key missing piece of the Eocene climate puzzle.

As evidence of this sort begins to accumulate, greenhouse gases other than carbon dioxide take on an altogether more profound significance for their involvement in producing a warm Eocene climate. By themselves, they are minor agents of warming but collectively they are a more potent climatic force. Not only had their high levels in the atmosphere been difficult to establish, but when added to the picture they also generated a pattern of warming more consistent with that suggested by the exotic fossil floras and faunas. Tyndall pointed us to this possibility over a century and a half ago. Yet the necessary development in scientific knowledge and technical advances in drilling and analysing ice-cores, as well as computing, took time to catch up with his pioneering ideas. Who knows what he would have made of the remarkable ice-core records of past global change and efforts at modelling the Earth system. I hope he would have been fascinated and delighted by them. For it was Tyndall's observations, and later those of Samuel Langley (1834–1906) on the nature of light emitted from the Sun, that permitted the Swedish chemist Svante Arrhenius (1859–1927) to develop a simplified global climate model that recognized water vapour and carbon dioxide as the dominant greenhouse gases. Arrhenius also took the pioneering step of calculating with his model the effect on the global average temperature of doubling the concentration of carbon dioxide. He estimated the warming to be 5–6 °C, a range remarkably close to that obtained with modern sophisticated climate models (1.5–4.5 °C).[61]

Undoubtedly, an atmosphere rich in greenhouse gases created by microbes, forests, and swamps is only part of the answer to the Eocene 'hothouse' climate enigma. Other greenhouse agents remain to be uncovered. For instance, as I pointed out in Chapter 6, we do not yet fully understand, or incorporate into the modelling, how the presence of polar forests themselves might influence high-latitude climates. One intriguing idea, following from the existence of a methane-rich atmosphere, is that polar forests stimulated the formation of thick clouds of frozen water vapour over the polar regions in the lower part of the stratosphere. It is a beguiling notion, because clouds of ice crystals trap outgoing energy Earth normally emits back into space and so could keep the planet's surface warmer in polar regions. The polar clouds are a high-altitude version of Tyndall's 'aqueous vapour blanket', which he believed necessary for 'vegetable life of England', only in this case it is not the vegetable life of England that is at stake but the warmth of the polar climate. In certain climate models, the presence of these clouds has a dramatic effect, warming up the high-latitude regions by a remarkable 20 °C.[62] Promising though it is, the idea leaves many scientists uneasy and has yet to gain general acceptance. Critics rightly point out that key properties of the clouds, like their extent and optical thickness, have to be prescribed, making it difficult to evaluate their climatic impacts objectively.[63]

In the lost Eocene paradise there is a certain pleasing symmetry reflecting the decisive role that plants had in identifying the climatic warmth and in offering a solution. The fossilized remains of subtropical floras excavated over 300 years ago initiated the debate, and later finds of fossil palms in Wyoming and elsewhere substantiated the findings. Three decades of scrutinizing the oceans failed to provide a decisive explanation for the hothouse world. Carbon dioxide levels, within the bounds indicated by the rocks and fossils, were also insufficient. Only when the activities of the terrestrial biosphere—plants and microbes—were considered did a warmer climate ensue, fuelled by a cascade of chemical reactions through the land and air. It could even be, though this is a long shot, that a methane greenhouse caused by swamplands triggered the formation of clouds of ice crystals in the stratosphere that warmed the polar regions.

Intriguing though such connections are, we have to be careful not to fall into a trap here. As I have tried to emphasize, greenhouse gases like methane, water vapour, and nitrous oxide are slaves to the climate system, by which I mean that a warmer or cooler climate raises or lowers their concentration in the atmosphere. The difficult question we face is, how did Earth's climate escape from the 'hothouse' condition and switch to something resembling today's state, with frozen polar regions?

The issue is clouded by controversy. The secrets locked up in sediment cores recovered from regions of the Pacific and Atlantic oceans can act as a sensitive 'dipstick' of ocean and atmospheric chemistry and offer some clues.[64] A possible course of events, read from the sediments with some skill, begins

50 million years ago with declining atmospheric carbon dioxide levels,[65] a decline that reflects its consumption by enhanced chemical weathering of fresh rock uplifted during the formation of the Himalayas.[66] As the carbon dioxide greenhouse effect slowly weakened the climate cooled, triggering a series of glaciations in both hemispheres, each locking up enough water to drop sea level by around 100 m. The Earth then shifted tentatively towards the 'icehouse' state of our modern world, and this sequence of glaciations was followed by one last abrupt drop in the atmospheric carbon dioxide level 30–40 million years ago. This time the climate flipped, creating an icehouse world that has existed ever since.

This neat explanation for Earth's escape from its former greenhouse state appears persuasive, but the problem with it has always been that the Arctic was thought to have frozen over much later, around 3 million years ago. If a declining carbon dioxide level really was the culprit for driving long-term climate cooling, how could Arctic cooling be delayed by some 30 million years? Only recently this major obstacle has been swept aside by the recovery of sediment cores from the bottom of the Arctic Ocean (see Chapter 6). The decisive evidence for ice recovered from these Arctic Ocean sediments comes in the form of pebbles and fine layers of sands that can only have been delivered by icebergs, and are dated to around 45 million years ago.[67] So, sea-ice and icebergs made an appearance in the Arctic much earlier than anyone had previously supposed and, indeed, the timing confirms that the Arctic froze over at about the same time as Antarctica. A synchronous freezing of the Arctic and Antarctic points the climatic finger at declining carbon dioxide as a major culprit of long-term cooling over the past 50 million years.

Chemical analyses of gases trapped in time in the bubbles of the polar ice-sheets unequivocally chart over two centuries rising levels of carbon dioxide, methane, and nitrous oxide.[68] Carbon dioxide levels have risen by a third since the onset of the Industrial Revolution in the mid-nineteenth century, due to the combustion of fossil fuels (mainly coal and oil) and deforestation. Methane levels have doubled over the same time with the spread of flooded agriculture, especially rice paddies in the tropics. Nitrous oxide levels have increased by 10% as the worldwide rise in nitrogen fertilizer applications has nourished the soil microbes.

The Earth's tropospheric ozone content has risen, too, although the rise cannot be gauged from ice-cores because the gas is too reactive to be faithfully preserved in the air bubbles trapped in ice-sheets. Instead, historical trends have been determined by comparing modern measurements with those that are thought to be the oldest reliable ozone datasets, made by Albert Lévy between 1876 and 1910 at the Observatoire de Montsouris, located on the outskirts of Paris.[69] For this purpose, Lévy designed a new instrument, replacing the earlier rather crude and unreliable approach of the German chemist Christian Schönbein (see Chapter 4), based on the colour change of starched paper impregnated with iodine. Lévy's new instrument worked by collecting the oxygen produced during the iodine-catalysed reaction of ozone with arsenate in a calibrated gas meter, and he used it to make thousands of ozone measurements at the Observatoire.

Andreas Volz and Dieter Kley, at the Institut für Chemie in Jürlich, brought the value of Lévy's forgotten measurements to light late in the twentieth century with an ingenious piece of historical science. The two researchers rediscovered Lévy's meticulous notes detailing exactly what he had been up to.

Recreating his apparatus to ensure its reliability, and comparing the measurements with those of a comparable rural European location today, Volz and Kley discovered that the background ozone level had more than doubled over the past century.[70] The doubling may have contributed to warming mean annual temperatures in the northern hemisphere by at least 0.4 °C.[71] It was also evident that the pattern of the seasonal ozone cycle had altered over the past century. In Lévy's day, ozone levels rose in the spring and fell during the summer, whereas now they rise in the summer. This distinctive difference points to a build-up of nitrogen oxides from cars and industrial processes as the culprit. Today, nitrogen oxides are abundant in the troposphere, and this causes ozone levels to rise in the summer as more sunlight is available to drive its photochemical production. However, the opposite happens when nitrogen oxides are scarce, with more sunlight destroying ozone in the summer—as seen a century ago. The springtime peak in the nineteenth century reflects the time of year when ozone is most efficiently drawn down from the stratosphere.

The recent historical trends in all major greenhouse gases highlight our capacity to drastically alter the chemistry of Earth's thin, fragile atmosphere. Comprehensive analyses indicate that our loading of the atmosphere with these gases is probably the primary reason why Earth's energy budget is presently out of balance, with more energy being absorbed from the Sun than is emitted from our planet out to space, to the tune of 0.85 watts per metre square.[72] Although it may not sound much, if it had been maintained for the last 10 000 years it would be sufficient to melt the polar ice-sheets, raising global sea-level by one kilometre, or to warm the surface oceans by 100 °C. Earth's climate has already warmed by an average of 0.8 °C between 1880 and 2003. Yet the total amount of forcing exerted by the rise in greenhouse gases is closer to 1.8 rather than the observed 0.85 watts per metre square. This means there is additional warming 'in the pipeline' that we do nothing about even if we hold greenhouse gases constant at present-day levels.

Rising levels of greenhouse gases and their possible climate consequences are worrying, and it is becoming increasingly clear that human activities are primarily responsible. An optimistic take on the situation is that it provides us with opportunities for mitigation. Set against the relentless rise in atmospheric carbon dioxide concentration, the rising levels of so-called trace greenhouse gases may seem inconsequential. But as James Hansen at NASA has pointed out, the increase in radiative forcing caused by the rise in all of them together (methane, nitrous oxide, and ozone, plus human-made black carbon aerosols and CFCs) is nearly equal to that caused by the rise in carbon dioxide itself in the past century.[73] Given projected trends in these gases over the next 50 years, one way to help 'defuse the global warming time bomb'[74] might be to focus on developing technological solutions to combat emissions of methane, tropospheric ozone, black carbon particles, and CFCs. Such solutions could allow carbon dioxide levels to stabilize at a higher value before they constitute a dangerous threat to the climate system.[75] The idea of this strategy is to offer a more realistic means of limiting future radiative forcing of climate, rather than relying solely on the more difficult-to-achieve reductions in carbon dioxide. Methane emissions might be lowered if production from ruminants was reduced through dietary adjustments and we sought to reduce methane leakage from natural gas distribution lines, landfill sites, coal mining, and oil drilling. Similarly, ozone levels might be controlled if we cleaned up emissions of its precursors from the primary sources like industry, power stations, and vehicle exhausts, and developed cleaner combustion technologies. Lowering ozone holds the additional benefit of alleviating its detrimental effects on global food production and human health.[76] Hansen also recommends we reduce emissions of black carbon from the combustion of diesel fuel and coal; black carbon reduces the reflectivity of aerosols and clouds and hence contributes to warming.

It remains to be seen if we will act quickly enough in this way to buy ourselves more time, and it is important to realize that limiting future increases in greenhouse gases other than carbon dioxide is not a substitute for curbing carbon dioxide emissions themselves. Middle-of-the-range scenarios of future increases in greenhouse gases forecast by the Intergovernmental Panel on Climate Change indicate that our atmosphere might, by the end of the twenty-first century, resemble that simulated for the Eocene.[77] Edward Jacob's fossils are a stark warning as to what the future climatic consequences of that drastic alteration to the chemistry of our atmosphere might be.

8

Nature's green revolution

Scientific progress and new technologies frequently go hand in hand and, by the late 1960s, mutually beneficial advances in both led to the striking discovery of a novel photosynthetic pathway. The new pathway involved an upgrade in the form of a solar-powered, carbon dioxide pump that gave the owners the ecological upper hand in hot, dry climates and under conditions of carbon dioxide starvation. This revelation was soon followed by the dramatic discovery that grasses with the newly recognized photosynthetic pathway transformed the subtropics, converting forests to grass-dominated savannas in a geological instant late in the Miocene, some 8 million years ago. A sudden bout of carbon dioxide starvation was widely regarded as an obvious trigger for the rise of savannas, but this view fell out of favour when records showed that the atmospheric content had dwindled millions of years before their ecological success. In the search for new explanations, fire science is the smouldering favourite to offer a solution.

The farther researches we make into things, the more beauty and harmony we see in them.

Stephen Hales (1727), *Vegetable staticks*

THE scientific revolution of the seventeenth and eighteenth centuries, if indeed it can be recognized as such, saw the foundations of modern science established. Developments by iconic figures, notably Francis Bacon (1561–1626), Galilei Galileo (1564–1642), Robert Boyle (1627–91), and Isaac Newton (1642–1727), among others, advanced the study of the natural world by moving it away from mystical concepts and grounding it firmly in the rational. Bacon outraged his intellectual contemporaries with the belief that scientific knowledge should be built on empirical observation and experimentation, and pursuing this theme is alleged to have done for him in the end, at the age of 65. According to Bacon's former secretary, the legend goes that Bacon was travelling in a coach towards London with one of the King's physicians on a snowy day in April 1626 when he decided to investigate whether meat could be preserved by ice. Seizing the opportunity for an experiment, Bacon purchased a chicken in Highgate, then a small village outside London, gutted it, and proceeded to stuff the carcass with snow to see if it delayed putrefaction. In his excitement he became oblivious to the cold, caught a chill, and took refuge in the Earl of Arundel's nearby house in Highgate, the Earl being away serving time in the Tower of London. Bacon died a few days later, probably from pneumonia, after being put up in a guest room with a damp bed disused for over a year, but not before penning a letter to the Earl communicating the success of the experiment.

This delightful story of Bacon's ultimate demise would have been fitting for his contribution to modern science, but is probably apocryphal. Surviving records indicate Bacon was already ill before the end of 1625, and inclined to inhale opiates and the vapours of chemical saltpetre (potassium nitrate) to improve his spirits and strengthen his ageing body. In those days, the saltpetre was impure, a mixture of potassium nitrate, sodium nitrate, and other compounds that may have given off toxic vapours. It seems possible, likely even, that Bacon overdosed on his inhalation of remedial substances to compensate for his ill health.[1]

As the scientific method became established, the world absorbed a deeper truth about humanity's place in the Universe, and the towering achievements of Bacon, Galileo, Boyle, Newton, and the other scientific greats set the agenda for the centuries ahead. The Cambridge historian Herbert Butterfield (1900–79) famously described the scientific revolution as being like 'putting on a new pair of spectacles', meaning it was central to defining our modern world.[2] But legitimate doubts remain about a version of history that claims the existence of a 'revolution'; suspicions linger that the concept is an artificial construct of historians.[3] We know, for example, that few people of the seventeenth century believed what scientific practitioners believed. In the days before communication highways like the Internet, the overwhelming majority of the population in England (and elsewhere) were completely unaware that the so-called scientific revolution was happening. Regardless of what we call it, knowledge of the natural world encompassing mathematics, physics, chemistry, and astronomy undoubtedly changed radically between the fifteenth and eighteenth centuries.

Lisa Jardine, a renowned historian at the University of London, persuasively argues for an alternative illuminating take on the development of 'classical' science.[4] Jardine wants us to recognize that the invention of instruments, and the pioneers who deployed them, critically underpin the scientific advances. Revolutions in technology went hand in hand with advancements in science. Technical instruments like microscopes, telescopes, pendulum clocks, balance-spring watches, and vacuum pumps catalysed the new science. These inventions gave the 'natural philosophers' a means of making original observations of the world around them from which theoretical foundations could emerge. This is how science progresses, and we should not find it surprising, when shifting our time horizon closer to the present, that significant recent discoveries also pivoted around new technologies and creative talents, much as the pursuit of science has always done.

Nowhere is this more apposite than in the sequence of momentous discoveries that eventually cracked the puzzle of how plants harness sunlight to synthesize biomass from carbon dioxide. The early roots of the puzzle date back to the Flemish natural philosopher, Jan Baptista Van Helmont (1577–1644) in the sixteenth century. Van Helmont conducted a pioneering experiment when he placed a willow cutting in dry soil and, after supplying it with no nutrients other than rainwater, recorded a 30-fold increase in mass.[5] He then rather spoilt it all by mistakenly arguing that the extra plant material—wood, bark, roots, and leaves—came from the water not air. This mistaken belief is quite common even today among producers of television gardening programmes in which the presenters exhort us to 'feed' our plants by adding food to the soil. Admittedly, the 'food' is required by plants to grow, but the basic building blocks for synthesizing biomass come from carbon dioxide extracted from the air.

Nevertheless, we should acknowledge Van Helmont's contribution for it called attention to the riddle of how plants grew. But it was not until the English clergyman Stephen Hales

(1677–1761) reported in his remarkably perceptive 1727 book *Vegetable staticks* that 'plants [were] very probably drawing thro' their leaves some part of the nourishment from the air' that it started to become clear what was happening. Hales drew this correct conclusion after recording that plants had reduced the volume of air in a closed vessel by around 15%. Although Hales could not really account for the observation, thankfully his thinking was on the right lines because the air volume was reduced by the removal of carbon dioxide. Hale went on to muse how air was 'imbibed into the substance of the plant' and noted 'may not light also, by freely entering the surfaces of leaves and flowers, contribute much to ennobling principles of vegetation'. And well he might, for plants achieve the feat by the remarkable process of photosynthesis. I say remarkable for good reason: a photon of light travels 93 million miles (150 million kilometres) to reach the Earth's surface in about eight minutes, yet it takes a plant seconds to capture its energy, process it, and store it in a chemical bond. Endlessly fascinating, photosynthesis is exceptionally complex and of considerable antiquity, dating back at least 2.5 billion years. It has played a pivotal role in the affairs of our planet, as a recent congress of scientific experts on photosynthesis highlighted with their telling statement: 'photosynthesis: the plant miracle that daily gives us bread and wine, the oxygen we breathe, and simply sustains all life as we know it'.[6] For all this, and perhaps unsurprisingly, its great elegance and sophistication only became clear when new technologies were brought to bear on the matter, some 300 years after Hales queried how it all worked.[7]

Foremost among the new technologies giving us the means to investigate the puzzle of photosynthesis was the cyclotron, a machine developed by the renowned nuclear physicist Ernest

Lawrence (1901–58) at the University of California's Berkeley Radiation Laboratory. Really a progenitor of modern particle accelerators, the cyclotron provided scientists with a means of studying the transformation of elements by disintegrating atoms with charged particles accelerated through millions of volts.[8] Its invention ushered in the beginnings of the modern nuclear age, and while nuclear physicists got excited by the new machines, others quickly appreciated that they had much to offer when it came to investigating how plants grew.

Two talented young researchers, Samuel Ruben (1913–43) and Martin Kamen (1913–2002), arriving at the right place (University of California, Berkeley) at the right time (late 1930s), quickly seized the opportunity to exploit the new technology of the physicists to investigate the riddle of how plants harvest sunlight and use its energy to make organic acids and carbohydrates (see Plate 15).[9] Recognizing that by the late 1930s traditional ideas about photosynthesis were crumbling, Ruben and Kamen began pioneering experiments using radioactive carbon atoms produced by the cyclotron. Their idea was to replace the carbon atom in the carbon dioxide molecule with a radioactive form, thereby 'labelling' it before it was taken up by the plant. It is a deceptively simple technique. It means that radioactive (labelled) compounds can be easily distinguished from unlabelled non-radioactive compounds and provides a novel way of tracing the progress of carbon dioxide through the plant as it is metabolized. Ingenious though this line of research was, the radioactive form of carbon then available to them (^{11}C) quickly lost its radioactivity, falling by a half in 21 minutes.[10] The short half-life of ^{11}C frustrated their efforts, and despite hundreds of experiments the original goal of establishing exactly how carbon dioxide was processed inside leaves remained tantalizingly out of reach.

To make progress, and give Ruben and Kamen a genuine chance of finding out how plants metabolized carbon dioxide, the pair urgently needed a radioactive form of carbon with a longer half-life. For that, their attention once again turned to the Radiation Laboratory's cyclotron. Kamen performed what turned out to be a crucial experiment on a rainy 19 February 1940, which involved bombarding graphite with protons in the cyclotron. Shortly before dawn, he scraped the disintegrated graphite off the probe into a vial and placed it on Ruben's desk for analysis, closed down the cyclotron, and headed home to get some sleep. Wearily stumbling home, Kamen was apprehended by the police on the lookout for suspects in a series of gruesome murders. The dishevelled scientist fitted the bill. However, the sole hysterical survivor of the massacre failed to identify the suspect, and he was released to crawl home and collapse into sleep, only to be awoken some 12 hours later by an excited Ruben. The sample Kamen had produced was giving off a faint trace of long-lived radioactivity, suggesting that it was the long-sought-after form of carbon. After exhaustive checks, Kamen and Ruben finally announced their momentous discovery on 27 February 1940.[11] They had discovered a new radioactive isotope of carbon (^{14}C) that decayed by a half not in minutes, but over millennia.

The breathtaking discovery of this supremely important isotope by Kamen and Ruben proved a watershed in the use of tracers in every area of biological and medical research. But exploiting it to understand the mysteries of photosynthesis had to wait until the Second World War was over and, by that time, neither of these forgotten heroes of photosynthesis research was in a position to do so. Ruben died tragically on 28 September 1943 while working on a National Defence Research Committee project concerning chemical warfare.

Sometime in September 1943 he had broken his right hand in a driving accident caused by falling asleep at the wheel. It was a seemingly minor event, but back in the laboratory the following Monday the injury made the usually straightforward task of plunging a small vial of phosgene gas into liquid air a difficult one. When the glass cracked unexpectedly, it released the deadly gas and Ruben inhaled a fatal dose. Kamen later summarized Ruben's outstanding but tragically short career:

> Ruben was responsible, almost single-handedly, for the growth of interest in tracer methodology ... His unique combination of experimental skills, energy, wide-ranging interests, and quick grasp of essentials when confronted with new and unfamiliar areas of science, provided a focus for an ever-increasing number of able investigators.[12]

Kamen, meanwhile, had been removed from the Berkeley laboratory after (unjust) persecution by the House of Un-American Activities Committee.[13] By July 1944, the Radiation Laboratory had been seconded into the war effort to produce radioisotopes for nuclear research in the Manhattan Project, an activity that intensified after the Japanese bombed Pearl Harbor on 7 December 1941. Following several years' secret research in this area at Berkeley, Kamen was declared a security risk, accused of being part of a 'spy ring' working for the Soviet Union. It left him deeply depressed and perplexed. A concerted campaign to clear his name in the courts followed and eventually saw his successful acquittal on all fronts. Kamen went on to enjoy a distinguished scientific career, receiving many honorary degrees and awards.[14]

By 1945, cyclotrons had become yesterday's technology, as the race to develop the atomic bomb spawned nuclear reactors. Nuclear reactors routinely produced large amounts of highly radioactive ^{14}C ideally suited for research on photosynthesis, making Kamen and Ruben's dream of discovering how photosynthesis works a reality. With Kamen and Ruben out of the picture, the opportunity to exploit the ready availability of ^{14}C fell to the team of scientists that Lawrence had newly installed at the Radiation Laboratory, Berkeley. The team was led by Melvin Calvin (1911–97), who oversaw an astoundingly productive 10-year period of research in collaboration with his colleagues Andrew Benson and James (Al) Bassham.[15] Driven by fierce competition with a rival group in Chicago, the team successfully elucidated the route by which plants produced their food from carbon dioxide and water.[16] Once again, the pattern of invention and discovery was repeated.

To get some flavour of the remarkable significance of the innovations and advances that have taken us to this point, it is worth noting that we have just breezed through the work of two Nobel Prizes, one awarded to Ernest Lawrence (Physics, 1939), the other to Melvin Calvin (Chemistry, 1961). The count rises to three if we include Ernest Rutherford (1871–1937) (Chemistry, 1908), whose concept of the atomic nucleus set the stage for Lawrence's ideas; four if we include the Prize awarded to John Cockcroft (1897–1967) and Ernest Walton (1903–95) (Physics, 1951), at the Cavendish Laboratory in Cambridge, for their pioneering linear particle accelerator that enabled them to split the atomic nucleus, one of the great scientific achievements of all time, and win the race against their American rivals.[17] Some time after Calvin received the 1961 Nobel Prize in Chemistry, he joined contemporary and former American Nobel Prize winners at the White House with President John F. Kennedy (1917–1963) and his wife. The event inspired Kennedy's famous quip: 'this is the most extraordinary collection of talent and human knowledge gathered at the White House, with the possible exception of when Thomas

Jefferson dined here alone'.[18] Kamen and Ruben are the two most significant omissions from this list of Nobel laureates. Colleagues declared that 'there is no doubt that Ruben and Kamen unequivocally earned a Nobel Prize for their discovery of long-lived radioactive carbon (^{14}C), which engendered a revolution in humanity's understanding of biology and medicine'.[19] Others thought the Nobel Committee might have been more generous in awarding the Prize jointly to Calvin and his principal collaborator Andy Benson, to recognize the importance of Benson's intellectual and experimental leadership in the research that saw the pathway of photosynthesis revealed.[20]

Calvin and Benson's team showed that plants converted carbon dioxide into a compound with a backbone of three carbon atoms, which subsequently feeds into a biochemical cycle to produce organic acids.[21] The cyclical conversion of carbon dioxide to organic acids and then sugars is catalysed by the enzyme Rubisco (see also Chapter 3, p. 49), of which we shall learn more later.[22] The details describe a universal process that explains how plants grow, and clarifies our understanding of the intermediate steps in the pathway; explaining and clarifying, two of the most satisfying attributes of a scientific discovery.[23] Afterwards, pretty much all plants were believed to process carbon dioxide in this way and became labelled 'C_3 plants', the C_3 terminology referring to the number of carbon atoms in the first molecule created when plants take up carbon dioxide. Thirty years on, though, it was becoming obvious that this simple view of the plant world was wide of the mark. The puzzle began when scientists at the Hawaiian Sugar Planters' Research Laboratory, Honolulu, noticed something odd when studying sugar cane: the radioactively labelled carbon turned up in organic acids with a four-carbon backbone instead of the expected three.[24] Other Russian and Australian scientists studying maize and salt marsh plants noticed the same oddity.[25] Indeed, the business of the four-carbon organic acid was so odd that the Russian scientist queried his experimental procedures.

None of this biochemical confusion escaped the attentions of our next two scientists, who went on to make an astonishing contribution. Ruminating over a few glasses of beer, that well-worn path to scientific progress, the Australian Hal Hatch and Briton Roger Slack resolved to take up the challenge and understand what, if anything, the four-carbon oddity meant. Pretty soon the duo uncovered something special. In a dazzlingly productive 5–6-year phase from 1965 onwards, they discovered that tropical grasses had evolved a revolutionary upgrade to the mode of photosynthesis recognized by Calvin, Benson, and their team.[26] The elaborate upgrade comes in the form of a solar-powered carbon dioxide pump boosting the carbon dioxide concentration around the enzyme—Rubisco—catalysing the synthesis of sugars.[27] In this mode, carbon dioxide is first captured and attached to a special carrier compound, which is then pumped into a special 'wreath-like' arrangement of cells surrounding the leaf veins. Inside these cells, the carrier's cargo is detached to create miniature carbon dioxide-rich greenhouses.[28] The arrangement boosts the concentration of carbon dioxide inside the cells to ten times that in the atmosphere and, bathed in this luxurious carbon dioxide-rich environment, Rubisco converts carbon dioxide into organic acids and then sugars with supreme efficiency, one unmatched by any other group of plants.[29] It is no surprise to learn, then, that C_4 plants rank amongst our most productive crops and worst weeds. The four-carbon compound puzzling the workers in Hawaii, Russia, and Australia turned out to be the carrier compound shuttling carbon dioxide into the special wreath-like cells. The discovery of the so-called C_4 photosynthetic pathway sealed the fame and honour of Hatch and Slack, who had given science a glimpse of the hidden internal workings of C_4 plants. When Hatch modestly revealed the circumstances surrounding the discovery of C_4 photosynthesis years later, we find they too echo Kamen and Ruben's discovery of ^{14}C:

> I think I am supposed, on occasions like this, to say how it is done. This reminds me of the situation when centenarians are asked what their secret is. Half say clean living and no drinking or smoking, the other half say the complete opposite. It may sound trite but it is true that luck is a most critical factor—just getting to be in the right place at the right time.[30]

As is so often the case with a new discovery, previously confusing observations fell into place. As early as 1884, the great German botanist Gottlieb Haberlandt (1854–1945) had bequeathed the term 'Kranz' to describe the special arrangement of cells inside the leaves of certain plant species, without appreciating their functional significance.[31] With the discovery of C_4 photosynthesis the arrangement made sense. The cells separate the two biochemical systems involved in processing carbon dioxide and its products spatially, a convenient division of labour within the leaf. Especially permeable to metabolites, this special arrangement of cells is remarkably impermeable to carbon dioxide. Kranz anatomy, or in modern parlance, the bundle sheath, is now recognized as a characteristic feature of the leaves of most species of C_4 plants.[32]

The C_4 photosynthetic pathway compensates for a deeply embedded evolutionary compromise in Rubisco, the enzyme at the heart of photosynthesis. The enzyme originally evolved in photosynthesizing micro-organisms called cyanobacteria nearly 3 billion years ago, when the atmosphere contained around 100 times more carbon dioxide than now and negligible amounts of oxygen.[33] Such a plentiful supply of carbon dioxide boosted photosynthesis for the entire biosphere, a planet-wide version of the C_4 pump allowing Rubisco to operate with maximum efficiency. When land plants inherited a version of Rubisco from cyanobacteria, who themselves ultimately became transformed in the critical photosynthetic organelles called chloroplasts,[34] the atmosphere contained only a tenth as much carbon dioxide. This shift in the atmospheric composition exposed a serious problem for Rubisco, the enzyme with an undeserved reputation for being slow and wasteful.[35] Starved of carbon dioxide, Rubisco splutters along, like an engine starved of fuel, often backfiring by capturing molecules of oxygen instead of carbon dioxide. When this happens, C_3 plants waste solar energy and lose valuable carbon dioxide. C_4 plants, equipped with their photosynthetic upgrade, avoid the pitfall but pay the price, if it can be considered as such, of generally being restricted to a subtropical existence where the warm climate is best suited to drive the pump. All of which points to the C_4 way of life as an adaptation to carbon dioxide starvation,[36] with a drop in the greenhouse gas millions of years ago having a decisive hand in the evolution of this novel photosynthetic pathway.[37]

Spanning three decades, the entwined chain of technological progress and discovery linking the beginnings of the nuclear age to the mysteries of Hawaiian sugar plantations saw nature's revolution—plants with the C_4 photosynthetic pathway— finally drawn to the attention of the scientific community. The current situation is that we recognize approximately 7500 species of C_4 plants, occupying a fifth of the vegetated land surface of our planet and accounting for 30% of the primary productivity of the terrestrial biosphere.[38] By far the majority are subtropical grasses, although some sedges and herbs have also benefited from the revolution.[39] Because their biochemical machinery operates most efficiently at high temperatures and in bright sunlight, it confines most C_4 plants to subtropical climates, where they dominate the grasslands and savannas. Only a single species of tree is known to use C_4 photosynthesis, *Chamaesyce forbesii*, and it occurs on Hawaii, an island where evolutionary pressures are altered by millions of years of isolation. Other than *C. forbesii*, the closest we have to a C_4 tree are a few woody shrubs that approach tree-like status with advanced age, notably black saxaul (*Haloxylon aphyllum*) in the hot, sandy deserts of central Asia.

It is a peculiar irony that C_4 plants only belatedly came to be recognized by modern science. For not only have we been exploiting them as part of our own agricultural revolution for the past 10 000 years, oblivious to their secret, but they also set the scene for the evolution of our ancestors.[40] Maize and sugar cane, the two blockbuster C_4 crops, had far-reaching impacts on society. Maize, domesticated from its wild ancestor teosinte somewhere in the western highlands of Mexico about 7500 years ago, spread across Mesoamerica, paving the way for the rise of complex pre-Columbian civilized societies in the Americas. Sugar cane was domesticated in New Guinea and has been known to history since the time of Alexander the Great (356–323 BC). It spread throughout the Caribbean in the seventeenth century, providing a cheap supply of sugar to the Western world and creating a revolution in human societies by altering our diets, social customs, and economies. Cultivating sugar cane is a labour-intensive business and large numbers of Africans were kidnapped and sold into slavery in the Caribbean to tend plantations, events that irreversibly changed the social fabric of the region. We may have tamed C_4 crops, but they have also tamed us by exerting a powerful influence on human social evolution and persuading us to cultivate them.

So, did chance alone create the elaborate and successful C_4 revolution in the plant world or is there more to it than that? Any account might sensibly begin with the question of when C_4 plants evolved. It is an obvious one to ask, but a difficult one to answer. The fossil record of grasses is pitiful. Rarely are the organic remains of grasses preserved as fossils because grasslands, by their very nature, occur in arid regions. The distinction of being the undisputed oldest known leaf fragments of a fossil C_4 grass goes to those recovered in 1978 from the appropriately named Last Chance Canyon in southern California.[41] Misdating of the sediments meant it was not recognized as the record holder until later, when the accepted age was established to be 12.5 million years old (Miocene age). Some scientists have tried to stake a claim for discovering the oldest ever C_4 grass, pushing back the date by 1.5 million years,[42] but their claim is based on tenuous evidence, the fragmentary remains of grass cuticles, and has been roundly rejected. Other than a fossil grass splinter with Kranz anatomy recovered from 5–7-million-year-old Miocene sediments in north-western Kansas,[43] this is the unsatisfactory sum of evidence the fossil record of leaf remains has revealed so far.

Fortunately, we can peer further down the historical corridor of C_4 plants to estimate the timing of their origins with molecular clocks. Molecular clocks are founded on the observation that the genetic code of all organisms—DNA—undergoes steady mutations over time in a fashion analogous to the ticking of a clock. These mutations are, in Charles Darwin's prescient phrase, 'neither beneficial or injurious'.[44] If the mutation rate is known, then if we know the number of mutations that separate the DNA sequences of species that once shared a common ancestor, it is possible to estimate the time that has elapsed since they diverged. For C_4 plants, grasses are the last common ancestor and their earliest unequivocal origins date to 50–65 million years ago.[45] These dates suggest that the (non-avian) dinosaurs did not eat grasses, but this idea was turned on its head only recently with the surprise discovery by a team of researchers from the Swedish Natural History Museum of microscopic silica structures characteristic of grasses in dinosaur droppings.[46] One group of grasses, the Poaceae, evidently originated and had already diversified in the Cretaceous, suggesting that the picture is not as clear-cut as originally thought. Still, molecular clocks of grasses calibrated against another ancient family of grasses, the Panicoideae, place the origins of C_4 plants at 25–32 million years ago, considerably younger than the palaeontological dates.[47]

At first sight, the enormous discrepancy between molecular and palaeontological dates for the origination of C_4 plants appears irreconcilable. Rapprochement, however, is closer than it seems. We should not expect the fossil record to provide the earliest date of C_4 plant origins. In all likelihood, C_4 photosynthesis evolved well before the 12.5-million-year-old Last Chance Canyon fossils; it is just that the plants using it were scarce in the ancient floras and the conditions for preservation as fossils were unlikely to be common in arid environments, making the fossils rare. Molecular clocks, too, are not without error. Uncertainties in the rates of mutations bedevil their timekeeping properties. All things considered, though, the molecular dates are probably the best indication we have for when C_4 plants originated. If we accept a date of 30 million years ago, it means the revolution sparked into life comparatively recently; compressing the entire time plants have occupied our planet into a single 24-hour day, C_4 plants arrive at the party around ten-thirty in the evening.

The story moved on when, throughout the 1980s and 1990s, a team of researchers at the University of Utah reported remarkable results from analyses of fossil teeth.[48] The researchers focused on the teeth of herbivorous mammals because their isotopic composition reflects that of the plants they consume. As Hatch and Slack showed, C_4 plants fix carbon dioxide differently compared to C_3 plants, and it later transpired that this results in them containing a different abundance of the heavy and light stable isotopes of carbon, a difference that is passed on to carbon-containing tissues, like the teeth and bones of the animals that eat them.[49] On the Athi Plains in Kenya, for example, where C_3 and C_4 plants coexist, grazing zebras preferentially select C_4 grasses and have teeth with an isotopic composition distinct from that of giraffes browsing C_3 trees and shrubs.[50] From the isotopic perspective, at least, we can see that fossil teeth speak to us about the dietary habits of their long-since-departed owners, who wandered the plains of ancient continents millions of years ago.

What the scientists from Utah learned, straight from the horse's mouth, as it were, was that the rise of C_4 grasses had been nothing less than meteoric (Fig. 13). Before 8 million years ago, C_3 trees and shrubs dominated the diets of large herbivorous mammals like zebras and horses in Africa and the Indian subcontinent. Within about a million years, however, the fossil teeth revealed a dramatic near-synchronous worldwide switch to diets consisting entirely of C_4 grasslands.[51] The switch implied that C_4 grasses exploded onto the ecological stage about 8 million years ago to replace former forests with C_4 savannas and grasslands. In the northern hemisphere, C_4 savannas originated in a latitudinal wave sweeping around the Earth from the hot tropics of eastern Africa to the cooler climes of North America. The rise of C_4 plants remains one of the most profound transformations of global ecosystems yet discovered in the fossil record, and its discovery ranks as one of the great

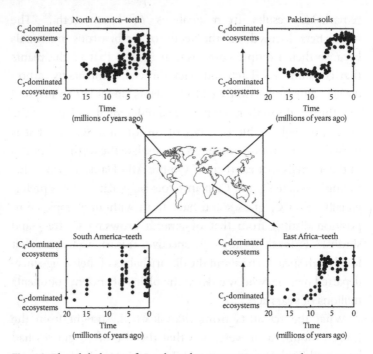

Fig. 13 The global rise of C_4 plant dominance in terrestrial ecosystems inferred from the isotopic composition of fossil tooth enamel and soils.

triumphs in the field of modern isotope geochemistry. It also suggests the cradle of C_4 photosynthesis lies deeply hidden in the tropics. The S-shaped curves marked out by the painstakingly acquired isotope records from four continents (Fig. 13) suggest we have ratcheted irreversibly towards the modern condition—a C_4 world.

Such profound changes in world vegetation demand explanation, and in the search for some answers attention focused on the fact that C_4 grasses had risen to dominate in the subtropics nearly synchronously in diverse geographic regions. This is strongly suggestive of a global trigger for the phenomenon and an obvious candidate is the carbon dioxide content of the atmosphere. So it was no surprise that researchers came up with a bold explanation, in the form of what we can call the 'carbon dioxide starvation' hypothesis.[52] According to this, dwindling carbon dioxide levels spelled double trouble for C_3 plants, starving them of carbon dioxide and water. This is because to combat carbon dioxide starvation they keep the microscopic stomatal pores on their leaves open wider for longer, causing more water to be lost by evaporation. Falling carbon dioxide in other words is equivalent to falling rainfall. C_4 plants, on the other hand, enjoyed immunity to both effects. Their superior photosynthetic pathway allows them to be productive under conditions of carbon dioxide 'starvation', and keep their stomatal pores closed when necessary to conserve precious water supplies. The Utah team argued that as the carbon dioxide content of the atmosphere breached some threshold value, it handed the ecological initiative to C_4 plants in hot climates. They concluded that

> the persistence of significant C_4 biomass beginning about 6–8 Myr [million years] ago and continuing to the present is compatible with atmospheric CO_2 levels in the Miocene declining below some 'crossover' point where C_4 grasses are favoured over C_3 grasses or other plants.[53]

It is a beguiling hypothesis, and understandably soon gained widespread acceptance as the leading explanation for the dramatic expansion of C_4 grasslands. It quickly found support from investigations revealing how tropical vegetation responded to the much more recent episode of carbon dioxide starvation that took place during the last ice age, a mere 20 000 years ago. The evidence, fossilized molecules of C_3 and C_4 plants, is preserved in lake sediments on tropical mountains in eastern Africa and in soils from the Chihuahuan Desert, New Mexico. Tracing the history of C_4 grasses in tropical Africa back to the last ice age reveals that they dominated mountain tops when the atmospheric carbon dioxide concentration was 50% lower than today, but as carbon dioxide levels rose they became replaced by C_3 forests.[54] It was a similar story in the Chihuahuan Desert: ice age, C_4 grasslands were abruptly replaced by C_3 desert shrubs about 9000 years ago, the time when carbon dioxide levels rose.[55] So, the 'carbon dioxide starvation' hypothesis scored two successes, elegantly explaining changes in tropical vegetation since the last ice age. Moreover, if these changes are achieved in a few thousand years, what is the capacity of carbon dioxide to restructure Earth's vegetation over millions of years?

Attractive though the 'carbon dioxide starvation' hypothesis is, it soon encountered a major obstacle. Serious difficulties surfaced with the publication of a carbon dioxide record in the international journal *Science* based on chemical analysis of the remains of marine phytoplankton in sediment cores drilled from the floor of the south-west Pacific Ocean.[56] According to the record, carbon dioxide levels had remained low geologically speaking (that is, similar to modern levels) between 5 and 16 million years ago. Later, two other approaches gauging the carbon dioxide content of the ancient atmosphere supported the same general picture.[57] All the evidence pointed to the onset of low carbon dioxide conditions 11 million years before the C_4 grasslands expanded, and a mismatch of this duration is not easily explained away.

If the atmospheric carbon dioxide records generated in the past decade render obsolete the idea that carbon dioxide starvation triggered the ecological success of C_4 grassland, it reinstates the search for other possible explanations. Our main clue to identifying what might have happened lies with the realization that the teeth and other isotopic records from the Old World are actually revealing a shift from subtropical C_3 forest or woodland to open savanna-style C_4 grasslands.[58] If forests are present, then they effectively exclude C_4 grasses, which cannot tolerate the cooler, shadier conditions. Deforestation, though, by whatever means, creates an open habitat, which if the climate is hot enough, allows the invasion of photosynthetically superior C_4 grasses. So what we are really searching for is a natural means of deforestation, a process that can only be brought about by climate change, and disturbance by fire and grazing animals.

Climate change over the long term (i.e. millions of years) is often driven by slow tectonic processes like the uplift of mountain ranges, and changing continental configurations that alter atmospheric and ocean circulation patterns governing the Earth's environment. We can glimpse the consequences of climate change for forests and grasslands with a pilgrimage to the Siwalik formation, a thick band of sediment and rock much beloved by geologists that stretches through Pakistan, northern India, and Nepal. South of the Siwalik sediments in Pakistan, in the province of Baluchistan, lie the famed Bugti Bone Beds. The Bugti Bone Beds are older than the Siwaliks, and have inspired the imaginations of palaeontologists for over 200 years since yielding up the bones of the largest land mammal of all times, a giant rhinoceros called *Paracerathium*; shaped like a giraffe, it stood 5 m tall at the shoulder.[59] Compacted into the younger Siwaliks further north is evidence for changes in the vegetation, fauna, and climate of the region over the past 18 million years. The richly detailed sediments are a remarkable repository of Earth history and give us an insight into climate changes during the C_4 revolution in that part of the world.

We find that as the revolution gathered pace and the C_4 savannas expanded, the Indian monsoon intensified causing the climate to dry and the once plentiful wintertime rains fail to appear.[60] Why the monsoon changed abruptly is still the subject of vigorous debate. On the one hand, uplift of the Tibetan Plateau around 8 million years ago has been found to do the job admirably in climate models.[61] On the other hand, some geologists believe the Plateau was already in place long before this.[62] Whatever the cause, with rains only appearing in the summer tree populations declined as recruitment of seedlings unable to survive the warm, dry winters failed. Forest deterioration inevitably followed, giving C_4 grasses the opportunity to take hold in the foothills of the Himalayas and the floodplains of the Ganges.

History tells us, then, that climate change can cause the decline and fall of C_3 forests and the establishment of C_4 grasses, but this is unlikely to be the whole story. There are other important influences at work that we are only beginning to appreciate. Exciting new detail was added to the picture when the South African workers William Bond and Guy Midgley realized they could improve upon the 'carbon dioxide starvation' hypothesis.[63] Thinking radically, they realized there was another way of looking at the problem, one integrating roles for climate change and wildfire. Instead of relying on a drop in carbon dioxide, their idea recognized that gaps in the forest, created by the death of trees during drought, allowed the establishment of patches of C_4 grasses. It is a seemingly small victory for the grasses. Yet once the establishment of these apparently innocuous patches of grasses got underway, it supplied highly combustible fuel loads in the dry season that increased the flammability of the ecosystem. More frequent fires are more likely to kill more trees, which, in turn, allow the further ingress of C_4 grasses better adapted to recover rapidly after a burn by sprouting from underground rhizomes. In Bond and Midgley's scheme of things, a self-reinforcing fire cycle accelerates deforestation and promotes grassland expansion.

The question raised by this speculation is whether invasion by C_4 grasses really produces the dramatic alterations to the fire regime of the landscape they invoke. Certainly, it is hard to overestimate the role of fire in shaping the world's vegetation patterns. In today's world, fire maintains more than half the land surface currently classified as savanna.[64] Without regular wildfires, modern forests could double in size, shrinking the extent of our tropical grasslands and savannas, like those in South America and South Africa, by a half. Evidence for C_4 grasses promoting fire activity and retarding forest development is found in Hawaii and the South Pacific island of New Caledonia. In Hawaii, the ecological drama plays out in the woodlands of the Volcanoes National Park, where invasion of two introduced species of C_4 perennial grasses (*Schizachyrium condensatum* and *Melinus minutiflora*) began in the late 1960s.[65] Before grasslands invaded, minor fires broke out once every other year. Twenty years later and the fires are three times more frequent and 50 times as large. Grass invasion apparently set in motion a positive feedback cycle efficiently converting non-flammable native woodland into highly flammable C_4 grasslands, eliminating endangered plant species in the process. Similar trends are being observed in New Caledonia, where deliberate fires on the drier side of the island have allowed alien C_4 grasses to invade the tropical rainforests.[66] The situations in Hawaii and New Caledonia represent an analogue, telescoped in time and space, for the more ancient spread of C_4 grasslands across the surface of the Earth. And these modern wars of grassland attrition are taking place against a background of rising atmospheric carbon dioxide concentrations, arguing for fire as the more important agent accelerating the conversion of C_3 forest to C_4 grassland.

Taking our cue from observations of this nature, it seems plausible that fire contributed to the ecological success of the C_4 grasslands millions of years ago.[67] However, observations by atmospheric physicists are revealing that even this hypothesis falls short of the complete picture because we are, in all likelihood, underestimating how powerfully fire could affect the spread of grasslands. Remarkable meteorological observations taken above the dense smoke produced by the hundreds of deforestation and agricultural wildfires that burn each year during the dry season in Amazonia are now revealing new connections between smoke, clouds, and climate. The fires burn at the boundary between tropical forests and C_4 grasslands and offer a unique insight into the consequences of fire-driven regime change millions of years ago. As we shall see, many of these findings shed new light on understanding how fire might reinforce the spread of C_4 grasslands by altering climate.

One of the first of the new connections between smoke and clouds was revealed by a young researcher, Ilan Koren, soon after he arrived at NASA's Goddard Space Flight Center in Maryland. Armed with a freshly minted doctorate from the University of Tel Aviv, Koren eagerly set about investigating the effects of tiny airborne particles produced by smoke on clouds using some of the world's most advanced satellite technology. He scrutinized satellite images taken above fires burning in the Amazon jungle during the dry season and quickly noticed that few, if any, showed clouds and smoke together—was smoke somehow suppressing cloud formation? Delving deeper into the phenomenon revealed why. Thick smoke plumes from fires travel for hundred of kilometres, shutting out sunlight and preventing it from reaching the ground. This effect slows evaporation on the ground and dries the air. Meanwhile the darker, smoky atmosphere above absorbs sunlight and warms up. Together, both effects reduce the flow and moisture content of the air needed to form certain types of clouds.[68] As Koren remarked, clouds really do seem to be 'nature's way of drawing in the sky the physics of what exactly is going on in the air'.

Clouds above the forest fires in Amazonia held further surprises for another team of researchers led by Meinrat Andreae at the Max Planck Institute for Chemistry in Mainz, Germany.[69] Andreae's team reported that smoke from the fires added huge numbers of aerosol particles to the atmosphere. An innocuous enough observation until it is realized that cloud droplets form when water vapour condenses around aerosol particles in the air. Thousands of droplets have to collide to form a drop large enough and heavy enough to fall as rain. Above the Amazon, the researchers found that smoke from fires increased the number of particles and dramatically reduced the size of the water droplets to the point where they were often not heavy enough to coalesce and fall as raindrops. Smoke from burning forests actually seemed to reduce rainfall.

From the trail-blazing work above the Amazon Basin came a new idea—wildfire can influence climate, and not just in the tropics. We usually think of a long hot spell of weather without rain as creating tinder-dry conditions that make natural and managed ecosystems prone to fire, placing agencies charged with fire management on a state of high alert. The reverse argument is that wildfire itself can influence climate to exacerbate the situation. In late April 1988, just such an effect was seen in the northern United States when the region experienced a severe drought, one of the driest of the twentieth century. Hundreds of thousands of acres of forest in the Yellowstone National Park burned in July of that year, adding vast amounts of black smoke to the atmosphere. One of the reasons for the severity of drought is believed to be the intense wildfires themselves. Smoke from the fires may have reduced cloud formation and disrupted atmospheric circulation patterns in the Midwest to an extent that this usually reliable and important source of precipitation for the region failed, exacerbating drought and priming the area for more fire.[70]

The expansion of C_4 grasslands millions of years ago, then, might be viewed as the Earth's switch to a flammable planet—a change accelerated by feedbacks causing climate change and more fire. The novel interactions between fire, trees, grasslands, clouds, smoke, climate, and carbon dioxide were more pervasive than at any time in Earth's history and form part of an intricate web of feedbacks (see Plate 16).[71] The complexity of the web can be analysed by representing each feature of interest as a node and connecting nodes in a linked chain of cause and effect by a series of arrows. The effect of one thing on another is examined and judged to be either positive or negative. A positive effect means more of one thing leads to more of another; a negative effect, indicated by a shaded mark, means more of one thing leads to less of the other. When a closed loop is identified by tracing out a pathway in a single direction from its start point to the end, the number of positives and negatives are summed to determine the final sign of the loop. This approach to breaking down a complex network into its component parts, systems analysis, is borrowed from information theory, widely used in social science, economics, chemical engineering, and circuit design.

The remarkable point to notice about the complex web that sums up the current state of affairs for C_4 plants is that by far the majority of the newly identified feedback loops are strongly self-reinforcing, creating a situation that has promoted and sustained C_4 grasslands for millions of years (positive). The implication is that once the system is pushed beyond some threshold point, perhaps by a change in climate, forest deterioration accelerates inexorably towards a new stable state of flammable C_4 grasslands. In this more complete view of how the world works, we can see that the apparent ratcheting towards a C_4 world implied by the records from fossil teeth appears to be inevitable.[72]

We should note that the proposed network does not rely on a drop in the carbon dioxide content of the atmosphere, but does require the observed conditions of carbon dioxide starvation to prime the system. Carbon starvation limits the growth of tree seedlings, preventing them from reaching the minimum height required to become fireproof. Put another way, when carbon dioxide is scarce, forest recruitment becomes extremely vulnerable to fire. Accepting this scenario for now means we can reprise the role for carbon dioxide starvation by viewing it as a necessary pre-condition for giving C_4 grasslands the ecological upper hand as they increase the flammability of ecosystems and wipe out the forests. Interestingly, C_4 plants may even fortuitously maintain this situation by enhancing the slow removal of carbon dioxide from the atmosphere by the chemical weathering of magnesium and calcium silicate rocks, at a time when the contribution by deteriorating forests is otherwise relaxed. More weathering helps ensure that an atmosphere impoverished in carbon dioxide continues, reinforcing the C_4 existence at the expense of C_3 trees.

It has to be said that as yet there is little evidence to support the idea of a set of self-reinforcing feedback loops amplifying the ecological success of C_4 grasslands in the Old World. Still, whatever the plausibility of the interpretation, the ecology and atmospheric physics underpinning it are sound. And one possible expression of the positive feedbacks is found on the floor of the western Pacific Ocean. It is, perhaps, hard to imagine a more unexpected destination for the combustion products of wildfire. Yet soot and charcoal are extremely resistant to chemical and microbial attack, and that transported on the trade winds from the Indian subcontinent to the Pacific has rained down through the depths to steadily accumulate on the seafloor for millions of years. Analysis of these deep ocean sediments reveals a striking thousand-fold increase in charcoal particles around 8 million years ago.[73] The charcoal contains charred fragments of grasses and wood, blackened remains testifying to the aggressive removal and replacement of one group of plants by the other. Although the record is patchy, it suggests that an increase in charcoal flux lags behind the conversion of forests to grasslands in India and Pakistan by about a million years. If the feedbacks operated in the sequence of climate change, vegetation response, and then fire, this lag is exactly what we would expect. It is too early to say if this is the smoking gun for regime change in the subtropics; the data are too sparse in the critical region of transition to make this out. The outlook is bright, however, because it is certainly not beyond the reach of modern geochemistry to bring the picture into sharper focus.

When astrophysicists started to unravel the mysteries of the Universe, they were struck by the long list of coincidences that suggested life depends very sensitively on the form of the physical laws and the values nature assigns to particle masses, force strengths, and the like.[74] The astrophysicist Fred Hoyle (1915–2001) was so impressed by what he called the 'monstrous series of accidents' that he commented that the Universe looked like a 'put-up job', meaning the laws of physics seemed finely tuned for life to appear.[75] As we unravel the hidden mysteries of C_4

plant success to reveal a web of feedbacks all leading in the same direction, we cannot help but wonder, in a less cosmic way, whether a C_4 world is also in Hoyle's words a 'put-up job'. If the carbon dioxide content of the atmosphere wasn't impoverished, if the climate hadn't changed, if grasses did not promote and tolerate fire, if C_4 grasses did not flourish under conditions of carbon dioxide starvation, if smoke did not beget drought and lightning. If only. The list of coincidences is impressive.

Impressive as it is, the question remains: why did C_4 plants leave it so late in the evolutionary day to appear on the ecological stage? Flowering plants evolved about 150 million years ago; C_4 plants failed to appear for another 120 million years. The obvious answer is that evolving the C_4 photosynthetic pathway is a difficult business. Complex changes in the expression of genes governing and regulating photosynthesis must be coordinated with those for other intricate metabolic processes and alterations to leaf anatomy.[76] Yet in spite of all this, we know that the C_4 photosynthetic pathway evolved independently from C_3 ancestors on at least 40 separate occasions.[77]

One of the reasons for multiple origins of C_4 plants may be that many C_3 plants possess components of the biochemical machinery necessary to conduct C_4 photosynthesis.[78] Celery (Apium graveolens) is an archetypal C_3 plant that has dimly lit green photosynthetic cells embedded inside the thick ridges of its stalks. In cut stems, these strange cells show up as green circles neatly aligned around the periphery. The explanation for the corridor of green through celery stalks is that it provides a means of recapturing carbon dioxide produced by respiring tissues (stems and roots) that might otherwise be lost as it leaked back to the atmosphere. In a surprise discovery, British researchers found that the clusters of green cells used the same enzyme as C_4 plants to strip the carbon dioxide from the four-carbon carrier compound found in the sap of C_4 leaves. The cells then convert it into organic acids with Rubisco through the pathway Calvin, Benson, and colleagues elucidated in the 1950s.[79] The working hypothesis here is that celery stalks perhaps function like an enlarged version of a C_4 leaf.

This exciting finding raises many questions. We don't yet know when C_3 plants evolved the capacity to manufacture C_4 plant enzymes and deployed them to scavenge 'waste' carbon dioxide. Could it be that recapturing carbon dioxide dissolved in sap represents some initial step on the road to C_4 plant evolution? Plant groups far more ancient than celery, like ferns and conifers, also have the capacity to do it, raising the possibility that C_4-type photosynthesis is much older than we think.[80] Could it be that plants 'learned' the trick millions of years ago and lost it, only to 'rediscover' it again more recently in their geological history? Our best guess about when it might have first been acquired is in the Carboniferous, 300 million years ago, when an atmosphere impoverished in carbon dioxide but enriched in oxygen prevailed for some 30 million years (see Chapter 3).[81] But so far no compelling evidence of C_4 photosynthesis in fossil plant remains dating to the Carboniferous and Permian has been unearthed.[82] If C_4 plants are 10 times older than we think and full-blown C_4 photosynthesis did arise so far back in time, it is well hidden indeed.

Getting to the bottom of questions like these is likely to be increasingly important in the coming decades. Thomas Malthus (1766–1834), the persistent pessimist who went on to become Britain's first professor of political economy, pointed out in his ground-breaking *Essay on the principle of population* of 1798 that the human population has the power to grow geometrically by repeatedly doubling over time. He concluded that the growth of food production would not keep up with that of the human population, writing 'the human species would increase in the ratio of 1, 2, 4, 8, 16 ... etc. and subsistence as 1, 2, 3, 4, 5 etc.' The spectre of population catastrophe loomed large but, of course, Malthus's forecast turned out to be wrong. Contraception eased birth rates, to slow the explosive rise of the human population from the 1970s onwards, and the green revolution in the late 1960s, engendered by shrewd genetics and the production of nitrogenous fertilizers, ensured that world cereal production trebled in the second half of the twentieth century.

Feeding the world in the second half of the twenty-first century could be a different matter. The global human population is predicted to reach 9 billion by 2050.[83] Meeting the escalating world food demand may call for a second green revolution, one that has its roots in nature's own C_4 uprising. The scope for engineering a better version of Rubisco in C_3 crops like wheat, barley, soybean, and potatoes to raise yields and meet the burgeoning demand is limited. An alternative solution may be to transfer the superior C_4 photosynthesis into our C_3 crops to achieve this end. With this ambitious goal in mind, the attention of the plant biotechnologists has focused on rice, the world's most important staple C_3 crop. Global production of rice has risen threefold over the past three decades, but yields are expected to peak shortly as the crop reaches its maximum efficiency in converting sunlight into food. Because the land area available for cultivation is limited, feeding the world with rice will mean boosting yields by growing a bigger crop with harder-working Rubisco.[84] One way to do this will be to re-engineer rice with the turbo-charged photosynthetic apparatus of C_4 plants.[85] Encouraged by the multiple origins of the C_4 pathway, the molecular geneticists believe it simply cannot be that difficult to do; the molecular engineering of C_4 rice is arguably the 'next frontier' in crop science, and a goal several decades away. But will success in this endeavour equate with popularity? The answer depends on how attitudes to genetically modified crops change as they become ever more widely grown in countries such as the USA, China, Argentina, and Brazil, and maybe even in Western Europe. If no adverse effects of the genetically modified materials are reported it may alter the public's attitude to, and demand for, genetically modified rice—a product that could substantially raise yields per hectare without requiring more nitrogen and water.

Regardless of how the second green revolution is engineered, nature's remarkable C_4 revolution continues and its momentum is likely to carry it forward for decades to come. The rapidly rising global carbon dioxide content of the atmosphere is not expected to halt or reverse its progress; the C_4 genie is out of the evolutionary lamp. True, carbon dioxide fertilization is expected to stimulate tree growth and favour the spread of woody plants into grasslands. But a far more significant issue is the magnitude and pace of human intensification of land use.[86] Since the seventeenth century, 12 million square kilometres of forest have been cleared for crop production, and tropical deforestation continues apace. In Amazonia, the conversion of forest to pasture is already increasing the accidental ignition and severity of fires to the point where fire now threatens the integrity of large areas of tropical forest by accelerating its conversion to C_4 savanna.[87] Thanks to humankind's intervention, the pace of nature's green revolution looks set to quicken.

9

Through a glass darkly

The argument of this book is twofold. First, that integrating plant physiology into palaeobotany allows us to recognize fossil plants as new entities: exquisite tachometers of Earth's history. Second, that plants themselves are significant geological forces of nature. It follows that decoding the wealth of information retained in the matrix of these fossil tachometers hinges on continued investigations of living plants. Integrating the information we glean in this way with cutting-edge knowledge from other research fields is a scientific frontier. It reveals how plants have shaped our planet's past. Like many frontiers, this one affords new opportunities for uncovering alternative perspectives on long-standing paradoxes from our past.

Hardly had we started when we came across signs that there were indeed wonders awaiting us.

Arthur Conan Doyle (1912), *The lost world*

THROUGHOUT this book we have encountered a varied cast of historical characters, who have pioneered the development of palaeobotanical thought over the past two centuries. Although the fascination of plant fossils has an exceptional pedigree, reaching back to at least the eleventh century,[1] Edward Jacob (Chapter 7) was the earliest of these 'searchers of scientific truth' introduced here. Jacob's eighteenth-century claim to fame lay in his descriptions of the fossilized remains of exotic subtropical floras and faunas in the crumbling sediments around the coastline of the Isle of Sheppey. Jacob was followed by the true palaeobotanical pioneers of the eighteenth and nineteenth centuries, who established the scientific basis for the anatomical and microscopic investigation of fossil plants. Through their synthesis of palaeontological knowledge, they established the study of fossils as technical and exacting, rather than a mere hobby.[2] And a common thread linking us with this 'golden age' of discovery and description is the notion that the fossils record some aspects of Earth's ancient climates. It's a telling reminder that curiosity compels us to ask why certain fossils are where they are and to speculate on what it means. Albert Seward, who examined Scott of the Antarctic's fossils (Chapter 6), codified this concept best with his celebrated and timely 1892 essay.[3] Seward's essay opened the eyes of devotees of fossil plants to possibilities beyond the traditional activities of description and classification.[4]

The argument of this book is that the emerging modern synthesis sees the dawning of a new era in the study of fossil plants. I am advocating that this modern synthesis arises out of the seamless integration of new knowledge concerning the physiological and ecological behaviour of living plants and ecosystems into the subject of palaeobotany. The promise of incorporating a powerful and exciting third strand, the science of the genetic pathways controlling the form of organisms— evolutionary developmental biology—into our thinking, is on the horizon. If this view is correct, then the study of living plants may be the driving force for unlocking greater riches from fossilized plant remains. Ultimately, the results from this endeavour will enrich our understanding of the Earth's history. Already, the underlying philosophy here is rejuvenating the field. It shows us that locked within the fossils is the story of the Earth over the past half billion years. Fossil plants are not, as is often assumed, 'silent witnesses to the passage of time', but exquisite tachometers of what has taken place throughout Earth history. And as we start to pay long overdue attention to them, a persuasive new field of research crystallizes, one that requires a reorientation of our thinking. It forces us to recognize that plants are a significant geological force of nature—a critical missing feedback linking biology, chemistry, and physics that has shaped our planet's history.

This chapter shows how new insights are proving enor- mously successful in transforming our efforts to recognize and decode nature's fossil tachometers. More often than not, insights spring from experiments, shifting the focus onto universal plant and ecosystem processes, rather than the packaging of those processes within discrete units we call species. This change in emphasis marks a significant departure from the traditional descriptive activities of the pioneers. It reflects how the definition of fossil species is a concept subject to change, as and when more specimens are examined, whereas those of the key processes are not. I reveal how the difficult business of unravelling the hidden role that plants have played in shaping Earth history is only likely to be achieved by the integration of information from a wide variety of disciplines. Occasionally, we find this integration offering a fresh and rewarding perspective on long-standing paradoxes from Earth history.

The recognition that fossils chronicle different facets of Earth's history stems from continued advances in our understanding of how plants work and how the environment shapes their form and function. It is also the case that plants shape their environment and, as we have seen, the challenge is identifying how they do so. Nevertheless, the point to emphasize here is that we have repeatedly seen that plant physiology opens our eyes to recognizing and unlocking the wealth of information coded into the fossilized organic remains of plants in three ways. First, we can use experiments to illuminate our understanding of the plant fossil record by providing a rational explanation for interpreting how the long-dead photosynthesizers might have functioned. By rational, I mean we can begin to appreciate the biological processes that attune the plant to the environment, extending in some instances right down to the underlying molecular details. Second, continued progress in defining the physiological details of how plants cope with a wide range of environmental conditions offers the prospect, at least, of recovering this same information from plant fossils. Third, experiments inject a healthy dose of empiricism into our thinking, by allowing direct testing of hypotheses that emerge from palaeontological studies.

A revealing example of the first point—how forging the link between physiology and fossils strengthens our interpretation of the fossil record—concerns the relationship between leaf shape and climate, a relationship palaeobotanists are fond of using as a 'thermometer' of past climatic conditions. The idea was first proposed by Harvard botanists Irving Bailey (1884–1967) and Edmund Sinnott (1888–1968) in their landmark study undertaken nearly a century ago.[5] With great insight, they recognized the potential in the observation that the leaves of tropical rainforest trees are large and lack teeth, whereas those of trees in the temperate forests tend to be small and toothed. In fact, the fraction of toothed species in floras rises in a very tight and orderly manner as we move from warm to cool climates.[6] The impressively strong relationship seen in modern floras offers a basis for inferring ancient climates from assemblages of fossil leaves, by looking at the proportion of toothed species in a fossil flora. However, the problem with this deceptively neat and simple arrangement that has dogged it for decades, is that we have little idea why it exists. Why should trees in colder climates have toothy leaves? If we cannot answer this salient question, what business do we have, nearly a century on from Bailey and Sinnott's study, in believing that leaves can act as reliable thermometers of ancient climates?

Some scientists have hypothesized that teeth offer protection from herbivores. Caterpillars, for instance, find it far more difficult to consume leaves adorned with teeth and spines than those with smooth margins.[7] But this is not generally believed to be the whole story and alternative insights into the secrets of this puzzle are emerging from experiments with modern plants. One team, led by Taylor Feild from Tulane University, New Orleans, decided to follow up on the well-known observation that water frequently 'leaks' through specialized valves located in the teeth at the tips of leaves.[8] The valves help release a build-up of hydrostatic pressure between the root and the leaf when soils are wet. In an elegant series of experiments, Feild's team showed that deliberately plugging the valves with wax inhibits the capacity of leaves to absorb carbon dioxide by photosynthesis. Instead of leaking out, water is 'forced' under pressure between and around the photosynthetic cells of the leaf, and the film of water greatly impedes the absorption of carbon dioxide. An intriguing twist to the story is that the cold, damp, misty mornings typical of spring days in the middle and high latitudes are the very conditions favouring water 'leakage' through teeth.

It is early days yet for this intriguing hypothesis; it must be tested on woody plants, the group conventionally used to infer ancient climates from fossil leaves. However, it may also be important that water loss through teeth studded along leaf margins improves the movement of water from the roots, through the stem, to the leaf by increasing the transpiration stream. Enhanced water flow could aid the refilling of the plumbing systems early in the spring when small bubbles (embolisms) form under freezing conditions, preventing the free-flowing passage of water through the plant.[9] If teeth really do act to help auto-repair freeze-damaged plumbing systems, the mechanism might further explain why floras in colder climates, experiencing freezing conditions more often, have a higher proportion of toothed species. These experimental studies of the functional significance of toothy leaves go hand in hand with the insights from molecular genetics. When taken together with breakthroughs in deciphering the nature of the genetic pathways that control leaf form,[10] we get a glimpse of the exciting opportunities for integrating palaeobotany, evolutionary developmental biology, and the geological history of our planet.

Interpreting ancient climates from plant fossils is a profoundly important endeavour, so it is not surprising to find it has been the daily concern of palaeontologists for more than 200 years. But the process of interpretation is not simple, as we have already seen, and occasionally the findings reported by plant physiologists provide a timely reminder of the need for caution in this concern. We saw in Chapter 7 that another pathway to reconstructing ancient climates from fossil plants relies on the observation that certain groups of modern plants, like palms, cycads, and gingers, are frost sensitive. The underlying assumption is that this frost intolerance has remained unchanged over millions of years, and that the scattered distribution of their fossil remains informs the debate by acting as biological indicators of where and when warm winters prevailed in the past. The question is, does this assumption stand up?

In July 1996, observations made during an outdoor experiment underway in Bungendore, south-eastern Australia, on the evergreen snow gum (*Eucalyptus pauciflora*) were to strike at the heart of that very question. The investigation focused on how this most freeze-tolerant of eucalypts, which is able to withstand temperatures as low as $-18\,^{\circ}$ C, might respond to future increases in the atmospheric carbon dioxide concentration. The opportunity arose quite by chance when, after an unusually hard, late frost in mid-spring, sharp-eyed botanists spotted that snow gums grown in a carbon dioxide-rich atmosphere had suffered greater damage than those grown in a normal carbon dioxide atmosphere.[11] How strange—elevated levels of carbon dioxide had altered the physiology of the plants, making them more susceptible to freezing temperatures. Strange or not, it is consistent with recent vegetation changes observed on the other side of the world, in the high northern latitudes. In Sweden, for instance, some species of woody shrubs are starting to migrate to lower altitudes, in spite of a warmer climate. The observed pattern of change—the removal of vegetation from colder regions—is at odds with the often-predicted migration of plants up mountains in a warmer climate, but is consistent with the theory that rising carbon dioxide levels enhance their susceptibility to freezing.[12]

Quite how carbon dioxide lowers the freezing tolerance of plants is unclear, but it appears to be related to its capacity to induce partial closure of the microscopic stomatal pores on the leaf surface.[13] This effect restricts the transpiration stream and limits evaporative cooling, causing leaf temperatures to rise (Chapter 2). It seems that the warmer daytime temperatures of leaves in a carbon dioxide-rich atmosphere delay the physiological adjustments necessary to confer cold tolerance, adjustments that are usually triggered by cooler temperatures.

Could it be that in the past, when palms, gingers, cycads, and the like enjoyed a carbon dioxide rich-atmosphere, they, too, were more sensitive to frost? Certainly, the evidence so far indicates this is the case for a range of ancient plant taxa, including frost-hardy palms.[14] The cautionary tale here is that the climatic tolerances of plants cannot be assumed to be fixed, immutable to the sands of time. Physiology matters, especially when carbon dioxide is brought into the equation, and the wise paint a sharper portrait of Earth's ancient climate by combining multiple lines of evidence.[15]

These two examples, the shape of leaves and frost intolerance of plants, illustrate how knowledge of plant physiological processes strengthens our interpretation of the plant fossil record and broadens the perspective it provides on Earth's history. Looking back at the narrative of earlier chapters, we can see how this familiar theme recurs. Chapter 2 explains why the widespread appearance of leaves, those pivotal light-harvesting organs, could have been delayed by up to 50 million years following the evolution of the earliest land plants. This showed how the remarkable discovery that the number of microscopic stomatal pores on leaves is controlled by the carbon dioxide content of the atmosphere paved the way for a new explanation about why.[16] Once carbon dioxide levels began to plummet late in the Palaeozoic, the evolution of leafy plants became a possibility. The risks of lethal overheating in the changing atmosphere afforded the opportunity for stepping up the production of stomatal pores to entrain a greater transpirational cooling stream. Leaves soon evolved in four major groups independently, through the steps recognized by Walter Zimmermann back in the 1930s and 1940s. Modern developmental genetic studies later showed how plants achieved it, through the use of a common molecular tool-kit that perhaps originated in primitive plants to regulate shoot branching.[17]

Investigations into the link between the carbon dioxide content of the atmosphere and climate are not of the same antiquity as the study of the fossil plants, extending back to the days of Joseph Fourier, John Tyndall, and Svante Arrhenius in the early nineteenth century. Nevertheless, a holy grail in this important field of study has been to find ways of assessing the strength of this linkage over geological time. Of course, the need to do so is made all the more pressing in the light of the current rapid increases in carbon dioxide. Fossil plants received special attention in this context with the discovery that the number of stomatal pores changes in tandem with the amount of carbon dioxide in the atmosphere. As with the evolution of leaves, this physiological adjustment of the plant to its atmospheric environment is underpinned by an understanding at the molecular level.[18] Inevitably, it raised the exciting possibility that fossil leaves might act as biosensors of the level of this important greenhouse gas in the ancient atmosphere. Only recently, the further potential of the plant kingdom in this regard has been revealed by physiological studies of the humble bryophytes (liverworts and mosses), whose fossil remains offer hope of another novel means for 'breathalyzing' the atmosphere.[19] Still, the point is that when the fossil leaf collections of the Reading University palaeobotanist Tom Harris were re-examined in the light of the stomatal–carbon dioxide link, they revealed previously hidden information (Chapter 5). They indicated large changes in atmospheric carbon dioxide levels across the Triassic–Jurassic boundary, some 200 million years ago. Soaring carbon dioxide and global warming were then deduced to have triggered a switch in the composition of the fossil floras from Greenland recorded by Harris some 60 years earlier. As the evidence from plant fossils mounts, it is reigniting what was becoming an increasingly sterile debate about the causes of the end-Triassic declines in animal biodiversity.

Discerning Earth's oxygen history has proved somewhat more problematic than has been the case for carbon dioxide. We do not yet have any equivalent means of inferring its past concentration from rocks and fossils and must instead infer it through a number of interpretive steps. These are based on the realization that the atmospheric oxygen content is controlled on a timescale of millions of years by the slow dance of carbon and sulfur cycles (Chapter 3).[20] Geochemists finally achieved what seemed to be a reasonable version of Earth's oxygen history over the past half billion years based on this reasoning, but whether it was the correct one or not was another matter. Only after it was recognized in laboratory growth experiments that an atmosphere enriched with oxygen shifted the carbon isotope composition of plants were geochemists handed the keys to smoothly unlock another version of Earth's oxygen history. A spin-off from this finding in the laboratory was that it provided a new means of interrogating the plant fossil record for evidence of a surge in oxygen levels during the Permo-Carboniferous, some 300 million years ago.[21] Once again, the findings of experiments facilitated the extraction of novel information from fossils.

Who knows what Seward would have made of these developments, a century on from penning his seminal essay on the use of fossil plants as 'tests of climate'. It is interesting to speculate that he might have been rather tickled by the contribution plant physiology made to the long-running transatlantic dispute about polar forests. The dispute began when he examined the fossils recovered from Antarctica by Scott's polar team early in the twentieth century (Chapter 6). The widely accepted scientific 'theory' was that ancient forests growing in the polar regions were deciduous, because shedding leaves in the autumn avoided carbon losses at a time when they could not photosynthesize in the darkness of warm polar winters. It seemed entirely sensible and, for a century, this beguiling notion held sway among the scientific community, even though it was long on speculation and short on hard facts. It was not until experimental trials cultivating trees in replicate ancient polar climates were brought to bear on the matter that it was exposed as entirely false. Modern plants shattered cherished beliefs about polar forests. Alternative explanations for the deciduous nature of polar forests must now be sought, with promising avenues of future investigation pointed up by real conifer forests at their northernmost limit, in eastern Siberia.

The story of leaf habit and the ecology of ancient polar forests is a case of living plants directly testing a palaeontological hypothesis. This is the third significant type of contribution that studies of living plants make to palaeobotany. As we have seen, this is also true for the zoological realm; in Chapter 3 the physiological studies of moths, flies, grasshoppers, and mammals yielded insights into evolutionary trends in body size revealed by the fossil record. In the botanical realm, however, the most controversial hypothesis—or 'crazy idea', depending on your viewpoint—I described was that mutated fossil spores in rocks dated to the great end-Permian mass extinction. Were they the product of excess ultraviolet-B radiation filtering through a tattered ozone layer damaged by massive volcanic eruptions in Siberia (Chapter 4)? Controversial it may be, but the important point is that it can readily be put to the experimental sword. Indeed, it has recently been reported that pollen produced by soybean (*Glycine max*) exposed to increased ultraviolet-B radiation took on a shrivelled appearance with a disfigured external ornamentation. Notably, like the mutated fossil spores, it also lacked the apertures through which the pollen tube germinates.[22] The catch is that high temperatures produced the same effects. This meagre experimental evidence is tantalizing but doesn't go far enough. Evaluating its credibility further will take more research. Only then will we know if this theorem is likely to stand the test of time or if its days are numbered.

The second argument presented in this book is that, as we strengthen our understanding of terrestrial plant activities, we gain a clearer picture of how they shape the environment and move towards a new view of Earth's history. This revised world-view recognizes plants themselves as significant geological forces of nature. We are familiar with a mental picture of Earth history shaped by mighty geological forces like plate tectonics, which build mountain ranges and create and destroy entire oceans over hundreds of millions of years. Yet over the same enormous, unimaginable lengths of time, plants, too, exert a cumulative and significant influence on the global environment. The formidable challenge of sorting out the geological consequences of terrestrial plant evolution and diversification is a second recurring theme of this book.

Nowhere is this theme better illustrated than by the profound global environmental consequences that flowed from the origination and colonization of the land by plants. In Chapter 2 we saw how the evolutionary appearance, diversification, and spread of forests across the ancient landscape transformed the nutrient, water, and energy circulation systems that maintain Earth's global environment. With the planet's surface clothed in forests recycling precipitation and leaking organic acids into the soils, continental erosion accelerated the removal of carbon dioxide from the atmosphere, sending it downstream as bicarbonate ions that were eventually to be buried at the bottom of the sea. A hundred million years of continental attrition by the activities of plants caused the carbon dioxide level to plummet and the climate to cool, bringing Earth to the brink of a catastrophic ice age. Global indigestion ensued as large amounts of dead organic debris became buried in swamplands, forcing atmospheric oxygen upwards and thus engendering evolutionary innovation—gigantism—in animals (Chapter 3).

Things might have been very different if plant life had not evolved. Without plants to accelerate the chemical weathering of rocks on the continents, the carbon dioxide content of the atmosphere would be 15 times the present-day concentration.[23] The resulting 'super-greenhouse' climate would have raised global temperatures by a sweltering 10°C, preventing the development of polar ice caps and leading to a sea level hundreds of metres higher than now, sufficient to make our modern coastlines unrecognizable. The situation for oxygen is in contrast with that of carbon dioxide. With the supply of oxygen to the atmosphere that was usually delivered by the slow burial of plant organic matter shut off, the oxygen content would be pinned to an asphyxiating 10%—equivalent to living in the thin atmosphere found at about 5.5 km above sea level. Meanwhile, in the hot climate and asphyxiating atmosphere, the barren terrestrial landscape of our hypothetical 'desert world', covered in a mixture of desert sands and thin algal crusts, would be unable to efficiently recycle water as modern tropical rainforests do, cutting rainfall on land by half.[24] The nature of rivers carrying the meagre rainfall out to sea would also be radically altered without plants to anchor soils and maintain the banks. Meandering rivers would be replaced by different kinds of watercourses, perhaps similar to the braided streams found on desert alluvial fans or in front of glaciers. Gradually, as the picture of our desert world planet comes into sharper focus, we can appreciate the enormous environmental legacy of plant life for our modern environment.

Plants, then, enhance the chemical erosion of continental rocks and produce organic debris that, over the eons, alters the global environment. It doesn't stop there, however, because in addition to these changes in the ebb and flow of materials that take millions of years to influence the climate and atmosphere, land plants also alter climate on shorter timescales. Accelerated warming in the Alaskan arctic, as shrubs have advanced northwards in the last decade, reminds us that even on human timescales vegetation can exert an influence on climate, in this case by changing the colour and hence reflectivity of the land surface (Chapter 6). Computer simulations of ancient climates are only starting to get to grips with evaluating the nature of this vegetation feedback.[25] What they tell us is that in the ancient 'greenhouse' world of the Cretaceous, the forests cloaking the

Arctic and Antarctica generated warmth by darkening the land surface and adding the greenhouse gas water vapour to the atmosphere from their transpiring canopies. Altered patterns of atmospheric circulation could then transport the warm air onto the continental mainland, and in doing so establish an atmospheric connection between polar forests and the climate of their lower-latitude temperate counterparts.

We can expect that through a wide range of activities, plant life has been silently, relentlessly, orchestrating global change over millions of years. Uncovering further manifold ways in which it does so remains a considerable scientific challenge, and the extent to which it can be achieved is uncertain. I suspect we have barely scratched the surface. Certainly, much more research is needed. Unfortunately, plants often alter the environment in ways that are imperceptible on the timescale of human lifetimes, which makes the changes they cause over geological timescales hard to uncover and difficult to investigate. New insights will require the application of knowledge from a wide variety of disciplines, and progress is likely to be underpinned by advances in technology.

I believe there are grounds for optimism. Take the case of the grasslands. We are beginning to uncover an intriguing network of linkages between this cosmopolitan biome, climate change, and evolution that has taken place since the end of the Eocene epoch. One school of thought, daring to challenge conventional wisdom, argues that the grasses themselves engineered the persistent long-term march towards cooler conditions over this time.[26] And they did this in numerous diverse ways. Grasslands liberate mineral nutrients from the young fertile soils they preferentially exploit and export unusually carbon-rich sediments into rivers. Both features sequester carbon from the atmosphere, one by supplying nutrients to enhance the productivity of marine ecosystems and the other by exporting bicarbonate ions to be buried at sea. If the capacity of grasslands to sequester carbon dioxide through these routes is sufficiently large, long-term climatic cooling will inevitably follow. The idea has also been mooted that the lighter colour of grasslands, compared to the darker forests they replaced, altered the reflectivity of the land surface to enhance regional aridity by reducing cloud cover.[27]

As the grassland machine rolled on, altering climate en route, it steered the evolutionary trajectory of animals, and even a group of phytoplankton called diatoms that require silica to produce their famously ornate shells. The fossil record beautifully documents the rise and evolutionary diversification of grazing animals (including horses, rhinos, and antelope) that appeared to exploit the newly available food source in tandem with the evolution and spread of grasses.[28] Grazing animals evolved high-crowned, enamel-edged teeth to deal with feeding on tough gritty grassland forage on the open plains. By feeding on grasses, the animals ingested cell walls built from tiny mineral deposits containing silica in a form that is twice as soluble as that in the mineral forms locked up in rocks. Before grasses appeared on the scene, silica was in short supply because of its slow release by the chemical breakdown of silicate rocks. Afterwards, grazers accelerated the delivery of silica to the oceans, greatly facilitating diatom success. It is more than mere coincidence that the diversity and ecological success of diatoms tracks that of the grasses.[29]

One more piece of this complex and incomplete grassland puzzle fell into place when it was appreciated that the astonishing geographical spread of grasses into virtually all parts of the world is strongly favoured by their ability to tolerate droughts. Drought tolerance in grasses is linked to the evolution of a special type of microscopic stomatal pore regulating the uptake of carbon dioxide for photosynthesis and the inevitable loss of water vapour. The unusual pore design comprises two long narrow cells with swollen ends aligned in parallel in a shape resembling dumb-bells.[30] It has the advantage over the kidney-shaped version possessed by most other groups of plants in that it requires smaller changes in the pressure inside the cells to cause the pore to open and close. This allows a speedier response to changes in the environment, especially light, and so helps conserve water and avoid the likely effects of drought. All of which leads us to the remarkable possibility that the evolution of a single type of stomatal pore is partly responsible for major climatic trends witnessed over the past 30 million years.

Of course, we do not know if this is really the case. The argument can be made, whether it can be made to stand up is another matter. The point is really to prompt us to notice the varied lines of evidence needed to uncover the role of plants in Earth's geological history. The pursuit of this controversial endeavour will require investigations embracing timescales ranging from milliseconds to millions of years, to take in ultra-fast biochemical reactions and the ultra-slow weathering of rocks, and spatial scales from micrometres to mega-kilometres to take in everything from individual cells to the entire terrestrial biosphere. These are the awesome transformations of scale likely to be necessary in our thinking to reveal the intimate connections between global-scale biology and Earth's history.

How wonderful that we have met with a paradox. Now we have some hope of making progress.

Niels Bohr, as quoted in Moore, R. (1967) *Niels Bohr, the man and the scientist.* Hodder & Stoughton, London (p.140).

In January 2005, the BBC broadcast a doom-laden documentary on the phenomenon of global dimming as part of its flagship science series *Horizon*.[31] The programme explained how society's industrial activities have added such vast quantities of tiny polluting aerosol particles to the atmosphere that it has been bouncing more incoming sunlight back into space.[32] Not only that, but the aerosols are altering the properties of clouds, making them brighter and causing them to act like giant mirrors.[33] The fear is that by loading the atmosphere with aerosols—a cooling pollutant—we have reduced the full strength of the warming due to rising greenhouse gas concentrations. It has been likened to driving a car with your foot on the accelerator and the brake. As the atmosphere becomes cleaner, the narrator Jack Fortune warned viewers 'temperatures will rise twice as fast as they [scientists] previously thought, with irreversible damage just twenty five years away'. Piling speculation on top of speculation Fortune, the prophet of doom, offered us an apocalyptic vision of our future world: 'Earth's climate would be spinning out of control, heading towards temperatures unseen in four billion years'.

The programme caused a sensation and understandably captured the public's imagination. 'Quite possibly the most disturbing programme any channel will show all year' cried the television critics.[34] Yet, shortly after the *Horizon* episode was transmitted, the leading international journal *Science* published several telling updates on the story offering both good and bad news. The good news is that the air we breathe has been getting cleaner and our skies brighter since 1990,[35] but the sting in the tail is that a cleaner atmosphere leaves us more vulnerable to the climatic consequences of a build-up of greenhouse gases.[36] As we ease our foot off the brake, will brighter skies really affect future warming as dramatically as the television producers would have us believe? It is an easy question to ask, but a hard one to answer. The climatic effects of aerosols are notoriously complex and difficult to unravel. Nevertheless, the policy implications are enormous so it is important that the pressing scientific issue of 'global dimming', with its potentially grave consequences for human society, is now on the climate change research agenda.[37] And the fact that it is on the table for discussion is thanks to the pioneering efforts of a handful of heroic researchers around the world, who have been diligently making routine measurements that established two paradoxes.

The first concerns the dramatic reductions in the amount of sunlight reaching the surface of the Earth first noticed by the British agricultural botanist Gerald Stanhill working in Israel, who coined the term 'global dimming'.[38] The paradox created by global dimming was received with fierce scepticism: the world was getting warmer whereas less sunlight meant it should be getting cooler. Why? The second paradox emerged with the discovery that the amount of water evaporating from pans placed outside around the world was decreasing.[39] To the surprise of many, pan evaporation actually decreased between 1950 and 1990 over large parts of the northern hemisphere. The same trend has also been seen in records extending back the last 50 years in Australia and New Zealand, suggesting it really is a widespread phenomenon.[40] How could pan evaporation go down when global temperatures were going up?

For many years, consensus on the reason for the unexpected historical trend in pan evaporation eluded the scientific community.[41] However, a few years ago, two straight-talking Australians, Michael Roderick and Graham Farquhar at the Australian National University in Canberra, put forward a valuable contribution to the controversy that is fast gaining traction.[42] They realized that, of the three main factors determining evaporation of water from a pan (solar energy, the amount of moisture in the air, and wind speed), solar energy is by far the most important. Reduce the amount of sunlight hitting the water and you reduce the amount of energy available to drive water molecules out of the pan. In the former Soviet Union, where pan evaporation and solar radiation datasets were available for the same 30-year period (1960–90), the arithmetic added up: the observed decline in pan evaporation matched that calculated from declines in solar radiation.

The Australian duo also brought another intriguing piece of evidence into play that lent further support to their arguments. If, they argued, global dimming is a genuine phenomenon resulting from more aerosol pollutants and more clouds, it should warm up the night-time temperatures by trapping outgoing heat—much as John Tyndall envisaged that water vapour prevented the 'warmth of our fields and gardens' from pouring itself 'unrequited into space'.[43] Warmer nights reduce the difference between the daily maximum and minimum temperatures, a difference known as the diurnal temperature range. Clouds were dramatically revealed to act in this way after all civilian aircraft were grounded for three days following the terrorist attacks on the United States on 11 September 2001. The grounding of aircraft reduced cloud cover produced by condensation trails, the trails of water vapour created in the wake of high-altitude aircraft. In response, the diurnal temperature range recorded by a network of weather stations across North America jumped by a factor of three.[44] Global dimming arising from cloudier skies could, then, be associated with widespread reductions in the diurnal temperature range, and indeed this is exactly what has been observed in the northern hemisphere, where aerosol pollution is greatest.[45]

The scientific developments leading up to the discovery of global dimming and all its wider ramifications for society are fascinating. Not least because a solution to two paradoxes—why global dimming and reductions in pan evaporation accompany a warming world—has been proposed by their unification. Very often, paradoxes are indicative of our inadequate insight into natural phenomena, rather than being features of the phenomena themselves. Consequently, rethinking our ideas when confronted with a paradox can often, as Niels Bohr was aware, lead to a deeper understanding of how the world works. Rarely is scientific progress marked with the exclamation of 'Eureka!'; more often it is the less euphoric 'hold on a minute' that signals something exciting is afoot, as the scientific endeavour advances, crab-like, through the iterative processes of conjecture and refutation.[46]

The challenge to orthodoxy can be mounted by revisiting timeless problems with new instruments, computer models, and observations made possible thanks to advances in technology and theory—a path reminiscent of that trod by the first 'natural philosophers' in the seventeenth and eighteenth centuries, as the scientific enterprise took shape.[47] Indeed, in this book, advances in all such areas have allowed us to integrate plants into the picture and cast new light on long-standing paradoxes from Earth history. We have encountered the case of oxygen and giant insects (Chapter 3), the extreme greenhouse world of the Eocene (Chapter 7), and the worldwide spread of grasses with the C_4 photosynthetic pathway in the face of a stable background level of carbon dioxide (Chapter 8).

As we have already seen, scientists had little idea of oxygen's atmospheric history over the period in which complex plants and animals evolved (Chapter 3). By the 1970s, the opinions of adherents to the Gaia hypothesis held sway. Leading the charge, James Lovelock argued a 'raging conflagration which would destroy tropical rainforests and arctic tundra alike',[48] was inevitable if atmospheric oxygen levels rose above 25%. According to these arguments, oxygen never strayed far from a baseline value of 21%, but if that was the case, how do we explain the spectacular fossil evidence for giant dragonflies, millipedes, scorpions, and the like that Charles Brongniart and his contemporaries began unearthing from 300-million-year-old rocks of European quarries?

It was not until a handful of geochemists showed that Earth's atmospheric oxygen content was regulated over millions of years by the slow birth and death of sedimentary rocks that this rigid picture started to change.[49] The birth of sedimentary rocks supplies oxygen to the atmosphere by burying organic matter and sulfur, and their death removes it by exposing both to oxidative destruction. Here was the critical advance in theory that provided the essential framework for constructing a scientifically robust history of oxygen's highs and lows over the past half billion years. As we have seen, the rest is history. The revised view shows oxygen levels reaching around 35% during the Permo-Carboniferous, a time that closely corresponds with the rise of the giant insects in the fossil record. Conversely, as oxygen levels subsequently plummeted, giant insects perished. Revisiting the paradox with a new theoretical foundation led to a more satisfying explanatory link between the evolution of Earth's atmosphere and animal life.[50]

In the case of Earth's early Eocene epoch, when alligators basked in the Arctic 50 million years ago, the paradox is why the climate was so much warmer than expected even if carbon dioxide levels were up to six times the present-day concentration (Chapter 7). Suspicion alighted on the fact that the state-of-the-art climate models were missing critical feedback loops that form part of the coupled Earth system, loops that were capable of amplifying the warming caused by carbon dioxide. One missing piece of the puzzle turned out to be the ancient swamps and forests that were emitting a cocktail of gases. When added into the models of the atmospheric chemists, they generated a range of potent greenhouse gases that had previously been completely overlooked. Collectively, the warming caused by the feedback of these gases on climate improved the capacity of the models to reproduce the global warmth indicated by the subtropical remains of fossil plants and animals in the Arctic. It had taken advances in computer models, new observations, and the erosion of strong boundaries between scientific disciplines, to recognize that swamps, forests, and microbes powerfully amplified carbon dioxide's effects on the Eocene tropical paradise.

Another challenge to entrenched orthodoxy followed hot on the heels of new records revealing a stable background level of carbon dioxide in the atmosphere at a time when grasses with a C_4 photosynthetic pathway transformed subtropical ecosystems worldwide in a geological instant 8 million years ago (Chapter 8). The C_4 grasses are supremely adapted to carbon dioxide starvation. So if carbon dioxide levels remained stable, the obvious explanation for this successful tropical takeover bid collapsed. The answer to how C_4 tropical grasses achieved near-synchronous worldwide success in the face of a stable background level of carbon dioxide remains elusive. But ecological detective work on the savannas of South Africa, meticulous research in the laboratories of plant physiologists, and the innovative aircraft and satellite measurement campaigns of atmospheric scientists combine to offer refreshing insights and novel avenues for further research. All the signs are that climate change and fire seem to have played a central role. Debate over the role of fire in the historical (and future) success of tropical grasslands and savannas is sure to smoulder on.

Arthur Conan Doyle (1859–1930) penned his classic tale *The lost world* in 1912. In it, he tells the story of an intrepid band of adventurers led by Professor Challenger exploring a mysterious plateau in the Amazonian rainforest. As the book unfolds, we find Challenger's initial claims for the continued existence of prehistoric life in central South America are understandably greeted by a sceptical reaction from members of the Zoological Institute. Such extraordinary claims demand extraordinary evidence and, in due course, a Committee of Investigation led by Challenger is dispatched to return with precisely what's required to win over the sceptics. The following year, at a specially convened meeting of the Institute in Regent Street, the adventurers report back. Wondrous tales of creatures new to science were recounted, but the evidence to corroborate them was slender, amounting to a few photographs of life on the plateau that may have been doctored. Eventually, provoked by the persistent dissenting voice of his life-long enemy Dr Illingworth, Challenger produced a large square wooden case and withdraws its sliding lid. To the astonishment of the hushed audience:

> an instant later, with a scratching, rattling sound, a most horrible and loathsome creature appeared from below and perched itself upon the side of the case. The face of the creature was like the wildest gargoyle that the imagination of a mad medieval builder could have conceived. It was malicious, horrible with two small red eyes as bright as points of burning red coal. Its long, savage mouth, which was held half-open, was full of a double row of shark-like teeth.

Challenger delivered the ultimate proof of prehistoric life on the plateau—the adventurers returned to London with a 'devil's chick', a young pterodactyl, crushing the objections of the doubters. I am not claiming that any such startling evidence for plants in shaping Earth's history is likely to come to light; the vector of causation is probably far too subtle. But remember the Wollemi Pine, one of the oldest and rarest living fossil species in the world, whose ancestry dates back to the time of the dinosaurs (Chapter 6). Living populations of this majestic conifer were discovered just 200 km from Sydney in Wollemi National Park, Australia in 1994. Wollemi Pine is the botanical equivalent of the pterodactyl chick. It may not have a beady red eye but it reminds us that, unlike on the plateau, in real life the 'ordinary laws of nature' are not suspended. As James Hutton and Charles Lyell recognized over two centuries ago, the laws of nature are immutable, and this leads to the crucial realization that observations of modern phenomena can be used to interpret the past. This uniformitarian principle, summarized by the maxim 'the present is the key to the past' (see also Chapter 5, p. 146), applies as equally to plants as it does to the environmental processes that slowly, imperceptibly, mould and recycle the landscape around us over millions of years. Consequently, the language chronicling the hidden geological history of our planet in fossil plants remains decipherable. As we stand on the threshold of a new era in this endeavour, it's a safe bet that the stock of these precious time capsules is set to rise. The opening epigraph of this chapter from Doyle's *The lost world* serves as a metaphor for the future of this exciting research frontier, for not only have we only just begun, but already there are indeed signs that 'wonders await us'.

NOTES

CHAPTER 1

Introduction

1 Darwin wrote six books devoted to plants. In *The ancestor's tale: a pilgrimage to the dawn of time* by Richard Dawkins (2004, Weidenfeld and Nicolson) plants got 11 out of 528 pages.

2 It would be churlish to list the geology textbooks that fall into this common trap, but for one that does recognize plants as an important element of the Earth sciences, see Kump, L.R., Keating, J.F., Crane, R.G., and Kasting, J.F. (2003) *The Earth system: an introduction to Earth system science*. Prentice Hall, London.

3 These uncertainties should not in any way trivialize or detract from the serious crisis facing our planet. The issues are discussed in Williams, P.D. (2005) Modelling climate change: the role of unresolved processes. *Philosophical Transactions of the Royal Society*, **A363**, 2931–46. For a discussion of the issues about making predictions of the biosphere's ecological future prospects, see Moorcroft, P.R. (2006) How close are we to a predictive science of the biosphere? *Trends in Ecology and Evolution*, **21**, 400–7. The other side of the argument is that the need to make ecological science more predictive has seen a welcome change in emphasis from the 'bean counting' of traditional ecological studies to those of underlying processes and mechanisms; see Pataki, D.E., Ellsworth, D.S., Evans, R.D. *et al.* (2003) Tracing changes in ecosystem function under elevated carbon dioxide concentrations. *Bioscience*, **53**, 805–18; Schlesinger, W.H. (2006) Global change ecology. *Trends in Ecology and Evolution*, **21**, 348–51.

4 Schellnhuber, H.J. (1999) 'Earth system' analysis and the second Copernican revolution. *Nature*, **402**, C19–C23.

5 Lovelock, J.E. (1988) *A biography of our living Earth*. Oxford University Press.

6 Lovelock, J.E. (1995) *The ages of Gaia*. W.W. Norton, New York.

7 Lovelock, *A biography of our living Earth* (above, n.5).

8 Arguments between proponents for and against the concept of Gaia continue. For a selection of these see Vol. 52, Issue 4 (2002) of the

academic journal *Climate Change*, published by Springer, which carries papers on the special theme of the Gaia hypothesis. Further responses, and responses to responses, were published in Vol. 58, Issues 1–2 (2003) in the same journal.

9 Lovelock, J.E. (2006) *The revenge of Gaia: why the Earth is fighting back – and how we can still save humanity*. Allen Lane, London.

10 All dates in this book use the timescale of Gradstein, F.M., Ogg, A.J. and Smith, A.G. *et al.* (2004) *A geologic timescale 2004*. Cambridge University Press.

11 Gribbin, J. (2003) *Science: a history. 1543–2001*. Penguin, London.

12 Crutzen, P.J. (2002) Geology of mankind. *Nature*, **415**, 23.

13 Houghton, J.T., Ding, Y., Griggs, D.J. *et al.* (2001) *Climate change 2001: The scientific basis*. Cambridge University Press. For a discussion of recent warming trends, see Hansen, J., Sato, M., Ruedy, R. *et al.* (2006) Global temperature change. *Proceedings of the National Academy of Sciences, USA*, **103**, 14288–93.

14 Pittock, A.B. (2006) Are scientists underestimating climate change? *EOS*, **87**, 340–1.

CHAPTER 2

Leaves, genes, and greenhouse gases

1 See Davies, P. (2003) *The origin of life*. Penguin.

2 Instrument measurements and their interpretation from the *Galileo* probe were reported by Sagan, C., Thompson, W.R., Carlson, R. *et al.* (1993) A search for life on Earth from the Galileo spacecraft. *Nature*, **365**, 715–21.

3 For persuasive discussions on why we might expect chlorophyll on an alien photosynthetic biosphere, see Wolstencroft, R.D. and Raven, J.A. (2002) Photosynthesis: likelihood of occurrence and possibility of detection in Earth-like planets. *Icarus*, **157**, 535–48, and Conway-Morris, S. (2003) *Life's solution: inevitable humans in a lonely universe*. Cambridge University Press. Calculations indicate that it would indeed be possible to detect 'exovegetation' on a terrestrial planet orbiting an M star (the commonest class of stars, containing red dwarfs; they have a diameter and mass less than one-third that of the Sun and a surface temperature of less than 3500 K),

although the strength of its 'biosignature' would be strongly dependent on the extent of the vegetated area, cloud cover, and viewing angle; see Tinetti, G., Rashby, S., and Young, Y.L. (2006) Detectability of red-edge shifted-vegetation on terrestrial planets orbiting M stars. *Astrophysical Journal*, **644**, L129–L132.

4 Stroeve, J.C., Serreze, M.C., Fetterer, F. *et al.* (2005) Tracking the Arctic's shrinking ice cover: another extreme September minimum in 2004. *Geophysical Research Letters*, **32**, doi:10.1029/2004GL021810.

5 On the west Antarctic ice sheet, ice shelf disintegration is documented by DeAngelis, H. and Skvarca, P. (2003) Glacier surge after ice shelf collapse. *Science*, **299**, 1560–2. A similar phenomenon has also been reported on Greenland by combining airborne laser altimetry with information from satellites by Joughin, I., Abdalati, W., and Fahnestock, M. (2004) Large fluctuations in speed on Greenland's Jakobshavn Isbræ glacier. *Nature*, **432**, 608–10.

6 Scambos, T.A., Bohlander, J.A., Shuman, C.A., and Skvarca, P. (2004) Glacier acceleration and thinning after ice shelf collapse in the Larsen B embayment, Antarctica. *Geophysical Research Letters*, **31**, doi:10.1029/2004GL020670.

7 Inductive reasoning is the process of generalization from a given set of facts rather than the sequential argument that characterizes deductive reasoning. The scientific endeavour uses both inductive and deductive logic, although hypotheses are usually framed with the former.

8 Cited in the French mathematician Jacques Hadamard's study of inspiration. See Hadamard, J. (1949) *The psychology of invention in the mathematical field*. Princeton University Press.

9 French, A.P. (1979) *Einstein: a centenary volume*. Heinemann, London.

10 See Tucker, C.J., Townshend, J.R.G., and Goff, T.E. (1985) African landcover classification using satellite data. *Science*, **277**, 369–74. The suggestion that satellites can be used to monitor terrestrial plant activity was reported by Tucker, C.J., Fung, I.Y., Keeling, C.D., and Gammon, R.H. (1986) Relationship between atmospheric CO_2 variations and a satellite-derived vegetation index. *Nature*, **319**, 195–9.

11 Myneni, R.B., Keeling, C.D., Tucker, C.J. *et al.* (1997) Increased plant growth in the northern high latitudes from 1981 to 1991. *Nature*, **386**, 698–702. See also the commentary by Fung, I. (1997) A greener North. *Nature*, **386**, 659–60.

12 A first attempt at integrating satellite data with models of plant growth to estimate marine and terrestrial primary productivity reported global net primary production was $104.9Pg$ C yr^- ($1Pg$ C $= 1 \times 10^{15}g$ C) with similar contributions from terrestrial ($.PgCyr^{-1}$) and marine ($.PgC$ yr^{-1}) plants. See Field, C.B., Behrenfeld, M.J., Randerson, J.T., and Falkowski, P. (1998) Primary production of the biosphere: integrating terrestrial and oceanic components. *Science*, **281**, 237–40.

13 Prentice, I.C, Farquhar, G.D., Fasham, M.J.R. *et al.* (2001) The carbon cycle and atmospheric carbon dioxide. In *Climate change 2001: the scientific basis* (ed. J.T. Houghton, Y. Ding, D.J. Griggs *et al.*), pp. 183–237. Cambridge University Press.

14 Cramer, W., Bondeau, A., Woodward, F.I. *et al.* (2001) Global response of terrestrial ecosystem structure and function to CO_2 and climate change: results from six dynamic global vegetation models. *Global Change Biology*, **7**, 357–73.

15 An extreme example of this positive feedback loop, whereby excessive carbon dioxide release from the Amazon amplifies future global warming, is reported by Cox, P.M., Betts, R.A., Jones, C.D. *et al.* (2000) Acceleration of global warming due to carbon-cycle feedbacks in a coupled climate model. *Nature*, **408**, 184–7. An intercomparison between elaborate coupled climate–carbon cycle models nevertheless indicates that an overall amplification of warming caused the anthropogenic addition of greenhouse gases involving the terrestrial carbon cycle is likely; see Friedlingstein, P., Cox, P., Betts, R. *et al.* (2006) Climate-carbon cycle feedback analysis, results from the C4MIP model intercomparison. *Journal of Climate*, **19**, 3337–53.

16 Jones, C.D. and Cox, P.M. (2005) On the significance of atmospheric CO_2 growth rate anomalies in 2002–2003. *Geophysical Research Letters*, **32**, doi: 10.1029/2005GL023027.

17 Leaves are classified into two main types, microphylls and megaphylls. Microphylls are tiny spine-like enations borne on the main stem, usually with a single vascular strand. These are found today in lycophytes (clubmosses, spikemosses, and quillworts) and were once common in some groups of extinct plants (lycopods). Megaphylls are the most common variety, characterized by a flat leaf blade and a network of branched veins, and are found in the

euphyllophytes (ferns, gymnosperms, and angiosperms). What concerns us throughout this chapter is the evolution of the megaphyll leaf.

18 Schulze, E.-D., Vygodskaya, N.N., Tchebakova, N.M. *et al.* (2002) The Eurosiberian transect: an introduction to the experimental region. *Tellus*, **54B**, 421–8.

19 Current thinking on land plant evolution, including palaeobotanical information and new discoveries on the relationships among land plants from molecular biology, is admirably set out by Kenrick. P. and Crane, P.R. (1997) The origin and early evolution of plants on land. *Nature*, **389**, 33–9. For a more comprehensive treatment, see also Kenrick, P. and Crane, P.R. (1997) *The origin and early diversification of land plants: a cladistic study*. Smithsonian Series in Comparative Evolutionary Biology, Smithsonian Institution Press, Washington. See also Niklas, K.J. (1992) *The evolutionary biology of plants*. University of Chicago Press.

20 Dawson achieved notoriety for the controversy sparked by his short paper published in 1859 describing two types of the strange fossil plants found on collecting trips in the Gaspé Peninsula. His description of one of these, *Psilophyton princeps*, sent ripples of confusion through the palaeobotanical community for more than a century (Dawson, J.W. (1859) On fossil plants from the Devonian rocks of Canada. *Quarterly Journal of the Geological Society*, **15**, 477–88). The problem was that in 1870 he published another, somewhat modified description of what he thought was a close relative of *P. princeps* that was unlike anything turned up by fossil hunters since (Dawson, J.W. (1870) The primitive vegetation of the earth. *Nature*, **2**, 85–8). Only after a further century of study did the mists of confusion clear. Dawson's reconstructions were derived from at least three separate specimens and at least two different species. In drawing his reconstruction Dawson had committed the cardinal sin in palaeontologists' eyes of not proving organic connection between the different parts of the plant, but instead following the dangerous route of merely inferring them by association.

21 Lang, W.H. (1937) IV. On the plant-remains from the Downtonian of England and Wales. *Philosophical Transactions of the Royal Society*, **B227**, 245–91.

22 Definitive evidence of vascular tissue in the axes of *Cooksonia* was presented by Edwards, D., Davies, K.L. and Axe, L. (1992) A vascular conducting strand in the early land plant *Cooksonia. Nature*, **357**, 683–5. See also the commentary, Hemsley, A.R. (1992) A vascular pipe dream, *Nature*, **357**, 641–2.

23 The absence of leaves in the earliest land-conquering vascular plants is perhaps understandable. Developing a leaf-like structure greatly increases the surface area of tissue losing water, a potentially ruinous situation for a plant without proper roots for taking up water to replenish its losses. Nevertheless, it remained deeply puzzling why plant life persisted in a leafless state for the ensuing 40 million years.

24 Kenrick and Crane, The origin and early evolution of plants on land (above, n.19).

25 Ibid.

26 Ibid.

27 Large fossil marine algae of Ordovician age are described by Fry, W.L. (1983) An algal flora from the upper Ordovician of the lake Winnipeg region, Manitoba, Canada. *Review of Palaeobotany and Palynology*, **39**, 313–41.

28 Many detailed photographs and drawings of *Eophyllophyton bellum* specimens are documented by Hao, D.G. and Beck, C.B. (1993) Further observations on *Eophyllophyton bellum* from the lower Devonian (Siegenian) of Yunnan, China. *Palaeontographica*, **B230**, 27–41. Further descriptions of specimens held in the Department of Geology, Peking University, were later reported in Hao, S., Beck, C.B., and Deming, W. (2003) Structure of the earliest leaves: adaptations to high concentrations of atmospheric CO_2. *International Journal of Plant Science*, **164**, 71–5.

29 The knotted homeobox gene family is abbreviated to KNOX. It gets its name after the appearance of 'knots' in leaves (similar to those seen in wood) of mutant plants in which the gene, normally only expressed in the shoot tip, was found to be expressed in leaves. See Vollbrecht, E., Veit, B., Sinha, N., and Hake, S. (1991) The development gene Knotted-1 is a member of a maize homeobox gene family. *Nature*, **350**, 241–3. For a recent round-up of KNOX gene functions, see Hake, S. and Ori, N. (2002) Plant morphogenesis and

KNOX genes. *Nature Genetics*, **31**, 121–2; Byrne, M.E. (2005) Networks in leaf development. *Current Opinion in Plant Biology*, **8**, 59–66.

30 Sano, R., Jurárez, C.M., Hass, B. *et al.* (2005) KNOX homeobox genes potentially have similar function in both diploid unicellular and multicellular meristems, but not haploid meristems. *Evolution and Development*, **7**, 69–78.

31 A common developmental mechanism has been identified to underpin the microphyll and megaphyll leaf formation; see Harrison, C.J., Corley, S.B., Moylan, E.C. *et al.* (2005) Independent recruitment of a conserved developmental mechanism during leaf evolution. *Nature*, **434**, 509–14. See also the review by Piazza, P., Jasinski, S., and Tsiantis, M. (2005) Evolution of leaf developmental mechanisms. *New Phytologist*, **167**, 693–710.

32 This remarkable discovery was reported by Floyd, S.K. and Bowman, J.L. (2004) Ancient microRNA target sequences in plants. *Nature*, **428**, 485–6.

33 Emery, J.F., Floyd, S.K., Alvarez, J. *et al.* (2003) Radial patterning of *Arabidopsis* shoots by class II HD-ZIP and KANADI genes. *Current Biology*, **13**, 1768–73.

34 Zimmermann's results were first published in book form (in German): Zimmermann, W. (1930) *Die Phylogenie der Pflanzen.* Gustav Fischer Verlag, Jena, and later summarized (in English) in the following paper: Zimmermann, W. (1932) Main results of the telome theory. *Palaeobotanist*, **1**, 456–70.

35 Two hundred years later, tests of Goethe's proposal have shown an aspect of it, by luck or good judgement, to be along the right lines. Developmental genetic studies by Garry Ditta and colleagues at the University of California, San Diego, showed that turning off a suite of gene regulators prevents the formation of flowers from whorls of sepals, petals, stamens, and carpels and leads instead to the formation of leaves. See Ditta, G., Pinyopich, A., Robles, P. *et al.* (2004) The *SEP4* gene of *Arabidopsis thaliana* functions in floral organ and meristem identity. *Current Biology*, **14**, 1935–40.

36 Kenrick and Crane, The origin and early evolution of plants on land (above, n.19).

37 Quote from Kenrick, P. (2002) The telome theory. In *Developmental genetics and plant evolution* (eds Q.C.B. Cronk, R.M. Bateman, and

J.A. Hawkins), pp. 365–87. Taylor and Francis, London. An effort to redress one criticism of Zimmermann's telome theory—that it lacks a molecular basis—was made by synthesizing advances in molecular biology with the fossil record; see Beerling, D.J. and Fleming, A.J. (2007) Zimmermann's telome theory of megaphyll leaf evolution: a molecular and cellular critique. *Current Opinion in Plant Biology*, **10**, 1–9.

38 Fossil soils retain in their fabric a record of atmospheric conditions at the time of their formation, which is recovered by combusting fossil soil carbonates back to gaseous carbon dioxide and measuring the abundance of light and heavy stable carbon isotopes. The abundance of the two isotopes broadly reflects the concentration of carbon dioxide in the atmosphere, once the effect of decomposing plant materials is corrected for. The technique yields estimates of the carbon dioxide content of the atmosphere with great uncertainty. Nevertheless, modern soils give the right answer so we assume it provides a reliable tool for probing the ancient atmosphere. Early evidence from soil carbonates suggesting atmospheric carbon dioxide levels were high in the Silurian and then dropped through the Devonian period was reported by Mora, C.I., Driese, S.G., and Colarusso, L.A. (1996) Middle to late Paleozoic atmospheric CO_2 levels from soil carbonate and organic matter. *Science*, **271**, 1105–7. Additional evidence from fossils is reported by Ekart, D.D., Cerling, T.E., Montanez, I.P., and Tabor, N.J. (1999) A 400 million year carbon isotope record of pedogenic carbonate: implications for paleoatmospheric carbon dioxide. *American Journal of Science*, **299**, 805–27.

39 See Berner, R.A. (2004) *The Phanerozoic carbon cycle: CO_2 and O_2*. Oxford University Press.

40 The discovery that recent rises in the carbon dioxide content of the atmosphere led to reductions in the stomatal density of trees in southern England was reported by Woodward, F.I. (1987) Stomatal numbers are sensitive to CO_2 increases from pre-industrial levels. *Nature*, **327**, 617–18. For commentary on the significance of this discovery, see Morison, J.I.L. (1987) Plant growth and CO_2 history. *Nature*, **327**, 560. In fact, Woodward was following up observations made in the late nineteenth century by botanists that showed that

stomatal density on leaves of alpine plant species increased with altitude. See Bonnier, G. (1895) Recherches expérimentales sur l'adaptation des plantes au climat alpin. *Annales des Science Naturelles, Botanique, Séries VII*, **20**, 217–360; Wagner, A. (1892) Zur Kenntniss des Blattbaues der Alpenflanzen und dessen biologischer Bedeutung. *Sitzungsberichte der Kaiserlichen Akademie der Wissenschaften, in Wien, Mathematisch-naturwissenschaftliche Klasse*, **100**, 487–547. Later experimental observations revealed that the response is due to the drop in the partial pressure of carbon dioxide that occurs with the fall in atmospheric pressure at altitude due to the monotonic fall in gravitational attraction towards the Earth's surface. See Woodward, F.I. and Bazzaz, F.A. (1988) The response of stomatal density to CO_2 partial pressure. *Journal of Experimental Botany*, **39**, 1771–81.

41 A fascinating account of Linnaeus's contribution to botany is given in Fara, P. (2003) *Sex, botany and the empire: the story of Carl Linnaeus and Joseph Banks.* Columbia University Press, New York.

42 Kohn, D., Murrell, G., Parker, J., and Whitehorn, M. (2005) What Henslow taught Darwin. *Nature*, **436**, 643–5.

43 Sulloway, F.J. (1982) Darwin and his finches: the evolution of a legend. *Journal of Historical Biology*, **15**, 1–52.

44 Woodward, F.I. (1987) Stomatal numbers are sensitive to CO_2 increases from pre-industrial levels (above, n.40).

45 Lake, J.A., Quick, W.P., Beerling, D.J., and Woodward, F.I. (2001) Plant development: signals from mature leaves. *Nature*, **411**, 154.

46 The *HIC* gene was one of the first genes identified that influences plant development under different carbon dioxide concentrations. Its discovery is reported in Gray, J.E., Holroyd, G.H., van der Lee, F.M. et al. (2000) The *HIC* signaling pathway links CO_2 perception to stomatal development. *Nature*, **408**, 713–16. For a commentary on this work, see Serna, L. and Fenoll, C. (2000) Coping with human CO_2 emissions. *Nature*, **408**, 656–7.

47 Further details on the nature of the signals are given by Bird, S.M. and Gray, J.E. (2003) Signals from the cuticle affect epidermal cell differentiation. *New Phytologist*, **157**, 9–23.

48 Edwards, D. (1998) Climate signals in Palaeozoic land plants. *Philosophical Transactions of the Royal Society*, **353**, 141–57.

49 *Archaeopteris* stomatal data are given in Osborne, C.P., Beerling, D.J., Lomax, B.H., and Chaloner, W.G. (2004) Biophysical constraints on the origin of leaves inferred from the fossil record. *Proceedings of the National Academy of Sciences, USA,* **101**, 10360–2.

50 The cooling effect of transpiration was demonstrated in classic early experiments by the German plant physiologist Otto Lange, at the Universät Würzburg, Germany. By preventing transpirational cooling in leaves of the desert plant *Citrullus colocynthsis,* he showed leaf temperatures rose by 20°C and resulted in severe heat injury. Lange, O.L. (1959) Untersuchungen über den- Wärmehaushalt und Hitzeresistenz mauretanischer Wüstern- under Savannenpflanzen. *Flora,* **147**, 595–651. However, not everyone agrees that transpiration has such a large effect on leaf temperatures. For a different perspective, see Tanner, W. (2001) Do drought-hardened plants suffer from fever? *Trends in Plant Science,* **6**, 507, and for counter-arguments, see Beerling, D.J., Osborne, C.P., and Chaloner, W.G. (2001) *Trends in Plant Science,* **6**, 507–8.

51 Smith, W.K. (1978) Temperatures of desert plants: another perspective on the adaptability of leaf size. *Science,* **201**, 614–16.

52 The hypothesis that the enlargement and diversification of leaves awaited a major drop in atmospheric carbon dioxide levels is reported in Beerling, D.J., Osborne, C.P., and Chaloner, W.G. (2001) Evolution of leaf-form in land plants linked to atmospheric CO_2 decline in the Late Palaeozoic Era. *Nature,* **410**, 352–4. A commentary on the idea, putting the hypothesis into the broader context of plant evolution, is given in Kenrick, P. (2001) Turning over a new leaf. *Nature,* **410**, 309–10.

53 For discussion of rooting system evolution, see Algeo, T.J. and Scheckler, S.E. (1998) Terrestrial-marine teleconnections in the Devonian: links between the evolution of land plants, weathering processes, and marine anoxic events. *Philosophical Transactions of the Royal Society,* **353**, 113–30.

54 Raven, J.A. and Edwards, D. (2001) Roots: evolutionary origins and biogeochemical significance. *Journal of Experimental Botany,* **52**, 381–401.

55 Osborne *et al.* Biophysical constraints on the origin of leaves inferred from the fossil record (above, n.49).

56 Ibid. The results of Osborne *et al.* are independently supported by those reported by Harvard University palaeontologists: Boyce, C.K. and Knoll, A.H. (2002) Evolution of developmental potential and the multiple independent origins of leaves in Paleozoic vascular plants. *Paleobiology*, **28**, 70–100; Boyce, C.K. (2005) Patterns of segregation and convergence in the evolution of fern and seed plant leaf morphologies. *Paleobiology*, **31**, 117–40.

57 Knoll, A.H. and Carroll, S.B. (1999) Early animal evolution: emerging views from comparative biology and geology. *Science*, **284**, 2129–37.

58 Carroll, S.B. (2000) Endless forms: the evolution of gene regulation and morphological diversity. *Cell*, **101**, 577–80.

59 See Duboule, D. and Wilkins, A.S. (1998) The evolution of 'bricolage'. *Trends in Genetics*, **14**, 54–9, and for a recent review of how these ideas fit into the evolution of animals, see Shubin, N.H. and Marshall, C.R. (2000) Fossils, genes, and origin of novelty. *Paleobiology (Supplement)*, **26**, 325–40.

60 Cronk, Q.C.B. (2001) Plant evolution and development in a postgenomic context. *Nature Genetics*, **2**, 607–19.

61 Harrison *et al.*, Independent recruitment of a conserved developmental mechanism during leaf evolution (above, n.31).

62 Maizel, A., Busch, M.A., Tanahashi, T. *et al.* (2005) The floral regulator LEAFY evolves by substitutions in the DNA binding domain. *Science*, **308**, 260–3.

63 For lively critical reviews of the controversial evidence, see Conway-Morris, S. (2000) Nipping the Cambrian 'explosion' in the bud? *BioEssays*, **22**, 1053–6; Conway-Morris, S. (2003) The Cambrian 'explosion' of metazoans and molecular biology: would Darwin be satisfied? *International Journal of Developmental Biology*, **47**, 505–15.

64 Carroll, S.B. (2005) *Endless forms most beautiful: the new science of Evo Devo and the making of the animal kingdom.* Weidenfeld and Nicolson, London.

65 Chaloner, W.G. and Sheerin, A. (1979) Devonian macrofloras. *Special Papers in Palaeontology*, **23**, 145–61.

66 Knoll and Carroll, Early animal evolution (above, n.57).

67 For a detailed treatment of all the relevant processes, see Berner, *Phanerozoic carbon cycle* (above, n.39).

68 It may even influence the life span of the biosphere. As the Sun burns ever brighter over time, the capacity of the long-term carbon cycle to remove carbon dioxide from the atmosphere and prevent Earth from overheating will be increasingly important. See Lovelock, J.E. and Whitfield, M. (1982) Life span of the biosphere. *Nature*, **296**, 561–3, and Caldeira, K. and Kasting, J.F. (1992) The life span of the biosphere revisited. *Nature*, **360**, 721–3.

69 Berner, R.A. (1998) The carbon cycle and CO_2 over Phanerozoic time: the role of land plants. *Philosophical Transactions of the Royal Society*, **353**, 75–82.

70 Temperature dependence of silicate rock weathering and its importance for long-term global climate were described by Walker, J.C.G., Hays, P.B., and Kasting, J.F. (1981) A negative feedback mechanism for the long-term stabilization of Earth surface temperature. *Journal of Geophysical Research*, **86**, 9776–82. Later, another group at Yale University led by Robert Berner took these ideas further by incorporating them into a mathematical model for simulating past variations in atmospheric carbon dioxide content and climate over millions of years. Their seminal work was published as Berner, R.A., Lasaga, A.C., and Garrels, R.M. (1983) The carbonate-silicate geochemical cycle and its effects on atmospheric carbon dioxide and climate. *American Journal of Science*, **283**, 641–83.

71 An accessible review of this topic is given by Kasting, J.F. and Catling, D. (2003) Evolution of a habitable planet. *Annual Review of Astronomy and Astrophysics*, **41**, 429–63.

72 Berner, E.K., Berner, R.A., and Moulton, K.L. (2003) Plants and mineral weathering: present and past. *Treatise in Geochemistry*, **5**, 169–88.

73 See ibid. for a thorough review of plants and weathering.

74 A general account of how plants influence the weathering of rocks and changes in atmospheric carbon dioxide levels over the past 600 million years is given by Berner, R.A. (1997) The rise of plants: their effect on weathering and atmospheric CO_2. *Science*, **276**, 544–6. See also Berner, *Phanerozoic carbon cycle* (above, n.39); Berner, The carbon cycle and CO_2 over Phanerozoic time (above, n.69); and Berner *et al.*, Plants and mineral weathering (above, n.72).

75 Beerling, D.J. and Berner, R.A. (2005) Feedbacks and the coevolution of plants and atmospheric CO_2. *Proceedings of the National Academy of Sciences, USA,* **102**, 1302–5.

76 Vigorous debate surrounds the notion that organisms modify their environment in a way that influences evolutionary processes. A flavour of the debate is to be found in the following short commentary: Laland, K.N., Odling-Smee, F.J., and Feldman, M.W. (2004) Causing a commotion. *Nature,* **429**, 609. The same protagonists set out a detailed case in their book: Odling-Smee, F.J., Laland, K.N., and Feldman, M.W. (2003) *Niche construction: the neglected process in evolution.* Princeton University Press. Some of the inflated claims of that book are exposed in a review by Laurent Keller (2003) Changing the world. *Nature,* **425**, 769–70. See also Dawkins, R. (2004) Extended phenotype—but not *too* extended. A reply to Laland, Turner and Jablonka. *Biology and Philosophy,* **19,** 377–96.

77 Beerling, D.J. (2005) Botanical briefing. Leaf evolution: gases, genes and geochemistry. *Annals of Botany,* **96**, 345–52.

CHAPTER 3

Oxygen and the lost world of giants

1 Priestley, J. (1775) *Experiments and observations on different kinds of air.* London: printed for J. Johnson.

2 Quotation from Gratzer, W. (2002) *Eurekas and euphorias: the Oxford book of scientific anecdotes.* Oxford University Press.

3 The episode, rediscovered by Gratzer (ibid.), is taken from Szabadváry, S. (1960) *History of analytical chemistry.* Gordon and Breach, London.

4 McKie, D. (1962) *Antoine Lavoisier: scientist, economist, social reformer.* Da Capo Press. See also Holmes, F.L. (1998) *Antoine Lavoisier: the crucial year.* Princeton University Press.

5 See Prinke, R.T. (1990) Michael Sendivogius and Christian Rosenkreutz. *Hemetic Journal,* 72–98. Further details (in English) are given by Szydlo, Z. (1994) *Water which does not wet hands: the alchemy of Michael Sendivogius.* Polish Academy of Sciences, Warsaw. Szydlo, Z. (1997) A new light on alchemy. *History Today,* **47**, 17–24.

6 Details of Drebbel's submarine, and many of his other discoveries including the perpetual motion machine, a microscope, dyes, and chemical technology can be found at the following websites: https://en.wikipedia.org/wiki/Cornelis_Drebbel

7 Boyle, R. (1660) *New experiments physico-mechanicall, touching the spring of air, and its effects.* Oxford.

8 For an introduction to the Commentry shale insects and a historical account of the finds made by Charles Brongniart, see Carpenter, F.M. (1943) Studies on Carboniferous insects from Commentry, France; Part I. Introduction and families Protagriidae, Meganeuridae, and Campylopteridaea. *Bulletin of the Geological Society of America,* **54**, 527–54.

9 Brongniart, C. (1894) Recherches pour servir à l'histoire des insectes fossiles des temps primaires. *Bulletin de la Société d'Industrie Minérale* **4**, 124–615.

10 Since Brongniart's day many of the giant insects have been re-described. See Carpenter, F.M. (1951) Studies on Carboniferous insects from Commentry, France. Part II. The Megasecoptera. *Journal of Paleontology,* **25**, 336–55.

11 Whyte, M.A. (2005) A gigantic fossil arthropod trackway. *Nature,* **438**, 576.

12 Carroll, R.L. (1988) *Vertebrate palaeontology and evolution.* Freeman, New York.

13 Harlé, E. and Harlé, A. (1911) Le vol de grands reptiles et insectes disparus semble indiquer une pression atmosphérique élevée. *Bulletin de la Société Géologique de France, 4th Series,* **11**, 118–21.

14 An important group of micro-organisms (nitrogen fixers) have evolved the knack of using the enzyme nitrogenase to convert nitrogen gas (N_2) into a metabolically useful form (NH_3). They ultimately make nitrogen available to all of the rest of life on Earth. Crucial though nitrogen fixers are to life, they have exerted no significant impact on the total amount of nitrogen gas in the atmospheric reservoir over the past half billion years. See Falkowski, P.G. (1997) Evolution of the nitrogen cycle and its influence on the biological sequestration of CO_2 in the ocean. *Nature,* **387**, 272–5.

15 Originally Earth's early atmosphere contained no oxygen. It was not until photosynthetic micro-organisms (cyanobacteria) evolved that

oxygen production added appreciable quantities, probably well before 2.5 billion years ago; see Summons, R., Jahnke, L.L., Hope, J.M., and Logan, G.A. (1999) 2-methylhopanoids as biomarkers for cyanobacterial oxygen photosynthesis, *Nature*, **400**, 554–7; Canfield, D.E. (1999) A Breath of fresh air. *Nature*, **400**, 503–4. Over several billion years, the activities of cyanobacteria added oxygen that was straightaway absorbed by the chemistry of the oceans and the crust. Eventually, as these reservoirs saturated, about 2.5 billion years ago, oxygen started to accumulate in the atmosphere. Controversy abounds about almost every aspect of events leading to the oxygenation of Earth's atmosphere; for brief overviews of recent developments, see Copley, J. (2001) The story of O. *Nature*, **410**, 862–4, and Kasting, J.F. (2001) The rise of atmospheric oxygen. *Science*, **293**, 819–20. Regardless, by the start of the Phanerozoic, Earth's atmosphere probably contained about 15% oxygen, mostly thanks to cyanobacteria, the microbial heroes of Earth history. The capacity of microbes and multicellular plant life to alter the oxygen content of the atmosphere from that point onwards is what concerns us here.

16 Ebelmen, J.J. (1845) Sur les produits de la décomposition des espèces minérales de la famile des silicates. *Annales des Mines*, **7**, 3–66.

17 Hunt, T.S. (1880) The chemical and geological relations of the atmosphere. *American Journal of Science*, **19**, 349–63. The significance of Ebelmen's long-forgotten work was rediscovered and finally brought to the attention of the wider scientific community to receive the recognition it deserves by Berner, R.A. and Maasch, K. A. (1996) Chemical weathering and controls on atmospheric O_2 and CO_2: fundamental principles were enunciated by J.J. Ebelmen in 1845. *Geochimica et Cosmochimica Acta*, **60**, 1633–7.

18 Ball's lecture, delivered at the evening meeting of the Geographical Society on 9 June 1879, was published as: Ball, J. (1879) On the origin of the flora from the European Alps. *Proceedings of the Royal Geographical Society*, **1**, 564–89.

19 Berner, R.A. (2006) The geological nitrogen cycle and atmospheric N_2 over Phanerozoic time. *Geology*, **34**, 413–15.

20 Cloud, P.E. (1965) Chairman's summary remarks. *Proceedings of the National Academy of Sciences, USA*, **53**, 1169–73.

21 Berkner, L.V. and Marshall, L.C. (1965) History of major atmospheric constituents. *Proceedings of the National Academy of Sciences, USA*, **53**, 1215–26. A more detailed treatment is given by Berkner, L.V. and Marshall, L.C. (1965) On the origin and rise of oxygen concentration in the Earth's atmosphere. *Journal of the Atmospheric Sciences*, **22**, 225–61.

22 Not all scientists accepted the proposed dominant role for the burial of organic matter produced by land plants in the oxygen cycle. Others believed oxygen production by microscopic marine plants—phytoplankton—to be the main factor controlling the oxygen content of the atmosphere. See Tappan, H. (1967) Primary production, isotopes, extinctions and the atmosphere. *Palaeogeography, Palaeoclimatology, Palaeoecology*, **4**, 187–210. Tappan proposed a link between oxygen and marine plants that implied that episodes of phytoplankton extinction, for example 65 million years ago when the dinosaurs died out suddenly, correspond to times of oxygen depletion. Conversely, times of phytoplankton diversification, perhaps due to increased nutrient supply with the shifting of the continents, boosted atmospheric oxygen levels.

23 Other groups were also actively engaged in the race to sort out the confusion, notably Preston Cloud at the University of Minnesota, Heinrich Holland at Harvard University, and James Walker at the University of Michigan. All wrote and published major contributions to the debate. See Cloud, P. (1972) A working model of primitive Earth. *American Journal of Science*, **272**, 537–48; Holland, H.D. (1984) *The chemical evolution of the atmosphere and the oceans*. Princeton Series in Geochemistry. Princeton University Press; Walker, J.C.G. (1974) Stability of atmospheric oxygen. *American Journal of Science*, **274**, 193–214.

24 Berner, R.A. (1992) Robert Minard Garrels. *Biographical Memoirs of the National Academy of Sciences*, **61**, 195–213. Canfield, D.E. (2005) The early history of atmospheric oxygen: homage to Robert M. Garrels. *Annual Review of Earth and Planetary Science*, **33**, 1–36.

25 Key influential papers include the following. Garrels, R.M. and Perry Jr, E.A. (1974) Cycling of carbon, sulphur, and oxygen through geologic time. In *The sea*, Vol. 5 (ed. E.D. Golberg), pp. 303–36. Wiley-Interscience, New York. Garrels, R.M., Lerman, A., and

Mackenzie, F.T. (1976) Controls of atmospheric O_2 and CO_2: past, present and future. *American Scientist*, **64**, 306–15. Garrels, R.M. and Lerman, A. (1981) Phanerozoic cycles of sedimentary carbon and sulfur. *Proceedings of the National Academy of Sciences, USA*, **78**, 4652–6. These ideas were developed further in Garrels, R.M. and Lerman, A. (1984) Coupling of the sedimentary sulphur and carbon cycles: an improved model. *American Journal of Science*, **284**, 989–1007.

26 Estimates of the carbon budget of the Amazonian river basin are reported by Richey, J.E., Melack, J.M., Aufdenkampe, A.K. *et al.* (2002) Outgassing from Amazonian rivers and wetlands as a large source of atmospheric CO_2. *Nature*, **416**, 617–20. See also the commentary: Grace, J. and Malhi, Y. (2002) Carbon dioxide goes with the flow. *Nature*, **416**, 594–5.

27 Forty years on from the mice helping Lavoisier to discover oxygen, they also had a serendipitous hand in uncovering the relationship between pyrite formation and organic matter. In his 1815 book, *Introduction to geology*, J. Bakewell describes how mice droppings accidentally got into a jar of iron sulfate. After they had been left to stand for some time, pyrite crystals covered the recovered droppings.

28 In skimming over the role of the sulfur cycle in the global oxygen cycle, we should note that there are those who believe seawater chemistry with basalts at mid-ocean ridges is important; for example, Hansen, K.W. and Wallman, C. (2002) Cretaceous and Cenozoic evolution of seawater composition, atmospheric O_2 and CO_2. *American Journal of Science*, **303**, 94–148. Strong objections to this view were raised on the grounds that these fluxes are trivial relative to those associated with pyrite burial and weathering; see Berner, R.A. (1999) Atmospheric oxygen over Phanerozoic time. *Proceedings of the National Academy of Sciences, USA*, **96**, 10955–7.

29 Averaged over the last half billion years, denudation of the continents by weathering has lowered their surface relief by a few tens of metres per million years. By contrast, the human activities related to construction and agriculture would transport enough sediment to lower continental relief by a few hundred metres per million years. See Wilkinson, B.H. (2005) Humans as geologic agents: a deep-time perspective. *Geology*, **33**, 161–4.

30 Evidence for microbial digestion of ancient shales is presented by Petsch, S.T., Eglington, T.I., and Edwards, K.J. (2001) ^{14}C-dead living biomass: evidence for microbial assimilation of ancient organic carbon during shale weathering. *Science*, **292**, 1127–31. See also the commentary by Pennisi, E. (2001) Shale-eating microbes recycle global carbon. *Science*, **292**, 1043.

31 High-precision measurements of the atmospheric oxygen are being made by Ralph Keeling at the Scripps Institution of Oceanography, University of California, San Diego. For reviews of this work and its significance for understanding the fate of fossil fuels, see Keeling, R. F. (1995) The atmospheric oxygen cycle: the oxygen isotopes of atmospheric CO_2 and O_2 and the O/N_2 ratio. *Reviews of Geophysics, Supplement*, 1253–62, and Bender, M.L., Battle, M., and Keeling, R.F. (1998) The O_2 balance of the atmosphere: a tool for studying the fate of fossil-fuel CO_2. *Annual Review of Energy and the Environment*, **23**, 207–23.

32 The atmosphere contains 210 000 parts per million of oxygen, and estimates (see Keeling, The atmospheric oxygen cycle (previous note)) indicate we are consuming it at a rate of 3 parts per million per year (210 000/3) = 70 000 years. For the same reason, chopping down all of the world's forest would not put our oxygen reserves as risk—the atmosphere contains such an enormous reserve of the gas that this loss, though catastrophic for biodiversity, would barely be noticed in terms of its impact on the atmospheric content.

33 Details of the 'rock abundance' approach to calculating the highs and lows of atmospheric oxygen during the last 600 million years are given in Berner, R.A. and Canfield, D.E. (1989) A new model of atmospheric oxygen over time. *American Journal of Science*, **289**, 333–61.

34 Evaporites are formed by the precipitation of salts in naturally evaporating marginal salt pans, including coastal lagoons. Calcium sulfate (gypsum) is very insoluble and therefore the first mineral to precipitate from seawater.

35 An isotope is an element that has an extra neutron, increasing its total mass without changing its charge.

36 This novel idea has two components. The first requires accepting that fossil charcoal is indeed the product of wildfire. Some workers

originally thought the black, opaque matter turning up in ancient sediments may represent the product of 'wet decay'. This view was largely overturned by Thomas Harris at Reading University; see Harris, T.M. (1958) Forest fire in the Mesozoic. *Journal of Ecology*, **46**, 447–53. The second requires accepting that the presence of fossil charcoal sets a minimum level of oxygen in the atmosphere. This idea was put forward by Cope, M.J. and Chaloner, W.G. (1980) Fossil charcoal as evidence of past atmospheric composition. *Nature*, **283**, 647–9. The question of whether charcoal really does reflect past oxygen levels was debated; see Clark, F.R.S. and Russell, D.A. (1981) Fossil charcoal and the palaeoatmosphere. *Nature*, **290**, 428, for objections and, on the same page, a robust response by Cope and Chaloner.

37 Since the work of Cope and Chaloner (above, n.36), the oldest wildfire got even older; see Glasspool, I.J., Edwards, D., and Axe, L. (2004) Charcoal in the Silurian as evidence for the earliest wildfire. *Geology*, **32**, 381–3.

38 Beerling, D.J., Woodward, F.I., Lomas, M.R. *et al.* (1998) The influence of Carboniferous palaeoatmospheres on plant function: an experimental and modelling assessment. *Philosophical Transactions of the Royal Society*, **B353**, 131–40.

39 Rubisco is an abbreviation of ribulose-1,5-bisphosphate carboxylase/oxygenase. The carboxylase/oxygenase ending reveals its dual function of catalysing the fixation of carbon dioxide (carboxylase) and oxygen (oxygenase) molecules.

40 Beerling, D.J., Lake, J.A., Berner, R.A. *et al.* (2002) Carbon isotope evidence implying high O/CO ratios in the Permo-Carboniferous atmosphere. *Geochimica et Cosmochimica Acta*, **66**, 3757–67.

41 Experiments with plants grown at different oxygen levels have revealed a wealth of further details; see Raven, J.A., Johnston, A.M., Parsons, R., and Kubler, J. (1994) The influence of natural and experimental high O_2 concentrations on O_2 evolving phototrophs. *Biological Reviews*, **69**, 61–94.

42 The 'atomic abundance' and 'rock abundance' methods for calculating ancient oxygen levels square with each other. See Berner, R.A., Petsch, S.T., Lake, J.A. *et al.* (2000) Isotope fractionation and atmospheric oxygen: implications for Phanerozoic O_2 evolution. *Science*,

287, 1630–3. Further details are given in Berner, R.A. (2001) Modeling atmospheric O_2 over Phanerozoic time. *Geochimica et Cosmochimica Acta*, **65**, 685–94.

43 Beerling, D.J., McGlashon, H., and Wellman, C. (2006) Palaeobotanical evidence for a late Palaeozoic rise in atmospheric oxygen. *Geobiology*, in preparation. See also Beerling *et al.*, Carbon isotope evidence implying high O/CO ratios in the Permo-Carboniferous atmosphere (above, n.40).

44 See Berner, R.A. and Landis, G.P. (1987) Chemical analysis of gaseous bubble inclusions in amber: composition of the ancient air? *American Journal of Science*, **287**, 757–62; Berner, R.A. and Landis, G.P. (1988) Gas bubbles in fossil amber as possible indicators of the major gas composition of ancient air. *Science*, **239**, 1406–9.

45 Five pages of technical concerns over the validity of amber bubbles as indicators of the ancient atmospheric composition were published; see, by various authors: (1988) Is air in amber ancient? *Science*, **241**, 717–21. Most of the criticisms relate to the possibility that amber does not provide an air-tight sealant of the gases inside the bubbles. Even though Landis and Berner resolutely defended their approach in the face of these criticisms (see Science, **241**, 721–4), doubts remain over amber as a source of useful information on the composition of the ancient atmosphere. Later, the problem of gaseous diffusion of air out of the bubbles was refuted, based on results of argon isotopes studies; see Landis, G.P. and Snee, L.W. (1991) $^{40}Ar/^{39}Ar$ systematics and argon diffusion in amber: implications for ancient earth atmospheres. *Palaeogeography, Palaeoclimatology, Palaeoecology (Global Change Section)*, **97**, 63–7.

46 See Robinson, J.M. (1990) Lignin, land plants, and fungi: biological evolution affecting Phanerozoic oxygen balance. *Geology*, **15**, 607–10, and Robinson, J.M. (1990) The burial of organic carbon as affected by the evolution of land plants. *Historical Biology*, **3**, 189–201.

47 See the references cited in the previous note.

48 Rayner, J.M.V. (2003) Gravity, the atmosphere and the evolution of animal locomotion. In *Evolution on planet Earth: the impact of the physical environment* (eds L.J. Rothschild and A.M. Lister), pp. 161–83. Academic Press, Amsterdam.

49 Westneat, M.W., Betz, O., Blob, R.W. *et al.* (2003) Tracheal respiration in insects visualized with synchrotron X-ray imaging. *Science*, **299**, 558–60.

50 Nevertheless, it is worth noting that flying insects achieve the highest rates of oxygen consumption of any animals and have spectacular endurance; see, for example, Harrison, J.F. and Roberts, S.P. (2000) Flight respiration and energetics. *Annual Review of Physiology*, **62**, 179–206.

51 This hypothesis was originally mooted by Graham, J.B., Dudley, R., Aguilar, N.M., and Gans, C. (1995) Implications of the late Palaeozoic oxygen pulse for physiology and evolution. *Nature*, **375**, 117–20. Further details are presented in Dudley, R. (1998) Atmospheric oxygen, giant Paleozoic insects and the evolution of aerial locomotor performance. *Journal of Experimental Biology*, **201**, 1043–50, and Harrison, J.F. and Lighton, J.R.B. (1998) Oxygen-sensitive flight metabolism in the dragonfly *Erythemis simplicicollis*. *Journal of Experimental Biology*, **201**, 1739–44.

52 Chapelle, G. and Peck, L.S. (1999) Polar gigantism dictated by oxygen availability. *Nature*, **399**, 114–15.

53 A minority remain unmoved. They argue that the amount of oxygen entering an animal from the surrounding water is not dictated by the absolute concentration. Instead, it is the gradient across the gas exchange surface. A given volume of polar seawater may contain more oxygen than the same volume of tropical water, but the partial pressure it exerts is similar. As a result, the gradient driving oxygen uptake is similar. See Spicer, J.I. and Gaston, K.J. (1999) Amphipod gigantism dictated by oxygen availability. *Ecology Letters*, **2**, 397–403. This technical point was shown to be flawed, however: see Peck, L.S. and Chapelle, G. (1999) Reply. *Ecology Letters*, **2**, 401–3. It was later holed below the waterline by a study of amphipods in the freshwater Lake Titicaca at an altitude of 3809 m above sea level: see Peck, L.S. and Chapelle, G. (2003) Reduced oxygen at high altitude limits maximum size. *Proceedings of the Royal Society of London (Supplement)*, **B270**, S166–S167.

54 Dudley's 'fly-pushing' experiment, and a review of several key topics (geochemistry, plants, insects, and fire) related to past variations in atmospheric oxygen is given in Berner, R.A., Beerling, D.J., Dudley,

R. *et al.* (2003) Phanerozoic atmospheric oxygen. *Annual Review of Earth and Planetary Sciences*, **31**, 105–34.

55 Another study showed female and male fruit flies grew to be 10 and 6% heavier, respectively, when grown at 40 rather than 21% oxygen under warm conditions; under cooler conditions the effect disappeared. See Frazier, M.R., Woods, H.A., and Harrison, J.F. (2001) Interactive effects of rearing temperature and oxygen on the development of *Drosophila melangaster*. *Physiological and Biochemical Zoology*, **74**, 641–50.

56 For an accessible popular account of oxygen's role in ageing, see Lane, N. (2002) *Oxygen: the molecule that made the world*. Oxford University Press. For a more technical briefing, see Halliwell, B. and Guttridge, J.M.C. (1999) *Free radicals in biology and medicine*. Oxford University Press.

57 The idea that insects primarily close off their tracheae to prevent the shedding of excess carbon dioxide, and so reduce oxygen toxicity, was advanced by Hetz, S.K. and Bradley, T.J. (2005) Insects breathe discontinuously to avoid oxygen toxicity. *Nature*, **433**, 516–19. See also the commentary by Burmester, T. (2005) A welcome shortage of breath. *Nature*, **433**, 471–2.

58 The tubes of tracheal systems in fruit flies, for example, become smaller in high-oxygen conditions and larger in low-oxygen conditions; see Henry, J.R. and Harrison, J.F. (2004) Plastic and evolved responses of tracheal dimensions to varied atmospheric oxygen content in *Drosophila melangaster*. *Journal of Experimental Biology*, **207**, 3559–67.

59 McAlester, A.L. (1970) Animal extinction, oxygen consumption, and atmospheric history. *Journal of Paleontology*, **44**, 405–09.

60 Greenlee, K.J. and Harrison, J.F. (2004) Development of respiratory function in the American locust *Schistocerca americana*. I. Across-instar effects. *Journal of Experimental Biology*, **207**, 497–508.

61 Huey, R.B. and Ward, P.D. (2005) Hypoxia, global warming and terrestrial late Permian extinctions. *Science*, **308**, 398–401. See also the commentary: Kerr, R.A. (2005) Gasping for air in Permian hard times. *Science*, **308**, 337.

62 Huey and Ward, Hypoxia, global warming and terrestrial late Permian extinctions (previous note).

63 The provocative idea that fire tightly regulates atmospheric oxygen to levels below 25% was put forward by the leading proponents of the Gaia hypothesis. The idea is based on ignition experiments in different atmospheres with paper tape. Watson, A., Lovelock, J.E., and Margulis, L. (1978) Methanogenesis, fires and the regulation of atmospheric oxygen. *Biosystems*, **10**, 293–8.

64 Lovelock, J.E. and Watson, A.J. (1982) The regulation of carbon dioxide and climate: Gaia or geochemistry? *Planetary Space Science*, **30**, 795–802.

65 Ibid.

66 Lovelock, J. (1979) *Gaia: A new look at life on Earth*. Oxford University Press.

67 Prehistoric forests may have been less combustible in a high-oxygen atmosphere than previously suggested. See Wildman, R.A., Hickey, L.J., Dickinson, M.B. *et al.* (2004) Burning of forest materials under late Palaeozoic high atmospheric oxygen levels. *Geology*, **32**, 457–60.

68 The negative feedback loop was published by Van Cappellen, P. and Ingall, E.D. (1996) Redox stabilization of the atmosphere and oceans by phosphorus-limited marine productivity. *Science*, **271**, 493–6. For a commentary, see Kump, L.R. and Mackenzie, F.T. (1996) Regulation of atmospheric O_2: feedback in the microbial feedbag. *Science*, **271**, 459–60.

69 Fire, phosphorous, and ocean feedbacks combined, the authors suggest, to dampen excursions in oxygen levels during the last half billion years. See Bergman, N.N., Lenton, T.M., and Watson, A.J. (2004) COPSE: a new model of biogeochemical cycling over Phanerozoic time. *American Journal of Science*, **304**, 397–437.

70 Kuhlbusch, T.A.J. and Crutzen, P.J. (1995) Toward a global estimate of black carbon residues of vegetation fires representing a sink of atmospheric CO_2 and a source of O_2. *Global Biogeochemical Cycles*, **4**, 491–501.

71 The evolution and diversification of fire prone terrestrial ecosystems is well documented from the fossil record through the rise in atmospheric oxygen during the late Palaeozoic. See Scott, A.C. and Glasspool, I.J. (2006) The diversification of Paleozoic fire systems and fluctuations in atmospheric oxygen concentration. *Proceedings of the National Academy of Sciences, USA*, **103**, 10861–5.

72 Owen, R. (1842) Report on British fossil reptiles. *Report of the British Association for the Advancement of Science 1841*, 60–204.

73 Vandenbrooks, J. (2004) The effect of varying pO_2 on vertebrate evolution. Abstract, 31–12. 2004 Geological Society of America Annual Meeting, Denver, Colorado.

74 Ebelmen, Sur les produits de la décomposition des espèces minérales de la famile des silicates (above, n.16).

CHAPTER 4

An ancient ozone catastrophe?

1 Cambridge University exams are still known as the Tripos, not because of the tripartite nature of the subject examined, but from the precarious three-legged stool on which examinees used to sit centuries ago.

2 Warwick, A. (2003) *Masters of theory: Cambridge and the rise of mathematical physics*. University of Chicago Press.

3 Grattan-Guiness, I. (1972) A mathematical union: William Henry Young and Grace Chisholm Young. *Annals of Science*, **29**, 105–86.

4 Egerton, A.C. (1949) Lord Rayleigh. *Obituary Notices of Fellows of the Royal Society*, **6**, 502–49.

5 Schönbein, C.F. (1840) Beobachtungen über den bei der Elektrolysation des Wassers und dem Ausströmen der gewöhnlichen Elektricität aus Spitzen sich entwickelnden Geruch. *Poggendorff's Annalen der Physik und Chemie*, **50**, 616–35. Schönbein, C.F. (1841) An account of researches in electrochemistry, in *Report of the British Association for the Advancement of Science for 1840*, pp. 209–15. Taylor, London. For a good and more accessible review of Schönbein's research, see Rubin, M.B. (2001) The history of ozone: the Schönbein period, 1839–1868. *Bulletin for the History of Chemistry*, **26**, 40–56.

6 Schönbein also invented the powerful explosive cellulose nitrate. The story goes that Schönbein mopped up spilled chemicals from a cotton apron and left them to dry outside. The apron subsequently exploded, causing him to analyse it for the chemicals responsible. Although he did not succeed in manufacturing cellulose nitrate, he did foresee the development of plastic explosives. See Williams, T.I. (ed.) (1982) *A biographical dictionary of scientists*. John Wiley, New York.

7 Stratospheric ozone absorbs ultraviolet (UV) radiation emitted from the Sun in the wavelength range 200–400 nm (nm = nanometre = ⁻ m). The UV part of the spectrum is divided into UV-A, UV-B and UV-C. UV-A lies in the range 315–400 nm and causes ageing and induces suntans; UV-B, the most biologically damaging radiation, has a wavelength of 280–315 nm; UV-C is absorbed in the atmosphere before it can reach the ground and has a wavelength of 200–280 nm.

8 Cornu, A. (1879) Sur l'absorption par l'atmosphère des radiations ultra-violettes. *Comptes Rendus*, **88**, 1285–90.

9 Hartley, W.N. (1881) On the absorption spectrum of ozone. *Journal of the Chemical Society*, **39**, 57–60.

10 The absorption bands were first observed by William Huggins and became known appropriately as the Huggins bands; see Huggins, W. (1890) On the limit of solar and stellar light in the ultra-violet part of the spectrum. *Proceedings of the Royal Society, Series A*, **48**, 133–6.

11 Quote from Egerton, Lord Rayleigh (above, n.4). Details are given in Fowler, A. and Strutt, R.J. (1917) Absorption bands of atmospheric ozone in the spectra of sun and stars. *Proceedings of the Royal Society, Series A*, **93**, 577–86.

12 Strutt, R.J. (1918) Ultra-violet transparency of the lower atmosphere, and its relative poverty in ozone. *Proceedings of the Royal Society, Series A*, **94**, 260–8.

13 Egerton, Lord Rayleigh (above, n.4).

14 The ingenious method for estimating the height of the ozone layer is reported in Götz, F.W.P., Meetham, A.R., and Dobson, G.M.B. (1934) The vertical distribution of ozone in the atmosphere. *Proceedings of the Royal Society, Series A*, **145**, 416–46. Before that, a French duo, Charles Fabry (1867–1945) and Henri Buisson (1873–1944), designed an instrument capable of measuring the absorption of sunlight at specific wavelengths and determining the total amount of ozone in the atmosphere (that is, from the surface of the Earth to the top of the atmosphere). They calculated that the total amount of ozone was equivalent to a layer of pure ozone only 5 mm thick at standard temperature and pressure—this has subsequently proved to be remarkably accurate, given the currently accepted value of ~3 mm. The authors correctly speculated that ozone was formed

by solar UV radiation and suggested that if this were the case, it would be mostly situated at a height of 40 km. See Fabry, C. and Buisson, H. (1913) L'absorption de l'ultraviolet par l'ozone et la limite du spectre solaire. *Journal de Physique*, **3 (Série 5)**, 196–206.

15 Chapman, S. (1930) A theory of upper-atmospheric ozone. *Memoirs of the Royal Meteorological Society*, 3, 103–25.

16 Teisserenc de Bort, L.P. (1902) Variations de la température de l'air libre dans la zona comprise entre 8 km et 13 km d'altitude. *Compres Rendus de l'Académie des Sciences de Paris*, **134**, 987–9.

17 Crutzen, P.J. (1970) The influence of nitrogen oxides on the atmospheric ozone content. *Quarterly Journal of the Royal Meteorological Society*, **96**, 320–5.

18 Pearce, F. (2003) High flyer. *New Scientist*, 5 July, pp. 44–7.

19 This important check was undertaken in two ways. First, the original instrument was replaced, and the new one gave similar readings the following spring. Second, the instrument that had reported the long-time series of data from Halley Bay on Antarctica was returned to Cambridge and its calibration rechecked and found to be accurate; see Farman, J.C., Gardiner, B.G., and Shanklin, J.D. (1985) Large losses of total ozone reveal seasonal ClO_x/NO_x interactions. *Nature*, **315**, 207–10.

20 Discovery reported in ibid. Later, two other rival groups attempted to claim priority for observing major Antarctic ozone depletion. A re-analysis of data from the French Antarctic station at Durmont d'Urville purportedly found low Antarctic ozone in 1958, but it involved an unproven technique and was inconsistent with all other credible measurements; for details of this episode see Newman, P.A. (1994) Antarctic total ozone in 1958. *Science*, **264**, 543–6. The second, more serious claim, arose from the Japanese group at the Antarctic station Syowa. A year earlier than the British announcement, the group reported at a conference anomalously low ozone readings at Syowa station during September and October 1982; the dataset was published as Chubachi, S. (1984) Preliminary result of ozone observation at Syowa Station from February 1982 to January 1983. *Memoirs of the National Institute of Polar Research*, Special Issue, **34**, 13–19. The consensus opinion, however, is that Chubachi's dataset was not in the same league as

those of the British: they were of a shorter time series, and, having been made at a lower latitude (Syowa Station is at 69°S compared to Halley Bay's 76°S), would have experienced more variability due to turbulent weather conditions associated with the margins of the Antarctic vortex. See also Chubachi, S. and Kajiwara, R. (1986) Total ozone variation. *Geophysical Research Letters*, **13**, 1197–8.

21 Benedick, R.E. (1991) *Ozone diplomacy*. Harvard University Press, Cambridge, MA.

22 Stolarski, R.S., Kruger, A.J., Schoeberl, M.R. *et al.* (1986) Nimbus-7 satellite measurements of the springtime Antarctic ozone decrease. *Nature*, **322**, 808–11.

23 One of the more exhaustive picks through the bones of history surrounding this affair is given in Christie, M. (2000) *The ozone layer: a philosophy of science perspective*. Cambridge University Press.

24 Concern originally surfaced over the possibility that exhaust gases of large fleets of supersonic aircraft might damage the ozone layer. By 1962, rapid technological progress since the Second World War had made the widespread use of supersonic aircraft for passenger transport a distinct possibility and with that there came, thanks to Crutzen's work ('The influence of nitrogen oxides on the atmospheric ozone content' (above, n.17)), a concern that the large quantities of nitrogen oxides they generated could seriously damage the ozone layer; see Johnston, H.S. (1971) Reduction of stratospheric ozone by nitrogen oxide catalysts from supersonic transport exhaust. *Science*, **173**, 517–22. Because large operating fleets of supersonic aircraft never materialized, this concern dissipated.

25 The key investigation into the causes of the Antarctic ozone depletion was primarily undertaken in 1986 by a team of American scientists in a project called National Ozone Experiment, #1 and in 1987 the National Ozone Experiment, #2. The famous 'smoking gun' result implicating chlorine was reported in Anderson, J.G., Brune, W.H., and Proffitt, M.H. (1989) Ozone destruction by chlorine radicals within the Antarctic vortex: the spatial and temporal evolution of ClO/O_3 anticorrelation based on *in situ* ER-2 data. *Journal of Geophysical Research*, **94**, 11465–79.

26 The ozone hole above the Arctic is less severe because the region lacks the ultra-cold temperatures that are far more widespread in

the Antarctic, due mainly to airflow over the Himalayas and the Rocky Mountains generating atmospheric waves—we usually experience these on the ground as the passage of storms. Some of the waves travel upwards to the stratosphere, mixing warm mid-latitude air with cold polar air, creating warmer winters and springs than in the south; see World Meteorological Organization (1991) *Scientific assessment of ozone depletion; 1991.* Report 44. WMO, Geneva.

27 Sherwood Rowland and Mario Molina proposed that they would decompose in the stratosphere to release chlorine atoms that would catalytically destroy ozone; see Molina, M.J. and Rowland, F.S. (1974) Stratospheric sinks for chlorofluorocarbons: chlorine catalyzed destruction of ozone. *Nature,* **249,** 810–14. Rowland and Molina's research paved the way for understanding the causes of the Antarctic ozone hole and the Montreal protocol banning CFCs. They shared the 1995 Nobel Prize for Chemistry with Crutzen; see http://nobelprize. org/nobel_prizes/ chemistry/laureates/1995/press.html. For a recent review, see Rowland, F.S. (2006) Stratospheric ozone depletion. *Philosophical Transactions of the Royal Society,* **361,** 769–90. The starting point for Molina and Rowland's work was James Lovelock's observations showing that CFCs—for which there is no natural source—were present in the air above rural Ireland, and later in air collected during a voyage on board the R.V. *Shackleton* from England to Antarctic, in amounts roughly equal to the total manufactured. See Lovelock, J.E. (1971) Atmospheric fluorine compounds as indicators of air movements. *Nature,* **230,** 379; Lovelock, J.E. (1973) Halogenated hydrocarbons in and over the Atlantic. *Nature,* **241,** 194–6. To his regret, Lovelock unfortunately remarked in his 1973 paper that 'the presence of these compounds [CFCs] constitutes no conceivable hazard'.

28 Solomon, S. (2004) The hole truth. *Nature,* **427,** 289–91. Weatherhead, E.C. and Andersen, S.B. (2006) The search for signs of recovery of the ozone layer. *Nature,* **441,** 39–45.

29 See above, note 7.

30 United Nations Environment Programme (1998) *Environmental effects of ozone depletion: 1998 Assessment.* UNEP, Nairobi, Kenya.

31 World Meteorological Organization (2002) *Scientific assessment of ozone depletion.* Global ozone research and monitoring report No. 47. WMO, Geneva.

32 Ruhland, C.T. and Day, T.A. (2000) Effects of ultraviolet-B radiation on leaf elongation, production and phenylpropanoid concentrations of *Deschampsia antarctica* and *Colobanthus quitensis* in Antarctica. *Plant Physiology*, **109**, 244–51. Xiong, F.S. and Day, T.A. (2001) Effect of solar ultraviolet-B radiation during springtime ozone depletion on photosynthesis and biomass production of Antarctic vascular plants. *Plant Physiology*, **125**, 738–51.

33 Smith, R.C., Prézelin, B.B., Baker, K.S. *et al.* (1992) Ozone depletion: ultraviolet radiation and phytoplankton biology in Antarctic waters. *Science*, **255**, 952–9.

34 Cockell, C.S. (1999) Crises and extinction in the fossil record: a role for ultraviolet radiation. *Paleobiology*, **25**, 212–25.

35 Marshall, H.T. (1928) Ultra-violet and extinction. *American Naturalist*, **62**, 165–87.

36 Various eyewitness reports and atmospheric effects are collated and reported in Whipple, F.J.W. (1930) The great Siberian meteor and the waves, seismic and aerial, which it produced. *Quarterly Journal of the Royal Meteorological Society*, **56**, 287–304. See also additional information in notes 37–9.

37 See Chyba, C.F., Thomas, P.J., and Zahnle, K.J. (1993) The 1908 Tunguska explosion: atmospheric disruption by a stony asteroid. *Nature*, **361**, 40–4.

38 See Turco, R.P., Toon, O.B., Park, C. *et al.* (1982) An analysis of the physical, chemical, optical and historical impacts of the 1908 Tunguska meteor fall. *Icarus*, **50**, 1–52.

39 Turco, R.P., Toon, O.B., Park, C. *et al.* (1981) Tunguska meteor fall of 1908: effects on stratospheric ozone. *Science*, **214**, 19–23.

40 Dobson, G.M.B. (1968) Forty years' research on atmospheric ozone at Oxford: a history. *Applied Optics*, **7**, 387–405.

41 Whipple, The great Siberian meteor and the waves, seismic and aerial, which it produced (above, n.36).

42 Ibid.

43 The celestial chunk of solar system debris that struck Earth and wiped out the dinosaurs 65 million years ago was approximately a million times larger than the Tunguska asteroid, and could have generated far more ozone-eating nitrogen compounds. But we have no evidence for the decimation of the ozone layer. It is even possible

that, because the target rock was rich in sulfate, the impact created a sulfate haze lasting long enough to help filter out the extra ultraviolet radiation penetrating a tattered ozone shield. Some scientists have speculated that this is why amphibians survived the expected 'ultraviolet spring' aftermath of the end-Cretaceous impact event; see Cockell, C.S. and Blaustein, A.R. (2000) 'Ultraviolet spring' and the ecological consequences of catastrophic impacts. *Ecology Letters*, **3**, 77–81. Amphibians are susceptible to small increases in ultraviolet radiation because they have a poor capacity to repair the DNA damage it causes. Who knows what might have happened if the asteroid had hit another part of our planet at the end of the Cretaceous?

44 McCormick, M.P., Thomason, L.W., and Trepte, C.R. (1995) Atmospheric effects of the Mt Pinatubo eruption. *Nature*, **373**, 399–404.

45 Newhall, C.G. and Punongbayan, R.S. (1996) *Fire and mud: eruptions and lahars of Mount Pinatubo, Philippines*. United States Geological Survey. Available free at http://pubs.usgs.gov/pinatubo/index.html

46 Brausseur, G. and Granier, C. (1992) Mount Pinatubo aerosols, chlorofluorocarbons, and ozone depletion. *Science*, **257**, 1239–42.

47 Key satellite evidence from a range of investigations is summarized in Solomon, S. (1999) Stratospheric ozone depletion: a review of concepts and history. *Review of Geophysics*, **37**, 275–316.

48 McCormick *et al.*, Atmospheric effects of the Mt Pinatubo eruption (above, n.44), and Newhall and Punongbayan, *Fire and mud* (above, n.45).

49 McCormick *et al.*, Atmospheric effects of the Mt Pinatubo eruption (above, n.44), and Newhall and Punongbayan, *Fire and mud* (above, n.45).

50 Vogelmann, A.M., Ackerman, T.P., and Turco, R.P. (1992) Enhancements of biologically effective ultraviolet radiation following volcanic eruptions. *Nature*, **359**, 47–9.

51 Courtilott, V. (1999) *Evolutionary catastrophes: the science of mass extinctions*. Cambridge University Press.

52 To be sure, the eruption had drastic local environmental consequences. It caused acid rain and crop failure, which led to people and livestock perishing through starvation. See Sigurdsson, H. (1982) Volcanic pollution and climate: the 1783 Laki eruption. *Eos*

Transactions, **63**, 601–2; Thordarson, T and Self, S. (2003) Atmospheric and environmental effects of the 1783–84 Laki eruption: a review and assessment. *Journal of Geophysical Research*, **108**, doi:10.1029/2001JD002042.

53 Johnston, D.A. (1980) Volcanic contribution of chlorine to the stratosphere: more significant to ozone than previously estimated? *Nature*, **209**, 491–3.

54 A recent summary of the end-Permian extinction is given by Benton, M.J. and Twitchett, R.J. (2003) How to kill (almost) all life: the end-Permian extinction event. *Trends in Ecology and Evolution*, **18**, 358–65. For a popular account, see Benton, M.J. (2003) *When life nearly died: the greatest mass extinction of all time*. Thames and Hudson, London.

55 For a global perspective on vegetation changes across the Permian–Triassic boundary, see Knoll, A.H. (1984) Patterns of extinction in the fossil record of vascular plants. In *Extinctions* (ed. M. Nitecki), pp. 21–69. University of Chicago Press; Rees, P.M. (2002) Land-plant diversity and the end-Permian mass extinction. *Geology*, **30**, 827–30.

56 The following papers describe the fossil record of terrestrial ecosystem devastation and recovery at and following the end-Permian. Looy, C.V., Brugman, W.A., Dilcher, D.L., and Visscher, H. (1999) The delayed resurgence of equatorial forests after the Permian–Triassic ecologic crisis. *Proceedings of the National Academy of Sciences, USA*, **96**, 13857–62; Looy, C.V., Twitchett, R.J., Dilcher, D.L. *et al.* (2001) Life in the end-Permian dead zone. *Proceedings of the National Academy of Sciences, USA*, **98**, 7879–83.

57 Visscher, H., Looy, C.V., Collinson, M.E. *et al.* (2004) Environmental mutagenesis during the end-Permian ecological crisis. *Proceedings of the National Academy of Sciences, USA*, **101**, 12952–6. See also the commentary: Pfefferkorn, H.W. (2004) The complexity of mass extinction. *Proceedings of the National Academy of Sciences, USA*, **101**, 12779–80. Further details of these unusual fossil spores are given in Looy, C.V., Collinson, M.E., Van Konijnenburg-Van Cittert *et al.* (2005) The ultrastructure and botanical affinity of end-Permian spore tetrads. *International Journal of Plant Science*, **166**, 875–87.

58 El Maâtaoui, M. and Pichot, C. (2001) Microsporogenesis in the endangered species *Cupressus dupreziana* A. Camus: evidence for

meiotic defects yielding unreduced and abortive pollen. *Planta*, **213**, 543–9.

59 DiMichele, W.A., Davis, J.I., and Ormstead, R.G. (1989) Origins of heterospory and the seed habit: the role of heterochrony. *Taxon*, **38**, 1–11.

60 Foster, C.B. and Afonin, S.A. (2005) Abnormal pollen grains: an outcome of deteriorating atmospheric conditions around the Permian-Triassic Boundary. *Journal of the Geological Society*, **162**, 653–9.

61 Schindewolf, O.H. (1954) Über die Faunenwende vom Paläozoikum zum Mesozoikum. *Zeitschrift der Deutschen Geologischen Gesellschaft*, **105**, 153–82.

62 See, for example, the respective suggestion that supernovae explosions should be considered to have caused mass extinctions, and even contributed to the extinction of the dinosaurs: Terry, K.D. and Tucker, W.H. (1968) Biologic effects of supernovae. *Science*, **159**, 421–3; Russell, D. and Tucker, W. (1971) Supernovae and the extinction of the dinosaurs. *Nature*, **229**, 553–4.

63 Technically, the so-called soft gamma-ray flashes detected by astronomers form a different class of bright gamma-ray flashes to those of gamma-ray bursts. Palmer, D.M., Barthelmy, S., Gehrels, N. *et al.* (2005) A giant γ-ray flare from the magnetar SGR 1806–20. *Nature*, **434**, 1107–9. See also the commentary: Lazzati, D. (2004) A certain flare. *Nature*, **434**, 1075–6.

64 Crutzen, P.J. and Bruhl, C. (1996) Mass extinctions and supernova explosions *Proceedings of the National Academy of Science, USA*, **93**, 1582–4.

65 Gehrels, N., Laird, C.M., Jackman, C.H. *et al.* (2003) Ozone depletion from nearby supernovae. *Astrophysical Journal*, **585**, 1169–76.

66 One team based at the University of Kansas propose that an ozone-destroying gamma-ray burst contributed to the mass extinction of marine animals at the end of the Ordovician 440 million years ago: see Thomas, B.C., Jackman, C.H., Melott, A.L. *et al.* (2005) Terrestrial ozone depletion due to a Milky Way gamma-ray burst. *Astrophysical Journal*, **622**, L153–L156; Melott, A.L., Lieberman, B.S., Laird, C.M. *et al.* (2004) Did a gamma-ray burst initiate the late Ordovician mass extinction? *International Journal of Astrobiology*, **3**, 55–61. The authors point out that marine faunal extinction patterns seen in the fossil

record are dependent on where the animals lived in the water column. Water strongly attenuates ultraviolet radiation and the speculative suggestion is that those reproduced with surface-dwelling planktonic larvae suffered worst. A major question, though, is whether a few years of partial ozone depletion by such an exotic mechanism is anything like sufficient to cause a mass extinction. After all, as we have seen, half of the ozone above Antarctica is lost every spring without devastating loss of life. Winning over the sceptics is likely to be difficult, not least because it is hard to falsify the hypothesis with firm empirical data. Some geochemical aspects of the hypothesis might be falsifiable, but the authors concede it unlikely that little, if any, evidence will be preserved.

67 Pavlov, A.A., Pavlov, A.K., Mills, M.J. et al. (2005) Catastrophic ozone loss during passage of the Solar system through an interstellar cloud. *Geophysical Research Letters*, **32**, L01815, doi: 10/1029/2004GL021601.

68 Solar winds have perhaps been suppressed a hundred times or more in the last 4 billion years as the solar system encounters a high-density interstellar dust cloud on its passage through the Milky Way. Earth's magnetic field drops when polarity reversals occasionally take place due to the natural dynamics of the molten core.

69 Pavlov et al., Catastrophic ozone loss during passage of the Solar system through an interstellar cloud (above, n.67).

70 Reid, G.C., Isaksen, I.S.A., Holzer, T.E., and Crutzen, P.J. (1976) Influence of ancient solar-proton events on the evolution of life. *Nature*, **259**, 177–9; Reid, G.C., McAfee, J.R.M., and Crutzen, P.J. (1978) Effects of intense stratospheric ionisation events. *Nature*, **275**, 489–92.

71 Wignall, P.B. (2001) Large igneous provinces and mass extinctions. *Earth-Science Reviews*, **53**, 1–33.

72 The dating of the extinction was facilitated by dating new rock sections with new radiometric techniques; see Bowring, S.A., Erwin, D.H., Jin, Y.G. et al. (1998) U/Pb zircon geochronology and tempo of the end-Permian mass extinction. *Science*, **280**, 1039–45; Reichow, M.K. et al. (2002) $^{40}Ar/^{39}Ar$ dates from the West Siberian Basin: Siberian flood basalt province doubled. *Science*, **296**, 1846–9.

73 See, for example, Czamanske, G.K., Gurevitch, A.B., Fedorenko, V., and Simonov, O. (1998) Demise of the Siberian plume: paleogeographic and paleotectonic reconstruction from the prevolcanic and volcanic record, North-Central Siberia. *International Geology Review*, **40**, 95–113.

74 Melnikov, N.V., Khomenko, A.V., Kuznetsova, E.N., and Zhidkova, L.V. (1997) The effect of traps on salt redistribution in the lower Cambrian of the western Siberian platform. *Russian Geology and Geophysics*, **38**, 1378–84.

75 Kontorovich, A.E., Khomenko, A.V., Burshtein, L.M. *et al.* (1997) Intense basic magmatism in the Tunguska petroleum basin, eastern Siberia, Russia. *Petroleum Geoscience*, **3**, 359–69.

76 Chemical analyses of organic coals and oils retrieved from boreholes sunk into Siberian rocks confirm they have indeed been subjected to great heating in this way and not after burial; see ibid., and Al'Mukhamedov, A.I., Medvedev, A.Y., and Zolotukhin, V.V. (2004) Chemical evolution of the Permian–Triassic basalts of the Siberian platform in space and time. *Petrology*, **12**, 339–53.

77 Proposed by Visscher *et al.*; see Environmental mutagenesis during the end-Permian ecological crisis (above, n. 57).

78 Campbell, I.H., Czamanske, G.K., Fedorenko, V.A. *et al.* (1992) Synchronism of the Siberian Traps and the Permian–Triassic boundary. *Science*, **258**, 1760–3.

79 Kamo, S.L., Czamanske, G.K., Amelin, Y. *et al.* (2003) Rapid eruption of Siberian flood-volcanic rocks and evidence for coincidence with the Permian–Triassic boundary and mass extinction at 251 Ma. *Earth and Planetary Science Letters*, **214**, 75–91.

80 Thordarson, Th. and Self, S. (1996) Sulfur, chlorine and fluorine degassing and atmospheric loading by the Roza eruption, Columbia River Basalt Group, Washington, USA. *Journal of Volcanology and Geothermal Research*, **74**, 49–73.

81 Beerling, D.J., Harfoot, M., Lomax, B. and Pyle, J.A. (2007) The stability of the stratospheric ozone layer during the end-Permian eruption of the Siberian Traps. *Philosophical Transactions of the Royal Society*, **A365**, 1843–1866. Hydrochloric acid is the most chemically important halogenated compound released. Due to its high solubility in water, it is suggested the proportion reaching the stratosphere

within an explosive volcanic eruption plume would be around 25%, see Textor, C., Graf, Hans-F., Herzog, M., and Oberhuber, J.M. (2003) Injection of gases into the stratosphere by explosive volcanic eruptions. *Journal of Geophysical Research*, **108,** doi:10.1029/2002JD002987. However, this value is relevant for a tropical (i.e., low-latitude) eruption, and the Siberian Traps erupted at a much higher latitude. Recent measurement of hydrochloric acid made by aircraft flying through the volcanic plume during the 2002 eruption of Hekla Volcano on Iceland revises this figure upwards, suggesting 75% of the emitted hydrochloric acid reaches the stratosphere, see Rose, W.I., Millard, G.A., Mather, T.A. *et al.* (2006) Atmospheric chemistry of a 33–34 hour old volcanic cloud from Hekla Volcano (Iceland): insights from direct sampling and application of chemical box modeling. *Journal of Geophysical Research*, **III,** doi:10.1029/ 2005JD006872.

The eruption of the Siberian Traps also probably released large amounts of sulfur dioxide, above n.80. In polluted regions in Russia, high sulfur dioxide concentrations are also associated with mutated pollen in gymnosperms, see for example, Tretyakova, I.N. and Noskova, N.E. (2004) Scotch pine pollen under conditions of environmental stress. *Russian Journal of Ecology*, **35,** 20–6. However, sulfur dioxide is rapidly converted to sulfuric acid and washed out of the atmosphere in precipitation and simulations indicate very little from the Siberian Traps reaches the lower latitudes and none makes it into the southern hemisphere. Consequently, sulfur dioxide pollution cannot explain the global pattern of mutated fossil spores.

82 See Kamo *et al.*, Rapid eruption of Siberian flood-volcanic rocks and evidence for coincidence with the Permian–Triassic boundary and mass extinction at 251 Ma (above, n.79).

83 See Beerling *et al.*, Stability of the stratospheric ozone layer during the end-Permian eruption of the Siberian Traps (above, n.81).

84 Brewer, A.W. (1949) Evidence for a world circulation provided by the measurements of helium and water vapour distribution in the stratosphere. *Quarterly Journal of the Royal Meteorological Society*, **75,** 351–63. Dobson, G.M.B. (1956) Origin and distribution of the polyatomic molecules in the atmosphere. *Proceedings of the Royal Society,*

A236, 187–93. Brewer and Dobson's American colleagues doubted the idea for some time and only became convinced when radioactive fallout produced by atomic weapons testing in the tropical Pacific rained out in the mid- and high-latitudes, see Peirson, D.M. (1971) Worldwide deposition of long-lived fission products from nuclear explosions. *Nature,* **234,** 144–75.

85 Harfoot, M., Beerling, D.J., Lomax, B.H., and Pyle, J.A. (2007) A 2-D atmospheric chemistry modelling investigation of Earth's Phanerozoic O_3 and near-surface ultraviolet radiation history. *Journal of Geophysical Research,* **112,** D07308.

86 Berner, R.A. (2005) The carbon and sulphur cycles and atmospheric oxygen from middle Permian to middle Triassic. *Geochimica et Cosmochimica Acta,* **69,** 3211–17.

87 See Beerling *et al.,* Stability of the stratospheric ozone layer during the end-Permian eruption of the Siberian Traps (above, n.81).

88 See Benton and Twitchett, How to kill (almost) all life, and Benton, *When life nearly died* (above, n.54).

89 Evidence for a possible impact crater was reported by Becker, L., Poreda, R.J., Basu, A.R. *et al.* (2004) Bedout: a possible end-Permian impact crater offshore of Northwestern Australia. *Science,* **304,** 1469–75. A host of technical objections followed shortly afterwards: see Wignall, P., Thomas, B., Willink, R., and Watling, J. (2004) Is Bedout an impact crater? Take 1. *Science,* **306,** 609; Renne, P.R., Melosh, H.J., Farley, K.A. *et al.* (2004) Is Bedout an impact crater? Take 2. *Science,* **306,** 610–11. Responses from Becker *et al.* follow each of these critiques.

90 The occurrence of extraterrestrial noble gases in carbon cages (fullerenes) was reported by Becker, L., Poreda, R.J., Hunt, A.G. *et al.* (2001) Impact event at the Permian–Triassic boundary: evidence from extraterrestrial noble gases in fullerenes. *Science,* **291,** 1530–3. It, too, later came under serious fire: see Farley, K.A. and Mukhopadhyay, S. (2001) An extraterrestrial impact at the Permian–Triassic boundary? *Science,* **293,** U1–U3.

91 See Benton and Twitchett, How to kill (almost) all life, and Benton, *When life nearly died* (above, n.54). A suite of modelling simulations putting the different elements of the killing model together is given in Berner, R.A. (2002) Examination of hypotheses for the

Permo–Triassic boundary extinction by carbon cycle modelling. *Proceedings of the National Academy of Sciences, USA*, **99**, 4172–7.

92 Ibid.

93 Lamarque, J., Kiehl, J.T., Shields, C., and Boville, B.A. (2005) Atmospheric chemistry response to changes in tropospheric methane concentration: application to the Permian–Triassic boundary. *EOS Transactions, American Geophysical Union Meeting, Supplement*, Abstract B41D-0227; Lamarque, J.F., Kiehl, J.T., Shields, C.A. *et al.* (2006) Modeling the response to changes in tropospheric methane concentration: application to the Permian–Triassic boundary. *Paleoceanography*, **21** doi:10.1029/2006PA001276.

94 Global oceanic anoxia was proposed by Wignall, P.B. and Twitchett, R.J. (1996) Oceanic anoxia and the end Permian mass extinction. *Science*, **272**, 1155–8. Since 1996, numerous papers have marshalled evidence in support of the idea.

95 Model simulations reporting how global warming at the end of the Permian triggers globally anoxic oceans through reduced dissolved oxygen amounts and a more sluggish circulation are reported in the following two papers: Hotinski, R.M., Bice, K.L., Kump, L.R. *et al.* (2001) Ocean stagnation and end-Permian anoxia. *Geology*, **29**, 7–10; Kiehl, J.T. and Shields, C.A. (2005) Climate simulation of the latest Permian: implications for mass extinction. *Geology*, **33**, 757–60.

96 The idea was proposed by Kump, L.R., Pavlov, A., and Arthur, M.A. (2005) Massive release of hydrogen sulphide to the surface ocean and atmosphere during intervals of oceanic anoxia. *Geology*, **33**, 397–400. The chemical mechanism involves the interaction of hydrogen sulfide with singlet oxygen, a reactive oxygen species caused by the splitting apart of molecular oxygen gas by sunlight. Oxygen singlets are required for the formation of ozone. In the model simulations reported by Kump *et al.*, massive hydrogen sulfide emissions mopped up oxygen singlets, causing a crash in ozone production.

97 Meinke, D.W., Cherry, J.M., Dean, C. *et al.* (1998) *Arabidopsis thaliana*: a model plant for genome analysis. *Science*, **282**, 662–82.

98 Preuss, D., Rhee, S.Y., and Davies, R.W. (1998) Tetrad analysis possible in *Arabidopsis* with mutation of the QUARTET (QRT) genes. *Science*, **264**, 1458–60.

316 NOTES TO CHAPTER 4

99 Rhee, S.Y. and Somerville, C.R. (1998) Tetrad pollen formation in *quartet* mutants of *Arabidopsis thaliana* is associated with persistence of pectic polysaccharides of the pollen mother cell wall. *Plant Journal*, **15**, 79–88.

100 Rousseaux, M.C., Ballaré, C.L., Giordano, C.V. *et al.* (1999) Ozone depletion and UVB radiation: impact on plant DNA damage in southern South America. *Proceedings of the National Academy of Sciences, USA*, **96**, 15310–15.

101 Ries, G., Heller, W., Puchta, H. *et al.* (2000) Elevated UV-B radiation reduces genome stability in plants. *Nature*, **406**, 98–101. See also the commentary: Britt, A.B. (2000) An unbearable beating by light? *Nature*, **406**, 30–1.

102 Møller, A.P. and Mousseau, T.A. (2006) Biological consequences of Chernobyl: 20 years on. *Trends in Ecology and Evolution*, **21**, 200–7.

103 Ibid.

104 Møller, A.P. (1993) Morphology and sexual selection in the barn swallow *Hirundo rustica* in Chernobyl, Ukraine. *Proceedings of the Royal Society*, **B252**, 51–7.

105 Shevchencko, V.A. *et al.* (1996) Genetic consequences of radioactive pollution of the environment caused by the Chernobyl accident for plants' populations. In *Consequences of the Chernobyl catastrophe: environmental health* (eds V.M. Zakharov and E.Y. Krysanov), pp. 112–26. Centre for Russian Environmental Policy; Sirenko, E.A. (2001) Palynological data from studies of bottom sediments in water bodies of 30-km Chernobyl zone. In *Proceedings of the first international seminar, pollen as indictor of environmental state and paleoecological reconstructions*, pp. 189–90. St Petersburgh [in Russian with English summary].

106 Rozema, J., Broekman, R.A., Blokker, P. *et al.* (2001) UV-B absorbance and UV-B absorbing compounds (*para*-coumaric acid) in pollen and sporopollenin: the perspective to track historic UV-B levels. *Journal of Photochemistry and Photobiology*, **62**, 108–17; Rozema, J., Noordijk, A.J., Broekman, R.A. *et al.* (2001) (Poly)phenolic compounds in pollen and spores of Antarctic plants as indicators of solar UV-B. *Plant Ecology*, **154**, 11–26; Rozema, J., van Geel, B., Björn, L.O. *et al.* (2002) Toward solving the UV puzzle. *Science*, **296**, 1621–2.

107 Blokker, P., Boelen, P., Broekman, R., and Rozema, J. (2006) The occurrence of p-coumaric acid and ferulic acid in fossil plant materials and their use as a UV-proxy. *Plant Ecology*, doi: 10.1007/ s11258–005–9026-y.

108 Blokker, P., Yeloff, D., Boelen, P. *et al.* (2005) Development of a proxy of past surface UV-B irradiation: a thermally assisted hydrolysis and methylation py-GC/MS method for the analysis of pollen and spores. *Analytical Chemistry*, **77**, 6026–31. Remarkable progress has also been made in recovering historical ultraviolet changes by analysing changes in the protective pigments of cyanobacteria preserved in sediments of shallow Antarctica lakes extending back nearly 40 000 years: see Hodgson, D.A., Vyverman, W., Verleyen E. *et al.* (2005) Late Pleistocene record of elevated UV radiation in an Antarctic lake. *Earth and Planetary Science Letters*, **236**, 765–72.

109 Lomax, B.L., Beerling, D.J., Callaghan, T.V. *et al.* (2005). The Siberian Traps, stratographic ozone, UV-B flux and mutagenesis. *GSA Speciality Meetings. Abstracts with Programs*, **1**, 37–3. Similar historical increases in UV-B-absorbing pigments have been obtained by analysing herbarium specimens of bryophytes from 1926 and 1996; see Huttunen, S., Lappalainen, N.M., and Turunen, J. (2005) UV-absorbing compounds in subarctic herbarium bryophytes. *Environmental Pollution*, **133**, 303–14.

110 Anon (1958) Innovations in physics. *Scientific American*, **199**, no. 3, September.

CHAPTER 5

Global warming ushers in the dinosaur era

1 Gordon, A.B. (1894) *The life and correspondence of William Buckland*. John Murray, London.

2 This great episode was recently unearthed by Walter Gratzer (*Eurekas and euphorias: the Oxford book of scientific anecdotes*, Oxford University Press, 2002) in Allen, D. (1978) *The naturalist in Britain*. Penguin, London.

3 Gordon, *The life and correspondence of William Buckland* (above, n.1).

4 We should note that the group Dinosauria is split into two, the Saurischia and the Ornithischia. Birds share a more recent com-

mon ancestry with the Saurischians, meaning the 'reptile-hipped', the group that includes *Tyrannosaurus*. Dinosaurs, then, cannot properly be regarded as extinct. It is more correct, but rather clumsy, to state that non-avian dinosaurs suffered extinction. I will use the word dinosaur in its vernacular form. The realization that the non-avian dinosaurs went extinct at the end of the Cretaceous emerged from studies largely made by Dale Russell, the curator of dinosaurs at the Royal Ontario Museum in Ottawa. See Russell, D.A. (1967) A census of dinosaur specimens collected in western Canada. *National Museum of Canada Natural History Paper*, **36**, 1–13; Russell, D.A. (1975) Reptilian diversity and the Cretaceous–Tertiary boundary in North America. *Geological Association of Canada Special Report*, **13**, 119–36.

5 See Sloan, R.E., Rigby, J.K, Van Valen, L.M., and Gabriel, D.L. (1986) Gradual dinosaur extinction and simultaneous ungulate radiation in the Hell Creek Formation. *Science*, **234**, 1173–5; Sheehan, P.M., Fastovsky, D.E., Hoffmann, R.G. *et al.* (1991) Sudden extinction of the dinosaurs: latest Cretaceous, Upper Great Plains, U.S.A. *Science*, **254**, 835–9. Technical concerns about the conclusion drawn by Sheehan *et al.* were raised by William Clemens and David Archibald and were swiftly rebutted: see Dinosaur diversity and extinction (1992). *Science*, **256**, 159–61.

6 See Fastovsky, D.E. and Sheehan, P.M. (2005) The extinction of the dinosaurs in North America. *GSA Today*, **15**, 4–19.

7 The debate concerns gradual environmental deterioration caused by the effects of Deccan Traps erupting in India, and the sudden catastrophic (for life) environmental effects of a large meteorite impact. The former viewpoint is championed by Vincent Courtillot: see Courtillot, V.E. (1990) A volcanic eruption. *Scientific American*, October, 53–60. The latter is espoused by father and son team Luis and Walter Alvarez: see Alvarez, W. and Asaro, F. (1990) An extraterrestrial impact. *Scientific American*, October, 44–52.

8 Alvarez, L.W., Alvarez, W., Asaro, F., and Michel, H.V. (1980) Extraterrestrial cause for the Cretaceous–Tertiary extinction. *Science*, **208**, 1095–108; Hsü, K.J., He, W.J., McKenzie, J.A. *et al.* (1982) Mass mortality and its environmental and evolutionary consequences. *Science*, **216**, 249–56.

9 Evidence for a 'raging wildfire' in the form of high soot in Cretaceous–Tertiary boundary sedimentary rocks has been seriously questioned; see Belcher, C.M., Collinson, M.E., Sweet, A.R. *et al.* (2003) Fireball passes and nothing burns—the role of thermal radiation in the Cretaceous–Tertiary event: evidence from the charcoal record of North America. *Geology*, **31**, 1061–4. The issue of whether the extraterrestrial impact had sufficient thermal power to actually ignite extensive wildfires has been questioned by the same team; see Belcher, C.M., Collinson, M.E., and Scott, A.C. (2005) Constraints on the thermal energy released from the Chicxulub impactor: new evidence from multi-method charcoal analysis. *Journal of the Geological Society, London*, **162**, 591–602.

10 Hildebrand, A.R., Penfield, G.T., Kring, D.A. *et al.* (1991) Chicxulub crater: a possible Cretaceous/Tertiary boundary impact crater on the Yucatán Peninsula, Mexico. *Geology*, **19**, 867–71.

11 See Belcher *et al.*, Fireball passes and nothing burns, and Belcher *et al.*, Constraints on the thermal energy released from the Chicxulub impactor (above, n.9).

12 I use the term mass extinction to mean a truly exceptional level of extinction, one statistically distinct from the continual background level that goes on as part of the usual business of evolution. Indeed 99% of all species that have ever lived are now extinct; species have an average lifetime of about a million years.

13 See, for example, Foote, M. (2003) Origination and extinction through the Phanerozoic: a new approach. *Journal of Geology*, **III**, 125–48.

14 For many reasons, Phillips' work might not accurately depict the history of Earth's biodiversity. His compilation is, for example, based on about 3000 fossils from British strata that did not necessarily reflect the global situation. He also drew curves of diversity without revealing the number of points in geological time used to shape them.

15 Detailed comparative analysis of the extinction of marine organisms over the last half billion years of Earth history is reported by Raup, D.M. and Sepkoski, J.J. (1982) Mass extinction in the marine fossil record. *Science*, **215**, 1501–3.

16 Sepkoski, J.J. (1993) Ten years in the library: new data confirm paleontological patterns. *Paleobiology*, **19**, 43–51. The original 1982 dataset received special scrutiny after publication of the highly controversial idea that mass extinctions occurred periodically about every 26 million years. See Raup, D.M. and Sepkoski, J.J. (1984) Periodicity of extinctions in the geologic past. *Proceedings of the National Academy of Sciences, USA*, **81**, 801–5; Raup, D.M. and Sepkoski, J.J. (1986) Periodic extinction of families and genera. *Science*, **231**, 833–6. For a popular account of mass extinctions and the controversy, see Raup, D.M. (1991) *Extinction: bad genes or bad luck?* Oxford University Press.

17 See Benton, M.J. (1995) Diversification and extinction in the history of life. *Science*, **268**, 52–8.

18 Newell, N.D. (1963) Crises in the history of life. *Scientific American*, **208**, 72–96. See also Valentine, J.W. (1969) Patterns of taxonomic and ecological structure of shelf benthos during Phanerozoic time. *Palaeontology*, **12**, 684–709.

19 Benton, Diversification and extinction in the history of life (above, n.17).

20 Ibid.

21 McGhee, G.R., Sheehan, P.M., Bottjer, D.J., and Droser, M.L. (2004) Ecological ranking of Phanerozoic biodiversity crises: ecological and taxonomic severities are decoupled. *Palaeogeography, Palaeoclimatology, Palaeoecology*, **211**, 289–97.

22 Recent opposition of this sort has been voiced most vociferously in Tanner, L.H., Lucas, S.G., and Chapman, M.G. (2004) Assessing the record and causes of Late Triassic extinctions. *Earth-Science Reviews*, **65**, 103–9. Earlier doubts were expressed by Hallam, A. (2002) How catastrophic was the end-Triassic mass extinction? *Lethaia*, **35**, 147–57. This paper represents a reversal of opinion for its author. A decade earlier, he argued the complete opposite, naming the end-Triassic as one of the five biggest extinctions in the last half billion years; see Hallam, A. (1990) The end-Triassic extinction event. *Geological Society of America Special Paper*, **247**, 577–83.

23 Bambach, R.K., Knoll, A.H., and Wang, S.C. (2004) Origination, extinction, and mass depletions of marine diversity. *Paleobiology*, **30**, 522–42.

24 Sereno, P.C. (1997) The origin and evolution of dinosaurs. *Annual Review of Earth and Planetary Science*, **25**, 435–89; Sereno, P.C. (1999) The evolution of dinosaurs. *Science*, **284**, 2137–47; Benton, M.J. (1983) Dinosaur success in the Triassic: a noncompetitive ecological model. *Quarterly Review of Biology*, **58**, 29–55.

25 The climate of Pangaea has been the subject of numerous numerical modelling studies; key contributions include: Crowley, T.J., Hyde, W.T., and Short, D.A. (1989) Seasonal cycle variations on the supercontinent of Pangaea. *Geology*, **17**, 457–60; Kutzbach, J.E. and Gallimore, R.G. (1989) Pangaean climates: megamonsoons of the megacontinent. *Journal of Geophysical Research*, **94**, 3341–57; Chandler, M.A., Rind, D., and Ruedy, R. (1992) Pangaean climate during the early Jurassic: GCM simulations and the sedimentary record of paleoclimate. *Geological Society of America Bulletin*, **104**, 543–59.

26 Loope, D.B., Rowe, C.M., and Joeckel, R.M. (2001) Annual monsoon rains recorded by Jurassic dunes. *Nature*, **412**, 64–6.

27 Knoll, A.H. (1984) Patterns of extinction in the fossil record of vascular plants. In *Extinctions* (ed. N.M Nitecki), pp. 21–68. University of Chicago Press. Niklas, K.J. (1997) *The evolutionary biology of plants*. University of Chicago Press. Wing, S.L. (2004) Mass extinctions and plant evolution. In *Extinctions in the history of life* (ed. P.D. Taylor), pp. 61–97. Cambridge University Press.

28 Officially a self-governing Danish territory, Greenland measures nearly 2735 km (1700 miles) long, has an average breadth of about 97 km (60 miles), and covers a total area of 2 175 600 square kilometres (840 004 square miles).

29 Chaloner, W.G. (1985) Thomas Maxwell Harris. *Biographical Memoirs of Fellows of the Royal Society*, **31**, 227–60.

30 Ibid.

31 Ibid.

32 See Woodward, F.I. (1987) Stomatal numbers are sensitive to CO_2 increases from pre-industrial levels. *Nature*, **327**, 617–18.

33 McElwain, J.C., Beerling, D.J., and Woodward, F.I. (1999) Fossil plants and global warming at the Triassic–Jurassic boundary. *Science*, **285**, 1386–90. See also commentaries in *New Scientist* (1999, 4 September, p. 16) and *Economist* (1999, 28 August, p. 71).

34 Mann, M., Bradley, R.S., and Hughes, M.K. (1999) Northern hemisphere temperatures during the past millennium: inferences, uncertainties, and limitations. *Geophysical Research Letters*, **26**, 759–62.

35 Summarized in Harris, T.M. (137) The fossil flora of Scoresby Sound, East Greenland. Part 5. Stratigraphic relations of the plant beds. *Meddelelser Om Grønland*, **112**, 1–114.

36 Chaloner, Thomas Maxwell Harris (above, n.29).

37 In today's climate and atmospheric carbon dioxide content, large-leaved tropical rainforest trees do not generally suffer thermal damage. This is because they possess among the highest stomatal densities of all species in the plant kingdom, ensuring adequate transpirational cooling.

38 Marzoli, A., Renne, P.R., Piccirillo, E.M. *et al.* (1999) Extensive 200-million-year-old continental flood basalts of the Central Atlantic Magmatic Province. *Science*, **284**, 616–18. See also commentary by Olsen, P. (1999) Giant lava flows, mass extinction, and mantle plumes. *Science*, **284**, 604–5. Later work established the eruptions as being synchronous with the extinctions: Marzoli, A., Bertrand, H., Knight, K.B. *et al.* (2004) Synchrony of the Central Atlantic magmatic province and the Triassic–Jurassic boundary climatic and biotic crisis. *Geology*, **32**, 973–6.

39 May, P.R. (1971) Pattern of Triassic–Jurassic diabase dikes around the North Atlantic in the context of predrift position of the continents. *Geological Society of American Bulletin*, **82**, 1285–91.

40 Knight, K.B., Nomade, S., Renne, P. *et al.* (2004) The Central Atlantic Magmatic Province at the Triassic–Jurassic boundary: paleomagnetic and $^{40}Ar/^{39}Ar$ evidence from Morocco for brief, episodic volcanism. *Earth and Planetary Science Letters*, **228**, 143–60.

41 See White, R.S. and McKenzie, D.P. (1989) Volcanism at rifts. *Scientific American*, July, 44–55.

42 Extensive fascinating documentation of the scientific and social side of the development of the theory of plate tectonics can be found in Oreskes, N. (ed.) (2001) *Plate tectonics: an insider's history of the modern theory of the earth*. Westview Press, Cambridge, MA. See also the following review, which includes suggestions for other books to consult on this topic: Morgan, J.P. (2002) When the Earth moved. *Nature*, **417**, 487–8.

43 The issue of whether the surfacing of deep-rooted mantle plumes is erratic and episodic, or a phenomenon directly linked with the break-up of a continent, is controversial. See White, R.S. and McKenzie, D. (1995) Mantle plumes and flood basalts. *Journal of Geophysical Research*, **100**, 17543–85; Courtillot, V., Jaupart, C., Manighetti, I. *et al.* (1999) On causal links between flood basalts and continental breakup. *Earth and Planetary Science Letters*, **166**, 177–95.

44 Carbon dioxide emission figures given in Gerlach, T.M. and Graeber, E.J. (1985) Volatile budget of Kilauea volcano. *Nature*, **313**, 273–7. Sulfur dioxide is produced from volcanoes in an amount equivalent, on average, to that of carbon dioxide: Symonds, R.B., Rose, W.I., Bluth, G.J.S., and Gerlach, T.M. (1994) Volcanic-gas studies: methods, results, and applications. *Reviews in Mineralogy*, **30**, 1–66.

45 Keppler, H., Wiedenbeck, M., and Shcheka, S.S. (2003) Carbon solubility in olivine and the mode of carbon storage in the Earth's mantle. *Nature*, **424**, 414–16.

46 The idea was originally advanced by Peter Vogt at the US National Oceanographic office; see Vogt, P.T. (1972) Evidence for global synchronism in mantle plume convections, and possible significance for geology. *Nature*, **240**, 338–42.

47 A flavour of this debate can be gained from Courtillot, V. (1999) *Evolutionary catastrophes: the science of mass extinctions.* Cambridge University Press. The following review does much to place the book in its proper context: Parker, W.C. (2000) *Palaios*, **15**, 582–3.

48 For a review of this topic, see Wignall, P.B. (2001) Large igneous provinces and mass extinctions. *Earth-Science Reviews*, **53**, 1–33.

49 See Tanner *et al.*, Assessing the record and causes of Late Triassic extinctions (above, n.22). The acid rain idea was repeated by Guex, J., Bartolini, A., Atudorei, V., and Taylor, D. (2004) High resolution ammonite and carbon isotope stratigraphy across the Triassic–Jurassic boundary at the New York Canyon (Nevada). *Earth and Planetary Science Letters*, **225**, 29–41. Puzzling suggestions of a global cooling at the Triassic–Jurassic boundary from changes in the abundance of fossil pollen of different 'warmth-loving' and 'cooling-loving' plant associations doesn't square with the rapid washing of sulfate aerosols out of the atmosphere by rain; see Hubbard, R. and Boulter, M.C. (1994) Phytogeography and paleoecology in western

Europe and eastern Greenland near the Triassic–Jurassic boundary. *Palaios*, **15**, 102–31.

50 See McElwain *et al.*, Fossil plants and global warming at the Triassic–Jurassic boundary (above, n.33); Guex *et al.*, High resolution ammonite and carbon isotope stratigraphy across the Triassic–Jurassic boundary at the New York Canyon (Nevada) (above, n.49); Ward, P. D., Haggart, J.W., Wilbur, D. *et al.* (2001) Sudden productivity collapse associated with the Triassic–Jurassic boundary mass extinction. *Science*, **292**, 1148–51; Hesselbo, S.P., Robinson, S.A., Surlyk, F., and Piasecki, S. (2002) Terrestrial and marine extinction at the Triassic–Jurassic boundary synchronized with major carbon cycle perturbation: a link to initiation of massive volcanism. *Geology*, **30**, 251–4; Ward, P.D., Garrison, G.H., Haggart, J.W. *et al.* (2004) Isotopic evidence bearing on Late Triassic extinction events, Queen Charlotte Islands, British Columbia, and implications for the duration and cause of the Triassic/Jurassic mass extinction. *Earth and Planetary Science Letters*, **234**, 589–600.

51 Pálfy, J., Demény, A., Haas, J. *et al.* (2001) Carbon isotope anomaly and other geochemical changes at the Triassic–Jurassic boundary from a marine section in Hungary. *Geology*, **29**, 1047–50.

52 A numerical evaluation of the effects of volcanic eruptions and methane releases on atmospheric carbon dioxide levels and the isotopic composition of organic matter is given by Beerling, D.J. and Berner, R.A. (2002) Biogeochemical constraints on the Triassic–Jurassic boundary carbon cycle event. *Global Biogeochemical Cycles*, **16**, doi: 10.1029/2001GB001637.

53 No means of interpreting the carbon dioxide content of the ancient atmosphere is without its uncertainties. For a technical review of the problems associated with the different techniques, see Royer, D.L., Berner, R.A., and Beerling, D.J. (2001) Phanerozoic atmospheric CO_2 change: evaluating geochemical and palaeobiological approaches. *Earth-Science Reviews*, **54**, 349–92. One issue for those techniques based on living organisms is selection by changing carbon dioxide levels. This raises the question of whether the responses of organisms to carbon dioxide we see today exactly mirror those of organisms tens of millions of years ago. Uncertainties of this sort transfer to uncertainties in the reconstructed carbon dioxide concentrations.

54 The mass of carbon stored as methane in gas hydrates is contentious and estimates range from a 'best guess' value of 5000 billion
tonnes to a lower range of 500–2500 billion tonnes; see, respectively, Buffet, B. and Archer, D. (2004) Global inventory of methane
clathrate: sensitivity to changes in the deep ocean. *Earth and Planetary
Science Letters*, **227**, 185–99, and Milkov, A.V. (2004) Global estimates
of hydrate-bound gas in marine sediments. *Earth-Science Reviews*, **66**,
183–97. For comparison, there are ∼ 1600 billion tonnes of carbon
in vegetation and soils globally; see Beerling, D.J. and Woodward. F.
I. (2001) *Vegetation and the terrestrial carbon cycle: modelling the first 400
million years*. Cambridge University Press.

55 Beerling and Berner, Biogeochemical constraints on the Triassic–
Jurassic boundary carbon cycle event (above, n.52).

56 Kennett, J.P. and Stott, L.D. (1991) Abrupt deep-sea warming, paleo-
ceanographic changes and benthic extinctions at the end of the
Palaeocene. *Nature*, **353**, 225–9. For evidence of a warming in the
tropical Pacific and Atlantic Oceans at the Palaeocene–Eocene
boundary, see Zachos, J.C., Wara, M.W., Bohaty, S. *et al.* (2003) A
transient rise in tropical sea surface temperature during the Palaeocene–Eocene thermal maximum. *Science*, **302**, 1551–4; Tripati, A.K.
and Elderfield, H. (2004) Abrupt hydrographic changes in the equatorial Pacific and subtropical Atlantic from foraminiferal Mg/Ca
indicate greenhouse origin for the thermal maximum at the
Paleocene–Eocene boundary. *Geochemistry, Geophysics, Geosystems*, **5**,
doi: 10.1029/2003GC000631.

57 Dickens, G.R., O'Neil, J.R., Rea, D.C., and Owen, R.M. (1995) Dissociation of oceanic methane hydrate as a cause of the carbon isotope
excursion at the end of the Paleocene. *Paleoceanography*, **10**, 965–71;
Dickens, G.R., Castillo, M.M., and Walker, J.C.G. (1997) A blast of gas
in the latest Paleocene: simulating first order effects of massive
dissociation of oceanic methane hydrate. *Geology*, **25**, 258–62.

58 Zachos, J.C., Rohl, U., Schellenberg, S.A. *et al.* (2005) Rapid
acidification of the ocean during the Paleocene–Eocene thermal
maximum. *Science*, **308**, 1611–15.

59 The effect of future carbon dioxide emissions on ocean acidity is
given in Caldeira, K. and Wickett, M.E. (2003) Anthropogenic carbon and ocean pH. *Nature*, **425**, 365.

60 The idea that carbon dioxide release during the eruption of the North Atlantic Volcanic Province caused gradual warming was mooted by Eldholm, O. and Thomas, E. (1993) Environmental impact of volcanic margin formation. *Earth and Planetary Science Letters*, **117**, 319–29. Later, detailed reconstruction of tropical ocean temperatures and circulation patterns supported the idea that changing ocean circulation may have been involved in triggering methane release from frozen gas hydrates. See Tripati, A. and Elderfield, J. (2005) Deep-sea temperature and circulation changes at the Paleocene–Eocene thermal maximum. *Science*, **308**, 1894–8. It should be emphasized that, even for the Palaeocene–Eocene thermal maximum, the case is far from open and shut when it comes to implicating methane release from gas hydrates: see Higgins, J.A. and Schrag, D.P. (2006) Beyond methane: towards a theory for the Paleocene–Eocene thermal maximum. *Earth and Planetary Science Letters*, **245**, 523–7.

61 Huynh, T.T. and Poulsen, C.J. (2005) Rising atmospheric CO_2 as a possible trigger for the end-Triassic mass extinction. *Palaeogeography, Palaeoclimatology, Palaeoecology*, **217**, 223–42.

62 Marzoli *et al.*, Extensive 200-million-year-old continental flood basalts of the Central Atlantic Magmatic Province (above, n.38), and Knight *et al.*, The Central Atlantic Magmatic Province at the Triassic–Jurassic boundary (above, n.40).

63 See Van de Schootbrugge, B., Tremoladi, F., Rosenthal, Y. *et al.* (2007) End-Triassic calcification crisis and blooms of organic-walled 'disaster species'. *Palaeogeography, Palaeoclimatology, Palaeoecology*, **244**, 126–141.

64 Huynh and Poulsen, Rising atmospheric CO_2 as a possible trigger for the end-Triassic mass extinction (above, n.61).

65 Pálfy, J., Mortensen, J.K., Carter, E.S. *et al.* (2004) Timing of the end-Triassic extinctions: first on land then in the sea? *Geology*, **28**, 39–42.

66 Wells, H.G. (1945). *Mind at the end of its tether*. Heinemann, London.

67 Cohen, A.S. and Cos, A.L. (2002) New geochemical evidence for the onset of volcanism in the Central Atlantic magmatic province and environmental change at the Triassic–Jurassic boundary. *Geology*, **30**, 267–70.

68 Evidence from fossil soils suggesting no change in the carbon dioxide content of the atmosphere across the Triassic–Jurassic

boundary was published as Tanner, L.H., Hubert, J.F., Coffey, B.P., and McInerney, D.P. (2001) Stability of atmospheric carbon dioxide levels across the Triassic–Jurassic boundary. *Nature*, **411**, 675–7. Tanner *et al.*'s conclusions have been challenged on scientific grounds. Correction of the technical concerns reveals the original data support a rise in atmospheric carbon dioxide levels: see Beerling, D.J. (2002) CO_2 and the end-Triassic mass extinction. *Nature*, **415**, 386–7. Only one of the original authors responded: see Tanner, L.H. (2002) Reply. *Nature*, **415**, 388.

69 Tanner *et al.*, Stability of atmospheric carbon dioxide levels across the Triassic–Jurassic boundary (above, n.67).

70 Beerling and Berner, Biogeochemical constraints on the Triassic–Jurassic boundary (above, n.52).

71 Was the rise of the dinosaurs related to a major impact event at the Triassic–Jurassic boundary? Controversial evidence is offered, but serious doubts remain: Olsen, P.E., Kent, D.V., Sues, H.D. *et al.* (2002) Ascent of dinosaurs linked to an iridium anomaly at the Triassic–Jurassic boundary. *Science*, **296**, 1305–7. See also the commentary, Kerr, R.A. (2002) Did an impact trigger the dinosaurs' rise? *Science*, **296**, 1215–16. Technical comments are raised in Thulborn, T. (2003) *Science*, **301**, 169b, and responded to: see Olsen, P.E. *et al.* (2003) *Science*, **301**, 169c.

72 The discovery was supported by later analyses from the rocks on Partridge Island, Nova Scotia. See Tanner, L.H. and Kyte, F.T. (2005) Anomalous iridium enrichment at the Triassic–Jurassic boundary, Blomidon Formation, Fundy basin, Canada. *Earth and Planetary Science Letters*, **240**, 634–41.

73 Fowell, S.J., Cornett, B., and Olsen, P.E. (1994) Geologically rapid late Triassic extinctions: palynology evidence from the Newark supergroup. *Geological Society of America Special Paper*, **288**, 197–206.

74 Shocked grains of quartz have been reported from a marine Triassic–Jurassic boundary site: Bice, D.M., Newton, C.R., McCauley, S. *et al.* (1992) Shocked quartz at the Triassic–Jurassic boundary in Italy. *Science*, **255**, 443–6. They have not yet been found in eastern North America despite a search by Mossman, D.J., Grantham, R.G., Langenhorst, F. (1998) A search for shocked quartz at the Triassic–Jurassic boundary in the Fundy and Newark basins of the Newark Supergroup. *Canadian Journal of Earth Science*, **35**, 101–9.

75 See review by Grieve, R.A.F. (1987) Terrestrial impact structures. *Annual Review of Earth and Planetary Sciences*, **15**, 245–70.

76 Walkden, G., Parker, J., and Kelley, S. (2002) A late Triassic impact ejector layer in southwest Britain. *Science*, **298**, 2185–8.

77 Simms, M.J. (2003) Uniquely extensive seismite from the latest Triassic of the United Kingdom: evidence for a bolide impact? *Geology*, **31**, 557–60.

78 Morgan, J.P., Reston, T.J., and Ranero, C.R. (2004) Contemporary mass extinctions, continental flood basalts, and 'impact signals': are mantle-induced lithospheric gas explosions the causal link? *Earth and Planetary Science Letters*, **217**, 263–84.

79 For discussion on this point, see Crowley, T.J. (1990) Are there any satisfactory geologic analogues for a future greenhouse world? *Journal of Climate*, **3**, 1282–92; Crowley, T.J. (1993) Geologic assessment of the greenhouse effect. *Bulletin of the American Meteorological Society*, **74**, 2362–73.

80 Houghton, J.T., Ding, Y., Griggs, D.J. *et al.* (2001) *Climate change 2001: the scientific basis*. Intergovernmental Panel on Climate Change. Cambridge University Press.

81 Mann *et al.*, Northern hemisphere temperatures during the past millennium (above, n.34).

82 Levitus, S., Antonov, J.I., Boyer, T.P., and Stephens, C. (2000) Warming of the world ocean. *Science*, **287**, 2225–9; Levitus, S., Antonov, J., and Boyer, T. (2005) Warming of the world ocean, 1955–2003. *Geophysical Research Letters*, **32**, doi: 10.1029/2004GL021592.

83 Archer, D. and Buffett, B. (2005) Time-dependent response of the global ocean clathrate reservoir to climatic and anthropogenic forcing. *Geochemistry, Geophysics, Geosystems*, **6**, doi: 10.1029/2004GC000854. See also Buffet and Archer, Global inventory of methane clathrate (above, n.54).

84 See the references cited in note 82.

CHAPTER 6

The flourishing forests of Antarctica

1 In fact, their survey data indicate that on the afternoon of 14 December 1911 they were at 89°56′ S. Further systematic reconnais-

sance of the area by the Norwegians ensured they passed within 100–600 m of the true geographic pole on 16 December 1911, the actual determination being extremely difficult to make owing to refraction of light by the cold atmosphere.

2 Scott, R.F. (1914) *Scott's last expedition: being the journals of Captain R.F. Scott, R.N., C.V.O.* 5th edn, 2 vols. Smith, Elder, London. Oxford University Press has reprinted Scott's journals with new introductions and extensive notes; the original publication was made after editing of passages that were felt to be too critical and unforgiving of his colleagues. See Scott, R.F. (2005) *Journals: Captain Scott's last expedition.* Oxford University Press.

3 The Antarctic plateau sits ∼ 2836 m above sea level. Wind circulation patterns around the continent further lower the pressure at the Pole, rendering its effective altitude even higher. In an effort to combat the difficulty of breathing oxygen in 'thin' air, the body responds by increasing heart and respiration rates to get more oxygen into the bloodstream. Higher rates of respiration, and the dry atmosphere, increase water loss, making dehydration a real problem. The human body requires about a gallon and half of water (∼ 7 litres) a day to combat dehydration in Antarctica; Scott and his party drank only a few glasses a day.

4 See Evans, E.R.G.R. (1921) *South with Scott.* Collins, London.

5 Wilson, E.A. (1972) *Diary of the Terra Nova expedition to the Antarctic, 1910–1912.* Blandford, Poole.

6 See Scott, *Scott's last expedition* (above, n.2).

7 See Evans, *South with Scott* (above, n.4).

8 See Scott, *Scott's last expedition* (above, n.2).

9 A fine case is made in Fiennes, R. (2003) *Captain Scott.* Hodder & Stoughton, London.

10 Ludlam, H. (1965) *Captain Scott.* Foulsham, London.

11 Solomon, S. (2001) *The coldest March: Scott's fatal Antarctic expedition.* Yale University Press, New Haven. See also Ranulph Fiennes biography for a valuable effort at debunking the myth: Fiennes, R. (2003) *Captain Scott.* Hodder & Stoughton, London.

12 See Scott, *Scott's last expedition* (above, n.2).

13 The current title is the Natural History Museum. The scientific report on the fossil plants collected by R.F. Scott and his party on the return

leg of their tragic expedition to the South Pole is give in Seward, A.C. (1914) Antarctic fossil plants: British Antarctic ('Terra Nova') Expedition, 1910. *British Museum of Natural History Report, Geology*, 1, 1–49.

14 See Rose, J. (1992) *Marie Stopes and the sexual revolution.* Faber, London.

15 Chaloner, W.G. (1995) Marie Stopes (1880–1958): the American connection. *Geological Society of America Memoir*, **158**, 127–34.

16 Chaloner, W.G. (2005) The palaeobotanical work of Marie Stopes. In *The history of palaeobotany: selected essays* (eds A.J. Bowden, C.V. Burek, and R. Wilding), pp. 127–35. Geological Society of London Special Publication, Vol. 241. The Geological Society, London.

17 Finds of fossil woods, presumed remnants of polar forests, have been extensively described by Scandinavian palaeobotanists. See Halle, T.G. (1913) The Mesozoic flora of Graham Land. *Wissenschaftliche Ergebnisse der Schwedischen Südpolar-Expedition 1901–03.* Bd. III, Lief. 14, 123 pp., Taf. 1–9; Nathorst, A.G. (1914) Nachträage zur Pälaozoischen Flora Spitsbergens. *Zur fossilen Flora der Polarländer.* Teil I. Lief. IV. Stockholm.

18 Guttridge, L.F. (2000) *Ghosts of Cape Sabine: the harrowing true story of the Greely Expedition.* Berkley Publishing Group, New York.

19 The 12 August 1884 edition of *The New York Times* reported under the heading 'A Horrible discovery', 'When their food gave out the unfortunate members of the colony, shivering and starving in their little tent on the bleak shore of Smith's Sound, were led by the horrible necessity to become cannibals. The complete history of their experience in that terrible Winter must be told, and the facts hitherto concealed will make the record of the Greely colony— already full of horrors—the most dreadful and repulsive chapter in the long annals of arctic exploration.'

20 From the early- to mid-nineteenth century onwards it was also becoming clear that the southern and northern polar forests had been populated by polar dinosaurs (see Rich, T.H., Vickers-Rich, P., and Gangloff, R.A. (2002) Polar dinosaurs. *Science*, **295**, 979–80). Polar dinosaurs probably needed sunlight to keep warm, and to stay alive when confronted with the dark winters they either had to migrate or hibernate. The smaller more mobile dinosaurs could do either, assuming seaways did not block their migration routes. Migration across long distances was a problem for the other, large

ones, which had slow top speeds (probably of about 2 km an hour), limiting their ability to reached sunnier lower latitude climes in time. Speculation concerns how the large ones managed to keep warm in the austral winters. One ingenious coping strategy they may have developed is similar to that found in giant leatherback turtles. The giant leatherback turtle, our largest living reptile, is able to swim in the seas from the tropics to north of the Arctic Circle. It achieves this remarkable feat by adjusting its circulation to regulate heat loss, and by having low metabolic rates and thick fat and skin layers for insulation. Dinosaurs, like turtles, could have possessed all of these traits with an effectiveness exaggerated by their large sizes. Gigantothermy, as it is known, may just have been sufficient to see them through the polar winters. See Benton, M.J. (1991) Polar dinosaurs and ancient climates. *Trends in Ecology and Evolution*, **6**, 28–30, and Paladino, F.V., O'Connor, M.P., and Spotila, J.R. (1990) Metabolism of leatherback turtles, gigantothermy, and thermoregulation of dinosaurs. *Nature*, **344**, 858–60.

21 Evidence, in the form of fossilized tree stumps some 20 cm high, that a Permian forest occupied Antarctica at a location 80–85°S. Taylor, E.J., Taylor, T.N., and Cuneo, R. (1992) The present is not the key to the past: a polar forest from the Permian of Antarctica. *Science*, **257**, 1675–7.

22 For an account that captures the excitement by one of the original team members, see Francis, J.E. (1991) Arctic Eden. *Natural History*, **1 (Jan)**, 57–63. Scientific descriptions of the forests can be found in Francis, J.E. (1991) The dynamics of polar fossil forests: tertiary fossil forests of Axel Heiberg Island, Canadian Archipelago. *Geological Survey of Canada Bulletin*, **403**, 29–38.

23 Spicer, R.A. and Chapman, J.L. (1990) Climate change and the evolution of high-latitude terrestrial vegetation and floras. *Trends in Ecology and Evolution*, **5**, 279–84.

24 A recent review is given in Grace, J., Berninger, F., and Nagy, L. (2002) Impacts of climate change on the tree line. *Annals of Botany*, **90**, 537–44.

25 Nathorst, A.G. (1890) Ueber die Reste eines Brotfruchtbaums *Artocarpus dicksoni* N. Sp. Aus den cenomanen Kreideablagerungen Grönlands. *Kongl. Svenska vetenskaps-akademiens handlingar* **24**, 2–9.

26 See Tarduno, J.A., Brinkman, D.B., Renne, P.R. *et al.* (1998) Evidence for extreme climatic warmth from late Cretaceous Arctic vertebrates. *Science*, **282**, 2241–4. See also a commentary by Huber, B.T. (1998) Tropical paradise at the Cretaceous poles? *Science*, **282**, 2199–200.

27 Prior to the discovery of champsosaur skeletal remains, only dinosaur and turtle remains had been found; see Estes, R. and Hutchinson, J. (1980) Eocene vertebrates from Ellesmere Island, Canadian Arctic Archipelago. *Palaeogeography, Palaeoclimatology, Palaeoecology*, **30**, 325–47. These are of limited significance because turtles could have escaped lethal freezing temperatures by hibernation in burrows and it has yet to be definitively established either way whether the dinosaurs were cold blooded or warm blooded.

28 For evidence of warm, high southern latitude Cretaceous oceans, see: Huber, B.T., Hodell, D.A., and Hamilton, C.P. (1995) Mid- to late-Cretaceous climate of the southern high latitudes: stable isotope evidence for minimal equator-to-pole thermal gradients. *Geological Society of America Bulletin*, **107**, 1164–91; Huber, B.T., Norris, R.D., and MacLeod, K.G. (2002) Deep-sea paleotemperature record of extreme warmth during the Cretaceous. *Geology*, **30**, 123–6. However, the evidence is controversial. It is based on measurements of the oxygen isotope composition of fossil shells of foraminifera, a feature reflecting that of the oceans. But ocean isotopic composition varies with factors other than temperature, particularly salinity and the presence of ice. After the organism dies, the shell is also subject to being dissolved and recrystallized, which can alter its isotopic composition. The controversy stems from the fact that it is not easy to rule out bias introduced from these two sources.

29 A recent calibration of Arctic Ocean temperatures during the Cretaceous was reported by Jenkyns, H.C., Forster, A., Schouten, S., and Sinninghe Damste, J.S. (2004) High temperatures in the Late Cretaceous Arctic Ocean. *Nature*, **432**, 888–92. Also see the accompanying commentary: Poulsen, C.J. (2004) A balmy Arctic. *Nature*, **432**, 814–15. The authors report a single estimate of ocean temperature inferred from the composition of membrane lipids of marine phytoplankton, which in part gets around the problems referred to in the previous note. However, the estimate is then used to calibrate

an oxygen isotope record giving very warm Arctic Ocean temperatures that are open to the same criticisms.

30 Brinkhuis, H., Schouten, S., Collinson, M.E. *et al.* (2006) Episodic freshwater in the Eocene Arctic Ocean. *Nature*, **441**, 606–9. See also the commentary, Stoll, H.M. (2006) The Arctic tells its story. *Nature*, **441**, 579–81.

31 Dutton, A.L., Lohmann, K.C., and Zinsmeister, W.J. (2002) Stable isotope and minor element proxies for Eocene climate of Seymour Island, Antarctica. *Paleoceanography*, **17**, doi: 10.1029/2000PA000593.

32 Larson, R.L. (1991) Geological consequences of superplumes. *Geology*, **19**, 963–6.

33 Extremely warm tropical surface ocean temperatures have been reconstructed from analyses of middle Cretaceous rocks in the proto-North Atlantic (32–36°C) and in the equatorial Pacific (27–32°C). See Schouten, S., Hopmans, E.C., Forster, A. *et al.* (2003) Extremely high sea-surface temperatures at low latitudes during the middle Cretaceous as revealed by archeal membrane lipids. *Geology*, **31**, 1069–72.

34 Creber, G.T. and Chaloner, W.G. (1984) Influence of environmental factors on the wood structure of living and fossil trees. *Botanical Review*, **50**, 357–448.

35 A similar signal can be read from the growth rings in the fossil bones of dinosaurs that inhabited the polar regions. Sections of bones of one group, the Hypsilophodontids, from southern Australia showed that growth continued throughout the wintertime, implying they were well adapted to the polar conditions. The same group also possessed enlarged optic lobes that may have afforded them greater visual acuity than their lower-latitude cousins to better deal with the winter darkness. See Rich, T.H. and Vickers-Rich, P. (2000) *Dinosaurs of darkness*. Indiana University Press.

36 Chaloner, W.G. and Creber, G.T. (1990) Do fossil plants give a climatic signal? *Journal of Geological Society*, **147**, 343–50.

37 Creber, G.T. and Chaloner, W.G. (1985) Tree growth in the Mesozoic and early Tertiary and the reconstruction of palaeoclimate. *Palaeogeography, Palaeoclimatology, Palaeoecology*, **52**, 35–60.

38 Siberian data for larch (*Larix dahurica*) from Von Middendorf, A.T. (1867) Dei Gewächse Sibiriens. In *Reise in den äussersten Norden und*

Osten Sibiriens. St Petersburg, 4, 1. Data for willow (*Salix arctica*) from Axel Heiberg Island (79°N): Beschel, R.E. and Webb, D. (1963) Growth ring studies on Arctic willows. In *Axel Heiberg Island. A preliminary report 1961–62*, pp. 189–98, McGill University, Montreal. Data for willow from Cornwallis Island (75°N): Warren Wilson, J. (1966) An analysis of plant growth and its control in arctic environments. *Annals of Botany*, **30**, 383–402.

39 Niklas, K.J. (1992) *Plant biomechanics: an engineering approach to plant form and function*. Chicago University Press.

40 See Williams, C.J., Johnson, A.H., LePage, B.A. *et al.* (2003) Reconstruction of Tertiary *Metasequoia* forests. I. Test of a method for biomass determination based on stem dimensions. *Paleobiology*, **29**, 256–70; Williams, C.J., Johnson, A.H., LePage, B.A. *et al.* (2003) Reconstruction of Tertiary *Metasequoia* forests. II. Structure, biomass, and productivity of Eocene floodplain forests in the Canadian Arctic. *Paleobiology*, **29**, 271–92.

41 Osborne, C.P. and Beerling, D.J. (2002) Sensitivity of tree growth to a high CO_2 environment: consequences for interpreting the characteristics of fossil woods from ancient 'greenhouse' worlds. *Palaeogeography, Palaeoclimatology, Palaeoecology*, **182**, 15–29.

42 Wolfe, J.A. (1985) The distribution of major vegetation types during the Tertiary. In *The carbon cycle and atmospheric CO_2: natural variations, Archean to present* (eds E.T. Sundquist and W.S. Broecker), pp. 357–75. Geophysical Monograph Series, Vol. 32. American Geophysical Union, Washington; Wolfe, J.A. (1987) Late Cretaceous-Cenozoic history of deciduousness and the terminal Cretaceous event. *Paleobiology*, **13**, 215–26.

43 Hickey, L.J. (1984) Eternal summer at 80 degrees North: fossil evidence for a warmer Arctic. *Discovery*, **17(1)**, 17–23.

44 Seward, Antarctic fossil plants (above, n.13).

45 Seward, A.C. (1926) II. The Cretaceous plant-bearing rocks of western Greenland. *Philosophical Transactions of the Royal Society*, **B215**, 57–172.

46 Andrews, H.N. (1980) *The fossil hunters: in search of ancient plants*. Cornell University Press, Ithaca.

47 The view that deciduous trees of ancient polar forests are an adaptation to the warm high-latitude environment is expressed in

Chaney, R.W. (1947) Tertiary centers and migration routes. *Ecological Monographs*, **17**, 140–8.

48 Mason, H.L. (1947) Evolution of certain floristic associations in western North America. *Ecological Monographs*, **17**, 201–10. Mason also initiated what became known as the 'obliquity debate', postulating that the inclination of Earth's rotational axis (obliquity) was substantially different in the past. It was a suggestion that he thought would eliminate long periods of winter darkness at high latitudes. A number of other workers misguidedly aligned themselves to the reduced obliquity argument; they wanted it lower by some 5–15°, far more than the well-understood periodic wobbles that have helped pace Earth's ice ages over the past 2 million years. It was a radical notion but one without a plausible mechanism. Quite apart from arguments concerning how to liquefy and redistribute large chunks of the mantle to affect it, a reduction in obliquity, while achieving the desired increase in sunlight during the polar winters, actually decreases it in summer. Climate model simulations later showed that these effects created substantial polar cooling incompatible with the absence of ice caps: see Barron, E.J. (1984) Climatic implication of the variable obliquity explanation of Cretaceous-Paleogene high-latitude floras. *Geology*, **12**, 595–8. Variable obliquity as an explanation for Cretaceous high-latitude forests simply has to be rejected.

49 Modern Siberian forests escape the same fate because winter temperatures fall to an icy −56°C which literally freezes their metabolism, preventing carbon losses by respiration. Climate information in northern Siberia is given in Schulze, E.-D., Vygodskaya, N.N., Tchebakova, N.M. *et al.* (2002) The Eurosiberian transect: an introduction to the experimental region. *Tellus*, **54B**, 421–8.

50 Hickey, Eternal summer at 80 degrees North (above, n.43).

51 Spicer, R.A. and Chapman, J.L. (1990) Climate change and the evolution of high-latitude terrestrial vegetation and floras. *Trends in Ecological Evolution*, **5**, 279–84.

52 Chaloner, W.G. and Creber, G.T. (1989) The phenomenon of forest growth in the Antarctic: a review. In *The origins and evolution of the Antarctic biota* (ed. J.A. Crame), pp. 85–8. Geological Society of London Special Publication, No. 47. The Geological Society, London.

53 Mooney, H.A. and Brayton, R. (1966) Field measurements of the metabolic responses of bristlecone pine and big sagebrush in the White Mountains. *Botanical Gazette*, **127**, 105–13.

54 Atkin, O.K. and Tjoelker, M.G. (2003) Thermal acclimation and the dynamic response of plant respiration to temperature. *Trends in Plant Science*, **8**, 343–51.

55 Read, J. and Francis, J. (1992) Responses of some Southern Hemisphere tree species to a prolonged dark period and their implications for high-latitude Cretaceous and Tertiary floras. *Palaeogeography, Palaeoclimatology, Palaeoecology*, **99**, 271–90.

56 Axelrod, D.I. (1984) An interpretation of Cretaceous and Tertiary biota in polar regions. *Palaeogeography, Palaeoclimatology, Palaeoecology*, **45**, 105–47.

57 Douglas, J.G. and Williams, G.E. (1982) Southern polar forests: the early Cretaceous floras of Victoria and their palaeoclimatic significance. *Palaeogeography, Palaeoclimatology, Palaeoecology*, **39**, 171–85; Wolfe, J.A. (1980) Tertiary climates and floristic relationships at high latitudes in the northern hemisphere. *Palaeogeography, Palaeoclimatology, Palaeoecology*, **30**, 313–23. To explain the fossil trees, the authors sought refuge in the obliquity debate (Mason, Evolution of certain floristic associations in western North America (above, n.48)), arguing that evergreens were credible evidence for a reduced obliquity, but they won few converts.

58 Spicer and Chapman, Climate change and the evolution of high-latitude terrestrial vegetation and floras (above, n.51).

59 Beerling, D.J. and Osborne, C.P. (2002) Physiological ecology of Mesozoic polar forests in a high CO_2 environment. *Annals of Botany*, **89**, 329–39.

60 The discovery of living dawn redwood trees in China is documented in Chaney, R.W. (1948) The bearing of the living *Metasequoia* on problems of Tertiary paleobotany. *Proceedings of the National Academy of Sciences, USA*, **34**, 503–15.

61 The most recent addition to the list of living fossils is that of Wollemi Pine (*Wollemia nobilis*). Known originally from fossils of Eocene-aged rocks, a small population was found growing in Wollemi National Park, north-west of Sydney in 1994: see Offord, C.A., Porter, C.L., Meagher, P.F., and Errington, G. (1999) Sexual

reproduction and early plant growth of the Wollemi Pine (*Wollemia nobilis*), a rare threatened Australian conifer. *Annals of Botany*, **84**, 1–9.

62 Tralau, H. (1968) Evolutionary trends in the genus *Ginkgo*. *Lethaia*, **1**, 63–101.

63 The following key papers debunking the myth of the deciduous view detail results from experiments and computer models tracking the effects of Cretaceous polar climates on the carbon balance of evergreen and deciduous trees: Royer, D.L., Osborne, C.P., and Beerling, D.J. (2003) Carbon loss by deciduous trees in a CO_2 rich ancient polar environment. *Nature*, **424**, 60–2; Royer, D.L., Osborne, C.P., and Beerling, D.J. (2005) Contrasting seasonal patterns of carbon gain in evergreen and deciduous trees of ancient polar forests. *Paleobiology*, **31**, 141–50.

64 A mathematical recipe for growing virtual forests using information on leaf lifespan, climate, and soils is given in Osborne, C.P. and Beerling, D.J. (2002) A process-based model of conifer forest structure and function with special emphasis on leaf lifespan. *Global Biogeochemical Cycles*, **16**, doi:10.1020/2001GB001467.

65 See the references cited in note 63.

66 For the issue of genetic drift to be significant, the carbon content of leaves would have to be 10 times less than now or respiration rates 10 times higher. Neither is likely.

67 Falcon-Lang, H.J. and Cantrill, D.J. (2001) Leaf phenology of some mid-Cretaceous polar forests, Alexander Island, Antarctica. *Geological Magazine*, **138**, 39–52; Parrish, J.T., Daniel, I.L., Kennedy, E.M., and Spicer, R.A. (1998) Palaeoclimatic significance of mid-Cretaceous floras from the middle Clarence Valley, New Zealand. *Palaios*, **13**, 149–54.

68 Differences in the photosynthetic strategies of the evergreen and deciduous trees and the underlying physiological mechanisms are reported in Osborne, C.P. and Beerling, D.J. (2003) The penalty of a long, hot summer: photosynthetic acclimation to high CO_2 and continuous light in 'living fossil' conifers. *Plant Physiology*, **133**, 803–12.

69 See the references cited in note 63.

70 Explaining the geographic patterns of evergreen and deciduous plants has long been a preoccupation of generations of ecologically minded scientists, the results of which have spawned a vast litera-

ture. Recent ideas concerned with explaining some of the oddities, and review of the extensive literature, are presented in Givnish, T.J. (2002) Adaptive significance of evergreen vs. deciduous leaves: solving the triple paradox. *Silva Fennica*, **36**, 703–43.

71 Betts, R.A. (2000) Offset of the potential carbon sink from boreal forestation by decreases in surface albedo. *Nature*, **408**, 187–90; Harding, R., Kuhry, P., Christensen, T. R. *et al.* (2002) Climatic feedbacks at the tundra-taiga interface. *Ambio*, Special Report **12**, 47–55.

72 See, for example, Sukachev, V.N. (1934) *Dendrology with basics of forest botany*. Goslesbumizdat, Moscow [in Russian], and Utkin, I.A. (1965) *Forests of Central Yakutia*. Nauka, Moscow [in Russian].

73 Oquist, G. and Huner, P.A. (2003) Photosynthesis of overwintering evergreen plants. *Annual Review of Plant Biology*, **54**, 329–55.

74 See Givnish, Adaptive significance of evergreen vs. deciduous leaves (above, n.70).

75 Aerts, R. (1995) The advantages of being evergreen. *Trends in Ecology and Evolution*, **10**, 402–7.

76 Falcon-Lang, H.J. (2000) A method to distinguish between woods produced by evergreen and deciduous coniferopsids on the basis of growth ring anatomy: a new palaeoecological tool. *Palaeontology*, **43**, 785–93; Falcon-Lang, H.J. (2000) The relationship between leaf longevity and growth ring markedness in modern conifer woods and its implications for palaeoclimatic studies. *Palaeogeography, Palaeoclimatology, Palaeoecology*, **160**, 317–28.

77 Brentnall, S.B., Beerling, D.J., Osborne, C.P. *et al.* (2005) Climatic and ecological determinants of leaf lifespan in polar forests of the high CO_2 Cretaceous 'greenhouse' world. *Global Change Biology*, **11**, 2177–95.

78 Groffmann, P.M., Driscoll, C.T., Fahey, T.J. *et al.* (2001) Colder soils in a warmer world: a snow manipulation study in a northern hardwood forest. *Biogeochemistry*, **56**, 135–60.

79 An accessible account of recent climate changes in the Arctic is given by Sturm, M., Perovich, D.K., and Serreze, M.C. (2003) Meltdown in the North. *Scientific American*, October, 60–7.

80 The northwards advance of trees and woody shrubs in northern Alaska has been taking place over the last three decades. See Sturm, M., Racine, C., and Tape, K. (2001) Increasing shrub abundance in

the Arctic. *Nature*, **411**, 546–7; Jia, G.J., Epstein, H.E., and Walkter, D.A. (2003) Greening of the Arctic Alaska: 1981–2001. *Geophysical Research Letters*, **30**, doi: 10.1029/2003GL018268; Hinzman, L.D., Bettez, N.D., Bolton, W.R. *et al.* (2005) Evidence and implications of recent climate change in Northern Alaska and other arctic regions. *Climatic Change*, **75**, 251–98. Its effects on land surface energy exchange are described in Sturm, M., Douglas, T., Racine, C., and Liston, G.E. (2005) Changing snow and shrub conditions affect albedo with global implications. *Journal of Geophysical Research*, **110**, doi: 10.1029/2005JG000013; Tape, K., Sturm, M., and Racine, C. (2006) The evidence for shrub expansion in Northern Alaska and the Pan-Arctic. *Global Change Biology*, **12**, 686–702.

81 Chapin, F.S., Sturm, M., Serreze, M.C. *et al.* (2005) Role of land-surface changes in Arctic summer warming. *Science*, **310**, 657–60. See also the commentary by Foley, J.A. (2005) Tipping points in the tundra. *Science*, **310**, 627–8.

CHAPTER 7

Paradise lost

1 Clouter, F., Mitchell, T., Rayner, D., and Rayner, M. (2000) *London Clay fossils of the Isle of Sheppey*. Medway Lapidary and Mineral Society, Northfleet.

2 Ibid.

3 A reprint of Bowerbank's 1840 letter, published in the March edition of the *Magazine of Natural History*, is bound into the front of the 1877 edition of Bowerbank's *A history of the fossil fruits and seeds of the London Clay*.

4 A brief history of the London Clay flora is provided in Collinson, M. E. (1983) *Fossil plants of the London Clay*. Palaeontological Association Field Guides to Fossils No. 1, London. The zoological equivalent of Bowerbank's work, documenting fossil carapaces and skulls of warm-water turtles, fish, and crocodiles, was published by Owen, R. and Bell, T. (1849) *Monograph of the fossil Reptilia of the London Clay. Part 1. Chelonia*. Palaeontological Society, London.

5 Reid, E.M. and Chandler, M.E.J. (1933) *The London Clay Flora*. British Natural History Museum, London. Even so, some doubted the fossil

floras really indicated subtropical climates in southern England: see Barghoorn, E.S. (1953) Evidence of climatic change in the geologic record of plant life. In *Climate Change* (ed. H. Shapley), pp. 235–48. Harvard University Press, Cambridge, MA; Van Steenis, C.G.G.J. (1962) The land-bridge theory in botany with particular reference to tropical plants. *Blumea*, **11**, 235–372; Daley, B. (1972) Some problems concerning the early Tertiary climate of southern Britain. *Palaeogeography, Palaeoclimatology, Palaeoecology*, **11**, 177–90.

6 Palaeoclimate reconstructions for the Eocene world from terrestrial fossil floras are reported by Greenwood, D.R. and Wing, S.L. (1995) Eocene continental climates and latitudinal temperature gradients. *Geology*, **23**, 1044–8. Debate over the statistical validity of the approach used by Greenwood and Wing followed, but the authors provided a robust defence. See Jordon, G.J. (1996) Eocene continental climates and latitudinal temperature gradients: comment. *Geology*, **23**, 1054, and Wing, S.L. and Greenwood, (1996) Reply. *Geology*, **23**, 1054–5. Later, the mild continental winter temperatures were shown to be consistent with an independent oxygen isotope-based approach to reconstructing palaeoclimates: see Fricke, H.C. and Wing, S.L. (2005) Oxygen isotope and palaeobotanical estimates of temperature and ^{18}O-latitude gradients over North America during the early Eocene. *American Journal of Science*, **304**, 612–35.

7 Greenwood and Wing, Eocene continental climates and latitudinal temperature gradients (above, n.6). Palms are vulnerable to freezing conditions because they possess a single growing tip which, if frosted, can kill them outright. This vulnerability compresses their modern-day distribution to a narrow latitudinal band within 10° of the equator. Fossils are therefore often taken to be indicative of mild, frost-free winters. However, palms are not all equally frost susceptible, as residents of Los Angeles learned in 1990 when an unusual deep freeze gripped the city for a week. The freeze wrought havoc on the tall elegant rows of Mexican fan palms (*Washingtonia robusta*) lining the streets, but left the compact and more cold-hardy Californian cousin (*Washingtonia filifera*), with its thicker, insulating trunks, unscathed.

8 Estes, R. and Hutchinson, J.H. (1980) Eocene lower vertebrates from Ellesmere Island, Canadian Arctic Archipelago. *Palaeogeography, Palaeoclimatology, Palaeoecology*, **30**, 325–47.

9 Pole, M.S. and Macphail, M.K. (1996) Eocene *Nypa* from Regatta Point, Tasmania. *Review of Palaeobotany and Palynology*, **92**, 55–67.

10 Reguero, M.A., Marenssi, S.A., and Santilla, S.N. (2002) Antarctic Peninsula and South America (Patagonia) Paleogene terrestrial faunas and environments: biogeographic relationships. *Palaeogeography, Palaeoclimatology, Palaeoecology*, **179**, 189–210.

11 Past ocean conditions are reconstructed from the oxygen isotope composition of the foraminifera shells and the ratio of the elements magnesium and calcium in the shells. Difficulties of interpretation with each approach notwithstanding, both reveal a consistent picture of climate change over the past 65 million years. Seminal papers reporting Earth's climate history from variations in the oxygen isotope composition of fossil foraminifera shells are: Shackleton, N.J. and Kennet, J.P. (1975) Palaeotemperature history of the Cenozoic and the initiation of Antarctic glaciation: oxygen and carbon isotope analyses in DSDP Sites 277, 279 and 281. *Initial Reports of the Deep Sea Drilling Project*, **29**, 743–55; Miller, K.G., Fairbanks, R.G., and Mountain, G.S. (1987) Tertiary oxygen isotope synthesis, sea level history, and continental margin erosion *Palaeoceanography*, **2**, 1–19; and Zachos, J.C., Pagani, M., Sloan, L. *et al.* (2001) Trends, rhythms, and aberrations in global climate 65Ma to present. *Science*, **292**, 686–93. Those reporting Earth's climate history from the magnesium/calcium ratios of fossil foraminifera shells are: Elderfield, H. and Ganssen, G. (2000) Past temperature and ^{18}O of surface ocean waters inferred from foraminiferal Mg/Ca ratios. *Nature*, **405**, 442–5, and Lear, C.H., Elderfield, H., and Wilson, P.A. (2000) Cenozoic deep-sea temperatures and global ice volumes from Mg/Ca in benthic foraminiferal calcite. *Science*, **287**, 269–72. For commentary on the Lear *et al.* paper, see Dwyer, G.S. (2000) Unravelling the signals of global climate change. *Science*, **287**, 246–7.

12 Part of the Eocene climate puzzle throughout much of the 1980s was that the tropical ocean temperatures were thought to be cooler than at present. How was it possible to have warm polar climates and cooler tropics? The problem became known as the 'cool tropics paradox'. It later transpired that there were issues with the reliability of some of the reconstructions of tropical ocean temperatures. We now believe tropical surface oceans were as warm, or slightly

warmer, than now in the Eocene. The problem is set out in Barron, E.J. (1987) Eocene equator-to-pole surface ocean temperatures: a significant climate problem. *Paleoceanography*, **2**, 729–39. See also Pearson, P.N., Ditchfield, P.W., Singano, J. *et al.* (2001) Warm tropical sea surface temperatures in the Late Cretaceous and Eocene epochs. *Nature*, **413**, 481–7.

13 The problem of explaining the Eocene climate is discussed by Sloan, L.C. and Barron, E.J. (1990) 'Equable' climates during Earth history? *Geology*, **18**, 489–92, and Sloan, L.C. and Barron, E.J. (1992) A comparison of Eocene climate model results to quantified paleoclimate interpretations. *Palaeogeography, Palaeoclimatology, Palaeoecology*, **93**, 183–202.

14 A succession of studies addressed the role of the oceans, all to no avail. An admirable review of all this is given in Sloan, L.C., Walker, J.C.G., and Moore, T.C. (1995) Possible role of oceanic heat transport in early Eocene climate. *Paleoceanography*, **10**, 347–56. More recently it has been argued that shifts in the configuration of the continents may also have altered the circulation of the oceans. Up until the Eocene, a westward-flowing ocean current spanned the circumference of the entire globe. Afterwards the gaps that separated Asia from Africa, and North America from South America gradually closed up, stopping the trans-global ocean gateway. Simulations with a simplified computer model suggest that a circum-global ocean circulation warming the high latitudes by pumping more heat up from the tropics makes a big difference to climate; closing it off shuts the heat pump down. However, this is a contentious finding not yet reproduced in more complex models; see Hotinski, R.M. and Toggweiler, J.R. (2003) Impact of a circumglobal passage on ocean heat transport and 'equable' climates. *Paleoceanography*, **18**, doi: 10.1029/2001PA000730.

15 Huber, M. and Caballero, R. (2003) Eocene El Niño: evidence for robust tropical dynamics in the 'hothouse'. *Science*, **299**, 877–81.

16 El Niño was originally recognized by fishermen off the coast of South America as the appearance of unusually warm water in the Pacific Ocean, occurring near the beginning of the year. El Niño is Spanish for The Little Boy or Christ child, indicating the tendency of the phenomenon to arrive around Christmas.

17 Trenberth, K.E., Caron, J.M., Stepaniak, D.P., and Worley, S. (2002) Evolution of El Niño-Southern Oscillation and global atmospheric surface temperatures. *Journal of Geophysical Research*, **107**, doi: 10.1029/2000JD000298.

18 Sun, D.Z. and Trenberth, K.E. (1998) Coordinated heat removal from the equatorial Pacific during the 1986–87 El Niño. *Geophysical Research Letters*, **25**, 2659–62.

19 Cane, M.A. (1998) A role for the tropical Pacific. *Science*, **282**, 59–61.

20 Wyoming records are reported by Ripepe, M., Roberts, L.T., and Fischer, A.G. (1991) ENSO and sunspot cycles in varved Eocene oil shales from image analysis. *Journal of Sedimentary Petrology*, **61**, 1155–63. German records are reported by Mingram, J. (1998) Laminated Eocene maar-lake sediments from Eckfield (Eifel region, Germany) and their short-term periodicities. *Palaeogeography, Palaeoclimatology, Palaeoecology*, **140**, 289–305.

21 Royer, D.L., Wing, S.C., Beerling, D.J. *et al.* (2001) Paleobotanical evidence for near present-day levels of atmospheric CO_2 during part of the Tertiary. *Science*, **292**, 2310–13. For further details, and possible problems with this work, see note 53, Chapter 5.

22 Estimates of the carbon dioxide content of the ancient atmosphere using the boron isotope composition of fossil foraminifera shells are reported by Pearson, P.N. and Palmer, M.R. (1999) Middle Eocene seawater pH and atmospheric carbon dioxide concentrations. *Science*, **284**, 1824–6, and Pearson, P.N. and Palmer, M.R. (2000) Atmospheric carbon dioxide concentrations over the past 60 million years. *Nature*, **406**, 695–9. The first paper came under fire: see Caldeira, K. and Berner, R.A. (1999) Seawater pH and atmospheric carbon dioxide. *Science*, **296**, 2043a, and the authors responded: see *Science*, **296**, 2043a–2043b. The second received criticism from a number of quarters. Some raised questions over the importance of other influences on the global boron isotope budget (see Lemarchand, D., Gaillardet, J., Lewin, É., and Allègre, C.J. (2000) The influence of rivers on marine boron isotopes and implications for reconstructing past ocean pH. *Nature*, **408**, 951–4; Pagani, M., Lemarchand, D., Spivack, A., and Gaillardet, J. (2005) A critical evaluation of the boron isotope-pH proxy: the accuracy of ocean pH estimates. *Geochimica et Cosmochimica Acta*, **69**, 953–61). Others showed that changes in seawater chemistry over the

past 60 million years strongly alter the reconstructions of atmospheric carbon dioxide: see Demicco, R.V., Lowenstien, T.K., and Hardie, L.A. (2003) Atmospheric pCO_2 since 60 Ma from records of seawater pH, calcium and primary carbonate mineralogy. *Geology*, **31**, 793–6.

23 Pagani, M., Zachos, J.C., Freeman, K.H. *et al.* (2005) Marked decline in atmospheric carbon dioxide concentrations during the Paleogene. *Science*, **309**, 600–3.

24 Sloan, L.C. and Rea, D.K. (1995) Atmospheric carbon dioxide and early Eocene climate: a general circulation modeling sensitivity study. *Palaeogeography, Palaeoclimatology, Palaeoecology*, **119**, 275–92.

25 Eocene tropical surface ocean temperatures were similar to, or only slightly warmer than, those of today; see Pearson *et al.*, Warm tropical sea surface temperatures in the Late Cretaceous and Eocene epochs (above, n.12). For a commentary on this work, see Kump, L. (2001) Chill taken out of the tropics. *Nature*, **413**, 470–1. Climate model simulations for the Eocene climate system response to carbon dioxide are reported by Shellito, C.J., Sloan, L.C., and Huber, M. (2003) Climate model sensitivity to atmospheric CO_2 levels in the early-middle Paleogene. *Palaeogeography, Palaeoclimatology, Palaeoecology*, **193**, 113–23.

26 Govan, F. (2003) Film stars ease their consciences with trees. *Sunday Telegraph*, 14 September 2003.

27 In November 2006, we drew attention to the importance of greenhouse gases other than carbon dioxide in the global change debate by organizing a discussion meeting at the Royal Society, London. The findings of that meeting are published in Beerling, D.J., Hewitt C.N., Pyle, J.A., and Raven, J.A. (2007) Critical issues in trace gas biogeochemistry and global change. *Philosophical Transactions of the Royal Society A* **365**, 1629–1954.

28 An Olympic-size swimming pool is 50 m long, 25 m wide, and has a uniform depth of 2 m, with a total volume of 2500 cubic metres. Given that 1 cubic metre is equal to 1000 litres, an Olympic swimming pool holds 2.5 million litres.

29 John Tyndall, Journal 8a, Wednesday, 18 May 1859. Tyndall Collection, Royal Institution of Great Britain, London.

30 A detailed and still highly readable account by John Tyndall of his experiments with gases is given in Tyndall, J. (1865) *Heat considered as*

a mode of motion. Second edn, with additions and illustrations. Longman Green, London.

31 The difference in the strength of the greenhouse effect between the gases arises because, for trace gases like methane, there is so much less of them in the atmosphere than carbon dioxide. The impact of an extra molecule of any greenhouse gas on the Earth's energy balance depends on its absorbing wavelengths, and the presence of other absorbers at those wavelengths. For carbon dioxide, which absorbs energy at a relatively cluttered part of the electromagnetic spectrum, the addition of each extra molecule leads to a slow (i.e. logarithmic) rise in its climatic importance. For other trace gases, like methane and nitrous oxide, which absorb energy in a less cluttered part of the electromagnetic spectrum, the climatic importance rises as their concentration increases in a more direct fashion than for carbon dioxide.

32 *Professor Tyndall's Lectures on Heat*. (1862) Lecture 12. 10 April. Tyndall Collection, Royal Institution of Great Britain, London.

33 Tyndall, *Heat considered as a mode of motion* (above, n.30).

34 Ibid.

35 Petit, J.R., Jouzel, J., Raynaud, D. *et al.* (1999) Climate and atmospheric history of the past 420,000 years from the Vostok ice core, Antarctica. *Nature*, **399**, 429–36.

36 See EPICA Community Members (2004) Eight glacial cycles from an Antarctic ice core. *Nature*, **429**, 623–8. For commentaries, see: McManus, J.F. (2004) A great grand-daddy of ice cores. *Nature*, **429**, 611–12, and Walker, G. (2004) Frozen time. *Nature*, **429**, 596–7. The long-term picture of climate and greenhouse gas variations provided by this exceptional core is reported in the following two papers: Siegenthaler, U., Stocker, T.F., Monnin, E. *et al.* (2005) Stable carbon cycle-climate relationship during the late Pleistocene. *Science*, **310**, 1313–17; Saphni, R., Chappellaz, J., Stocker, T.F. *et al.* (2005) Atmospheric methane and nitrous oxide of the late Pleistocene from Antarctic ice cores. *Science*, **310**, 1317–21.

37 Severinghaus, J.P. and Brook, E.J. (1999) Abrupt climate change at the end of the last glacial period inferred from trapped air in the polar ice. *Science*, **286**, 930–4.

38 Shackleton, N.J. (2000) The 100,000 yr ice-age cycle identified and found to lag temperature, carbon dioxide and orbital eccentricity. *Science*, **289**, 1897–902. See also the commentary: Kerr, R.A. (2000) Ice, mud point to CO_2 role in glacial cycle. *Science*, **289**, 1868.

39 In the natural wetlands of Scandinavia and elsewhere, a 2°C rise in summer temperatures can increase by fifty per cent the amount of methane emitted; see Christensen, T.R., Ekberg, A., and Strom, L. (2003) Factors controlling large scale variations in methane emissions from wetlands. *Geophysical Research Letters*, **30**, doi: 10.1029/2002GL016848.

40 Dacey, J.W.H., Drake, B.G., and Klug, M.J. (1994) Stimulation of methane emission by carbon dioxide enrichment of marsh vegetation. *Nature*, **370**, 47–9. Later reports on a range of wetlands with herbaceous and woody plants also show dramatic stimulation of methane emission when exposed to elevated carbon dioxide. For a review, see Vann, C.D. and Megonigal, J.P. (2003) Elevated CO_2 and water depth regulation of methane emissions: a comparison of woody and non-woody wetland plant species. *Biogeochemistry*, **63**, 117–34. However, conflicting results have been reported for oligotrophic (nutrient-poor) peatlands in Europe: see Silvola, J., Saarnio, S., Foot, J. *et al.* (2003) Effects of elevated CO_2 and N deposition on CH_4 emissions from European mires. *Global Biogeochemical Cycles*, **17**, doi: 10.1029/2002GB001886.

41 Lelieveld, J., Crutzen, P.J., and Dentener, F.J. (1998) Changing concentration, lifetime and climate forcing of atmospheric methane. *Tellus*, **50B**, 128–50.

42 Flückiger, J., Blunier, T., Stauffer, B. *et al.* (2004) NO and CH_4 variations during the last glacial epoch: insight into global processes. *Global Biogeochemical Cycles*, **18**; doi: 10.1029/2003GB002122.

43 For an overview of the marine nitrous oxide cycle, see Suntharalingham, P. and Sarmiento, J.L. (2000) Factors governing the oceanic nitrous oxide distribution: simulations with a general circulation model. *Global Biogeochemical Cycles*, **14**, 429–54.

44 Sloan and colleagues point the finger at a methane greenhouse during the early Eocene: Sloan, L.C., Walker, J.C.G., and Moore, T. C. (1992) Possible methane-induced polar warming in the early Eocene. *Nature*, **357**, 320–32.

45 Frankenberg, C., Meirink, J.F., van Weele M. *et al.* (2005) Assessing methane emissions from global space-borne observations. *Science*, **308**, 1010–14.

46 Sloan *et al.*, Possible methane-induced polar warming in the early Eocene (above, n.44).

47 The basic approach was first successfully evaluated by simulating the ice-age methane budget reported from analyses of air bubbles trapped in ice cores; see Valdes, P.J., Beerling, D.J., and Johnson, C.E. (2005) The ice-age methane budget. *Geophysical Research Letters*, **32**, doi: 10.1029/2004GL021004. We reported early results for the Eocene simulations at the annual fall meetings of the American Geophysical Union in San Francisco in 2002 and 2003: see Beerling, D.J. and Valdes, P.J. (2002) Feedback of atmospheric chemistry, via CH_4, on the Eocene climate. *EOS Transactions, AGU*, **83** (47), Fall Meet. Suppl., PP12B-01. American Geophysical Union, Washington; Beerling, D.J. and Valdes, P.J. (2003) Global warming in the early Eocene: was it driven by carbon dioxide? *EOS Transactions, AGU*, **84** (46), Fall Meet. Suppl., PP22B-04 (Invited). American Geophysical Union, Washington.

48 Ibid.

49 Haagen-Smit, A.J. (1952) Chemistry and the physiology of Los Angeles smog. *Industrial & Engineering Chemistry Research*, **44**, 1342–6. See also Haagen-Smit, A.J. (1970) A lesson from the smog capital of the world. *Proceedings of the National Academy of Sciences, USA*, **67**, 887–97.

50 Current thinking is set out in Lelieveld, J. and Dentner, F.J. (2000) What controls tropospheric ozone? *Journal of Geophysical Research*, **105**, 3531–51.

51 The 2003 blackout in North America offered the authors a serendipitous opportunity to make atmospheric chemistry measurements from a light aircraft, investigating the effects of a reduction in emissions of nitrogen oxides. Marufu, L.T., Taubman, B.F., Bloomer, B. *et al.* (2004) The 2003 North American electrical blackout: an accidental experiment in atmospheric chemistry. *Geophysical Research Letters*, **31**, doi: 10.1029/2004GL019771.

52 Guenther, A., Hewitt, C., Erikson, D. *et al.* (1995) A global model of natural volatile organic compound emissions. *Journal of Geophysical Research*, **100**, 8873–92.

53 Dreyfus, G.B., Schade, G.W., and Goldstein, A.H. (2002) Observational constraints on the contribution of isoprene oxidation to ozone production on the western slope of the Sierra Nevada, California. *Journal of Geophysical Research*, **107**, doi: 10.1029/2001JD001490.

54 Quoted in *Sierra* magazine, 10 September 1980. An important early report of the role of natural hydrocarbons in ozone formation in urban areas is Chameides, W.L., Lindsay, R.W., Richardson, J., and Kiang, C.S. (1988) The role of hydrocarbons in urban photochemical smog: Atlanta as a case study. *Science*, **241**, 1473–5.

55 See Went, W. (1960) Blue hazes in the atmosphere. *Nature*, **187**, 641–3. Further discussion is also given in Hayden, B.P. (1998) Ecosystem feedbacks on climate at the landscape scale. *Philosophical Transactions of the Royal Society*, **B353**, 5–18.

56 Wayne, R.P. (2000) *The chemistry of atmospheres*. Third edn. Oxford University Press.

57 Dunn, D.B. (1959) Some effects of air pollution on *Lupinus* in the Los Angeles area. *Ecology*, **40**, 621; Miller, P.R., Parmeter, J.R., Taylor, O.C., and Cardiff, E.A. (1963) Ozone injury to foliage of *Pinus ponderosa*. *Phytopathology*, **53**, 1072–6; Miller, P.R., McCutcheon, M.H., and Milligan, H.P. (1972) Oxidant air pollution in the Central Valley, Sierra Nevada foothills, and Mineral King Valley of California. *Atmospheric Environment*, **6**, 623–33. For a recent review of the thousands of papers published since these pioneering contributions, see Ashmore, M.R. (2005) Assessing the future global impacts of ozone on vegetation. *Plant, Cell and Environment*, **28**, 949–64.

58 Broadmeadow, M.S.J., Heath, J., and Randle, T.J. (1999) Environmental limitations to O_3 uptake: some key results from young trees growing at elevated CO_2 concentrations. *Water Air and Soil Pollution*, **116**, 299–310.

59 A summary of the experimental findings from this ongoing investigation is presented in Karnosky, D.F., Pregitzer, K.S., Zak, D.R. *et al.* (2005) Scaling ozone responses of forest trees to the ecosystem level in a changing climate. *Plant, Cell and Environment*, **28**, 965–81. Alternative ways of looking at carbon dioxide–ozone interactions are also being pursued. See, for example, Hanson, P.J., Wullschleger, S.D., Norby, R.J. *et al.* (2005) Importance of changing CO_2,

temperature, precipitation, and ozone on carbon and water cycles of an upland-oak forest: incorporating experimental results into model simulations. *Global Change Biology*, **11**, 1402–23.

60 Hansen, J. (2004) Defusing the global warming time bomb. *Scientific American*, March, 68–77.

61 Discussed in Fleming, J.R. (1998) *Historical perspectives on climate change.* Oxford University Press, New York.

62 The climatic effects of optically thick polar stratospheric clouds are reported by Sloan, L.C. and Pollard, D. (1998) Polar stratospheric clouds: a high latitude warming mechanism in an ancient greenhouse world. *Geophysical Research Letters*, **25**, 3517–20.

63 Another related idea is that the clouds form due to a change in the circulation of the stratosphere that may take place in a high carbon dioxide world: see Kirk-Davidoff, D.B., Schrag, D.P., and Anderson, J.G. (2002) On the feedback of stratospheric clouds on polar climate. *Geophysical Research Letters*, **29**, doi: 10.1029/2002GL014659.

64 Tripati, A., Backman, J., Elderfield, H., and Ferretti, P. (2005) Eocene bipolar glaciation associated with global carbon cycle changes. *Nature*, **436**, 341–6. See also the commentary: Kump, L.R. (2005) Foreshadowing the glacial era. *Nature*, **436**, 333–4.

65 Pagani *et al.*, Marked decline in atmospheric carbon dioxide concentrations during the Paleogene (above, n.23).

66 Suggested in Tripati *et al.*, Eocene bipolar glaciation associated with global carbon cycle changes (above, n.64), based on the large increase in the slope of the seawater strontium isotope curve around 40 million years ago reported in McArthur, J.M., Howarth, R.J., and Bailey, T.R. (2001) Strontium isotope stratigraphy; LOWESS Version 3; best fit to the marine Sr-isotope curve for 0–509 Ma and accompanying look-up table for deriving numerical age. *Journal of Geology*, **109**, 155–70.

67 Moran, K., Backman, J., Brinkhuis, H. *et al.* (2006) The Cenozoic palaeoenvironment of the Arctic Ocean. *Nature*, **441**, 601–5. See also the accompanying commentary article, Stoll, H.M. (2006) The Arctic tells its story. *Nature*, **441**, 579–81.

68 Friedli, H., Lötscher, H., Oeschger, H. *et al.* (1987) Ice core record of the $^{13}C/^{12}C$ ratio of atmospheric CO_2 in the past two centuries. *Nature*, **324**, 237–8; Etheridge, D.M., Pearman, G.I., and Fraser, P.J.

(1992) Changing tropospheric methane between 1841 and 1978 from a high accumulation-rate Antarctic ice core. *Tellus*, **44B**, 282–94; Machida, T., Nakazawa, T., Fujii, Y. *et al.* (1995) Increase in the atmospheric nitrous oxide concentration during the past 250 years. *Geophysical Research Letters*, **22**, 2921–4.

69 Albert Lévy (1878) *Annuaire de l'Observ. de Montsouris*, pp. 495–505. Gauthier-Villars, Paris.

70 Volz, A. and Kley, D. (1988) Evaluation of the Montsouris series of ozone measurements made in the nineteenth century. *Nature*, **332**, 240–2.

71 See Mickley, L.J., Jacob, D.J., Field, B.D., and Rind, D. (2004) Climate response to the increase in tropospheric ozone since preindustrial times: a comparison between ozone and equivalent CO_2 forcings. *Journal of Geophysical Research*, **109**, doi: 10.1029/2003JD003653. Note, though, that atmospheric chemistry models have a hard time reproducing the low tropospheric ozone values of the preindustrial atmosphere reported by Volz and Kley (Evaluation of the Montsouris series of ozone measurements made in the nineteenth century (above, n.70)) and in general overestimate it, with the result that radiative forcing and climatic warming may be underestimated. See Mickley, L.J., Jacob, D.J., and Rind, D. (2001) Uncertainty in preindustrial abundance of tropospheric ozone: implications for radiative forcing calculations. *Journal of Geophysical Research*, **106**, 3389–99.

72 Hansen, J., Nazarenko, L., Ruedy, R. *et al.* (2005) Earth's energy imbalance: confirmation and implications. *Science*, **308**, 1431–5.

73 Hansen, J. and Sato, M. (2004) Greenhouse gas growth rates. *Proceedings of the National Academy of Sciences, USA*, **101**, 16109–14; Hansen, J., Sato, M., Ruedy, R., and Oinas, V. (2000) Global warming in the twenty first century: an alternative scenario. *Proceedings of the National Academy of Sciences, USA*, **97**, 9875–80.

74 Hansen, Defusing the global warming time bomb (above, n.60).

75 Hansen and Sato, Greenhouse gas growth rates; Hansen *et al.*, Global warming in the twenty first century (above, n.73).

76 Long, S.P., Ainsworth, E.A., Leakey, A.D.B., and Morgan, P.B. (2005) Global food security: treatment of major food crops with elevated carbon dioxide or ozone under large-scale fully open-air conditions

suggests recent models may have overestimated future yields. *Philosophical Transactions of the Royal Society*, **B360**, 2011–20; Long, S.P., Ainsworth, E.A., Leakey, A.D.B. *et al.* (2006) Food for thought: lower-than-expected crop yield stimulation with rising CO_2 concentrations. *Science*, **312**, 1918–21.

77 Houghton, J.T., Ding, Y., Griggs, D.J. *et al.* (2001) *Climate change 2001: the scientific basis*. Intergovernmental Panel on Climate Change. Cambridge University Press.

CHAPTER 8
Nature's green revolution

1 The circumstances surrounding Bacon's death have been revisited by Jardine, L. and Stuart, A. (1998) *Hostage to fortune: the troubled life of Francis Bacon 1561–1626*. Phoenix, London.

2 Butterfield, H. (1965) *The origins of modern science*. Revised edn. Free Press, New York.

3 See Shapin, S. (2001) *The scientific revolution*. Second edn. University of Chicago Press, for a revisionist version of the scientific revolution. A more orthodox treatment is given in Henry, J. (1997) *The scientific revolution and the origins of modern science*. Palgrave, New York.

4 Jardine, L. (1999) *Ingenious pursuits: building the scientific revolution*. Abacus, London.

5 A contemporary account of Van Helmont's experiments is given by Boyle, R. (1661) *The sceptical chemist: or chymico-physical doubts and paradoxes*. Creek, London. Boyle valued observation and experiment at least as much as logical thinking in formulating accurate scientific understanding, and Van Helmont's approach fitted with this viewpoint.

6 Theme Statement; 12th International Congress on Photosynthesis, Brisbane, 2001.

7 For a brief review of the progress made by a succession of early thinkers, from Hales onwards, that also includes many more recent developments, see Govindjee and Gest, H. (2002) Celebrating the millennium: historical highlights of photosynthesis. *Photosynthesis Research*, **73**, 1–6.

8 The name cyclotron stems from the fact that the machine guides charged particles with a magnetic force along spiral paths, accelerating

them across a gap between two large semicircular magnets with an applied electrical field. The British scientists John Cockcroft and Ernest Walton invented a linear version which was used to win the race to split the atomic nucleus.

9 Ruben was employed as an instructor in the Department of Chemistry, and Kamen as a research fellow in the Radiation Laboratory headed by Lawrence. Both men initially worked together investigating the mysterious results of James Cork, a physicist from the University of Michigan visiting Lawrence in Berkeley, found to be in error by the great physicist Niels Bohr (1885–1962). Resolving where the technical problem in Cork's experiments lay was an arduous task for Kamen and Ruben, involving many continuous 18-hour stints in the lab. The experience drew them into a strong friendship and they forged an enduring partnership to exploit the Radiation Laboratory's cyclotron and the facilities in the Chemistry Department.

10 The half-life is the amount of time it takes for half of the atoms in a sample to decay. The half-life for a given isotope is always the same, and does not depend on how many atoms you have or on how long they've been sitting around.

11 The announcement of the discovery of radioactive carbon (^{14}C) was made in two papers. The first, in 1940, was a short note intended to establish priority to the discovery, the second, in 1941, is a fuller account fleshing out the nuclear physical details. Ruben, S. and Kamen, M.D. (1940) Radioactive carbon of the long half-life. *Physical Review*, **57**, 549; Ruben, S. and Kamen, M.D. (1941) Long-lived radio carbon: C^{14}. *Physical Review*, **59**, 349–54. A retrospective account of the discovery is given in Kamen, M.D. (1963) Early history of carbon-14. *Science*, **140**, 584–90. Radioactive ^{14}C decays to nitrogen-14 (^{14}N) with a half-life of 5730 years and is formed in nature by the bombardment of ^{14}N with cosmic rays. It is present in very small amounts in the atmosphere and in living organisms. When an organism dies, equilibration with the atmosphere ceases and its ^{14}C content begins to decay. This fact, and its slow decay time, makes ^{14}C useful for dating prehistoric fossilized plant and animal remains.

12 A summary of the progress Ruben made during those earliest experiments with the very short-lived isotope ^{11}C is given by Gest, H. (2004) Samuel Ruben's contribution to research on photosynthesis and bacterial metabolism with radioactive carbon. *Photosynthesis Research*, **80**, 77–83. Quotation from Kamen, M.D. (1985) *Radiant science, dark politics*. University of California Press, Berkeley.

13 A poignant account of Martin Kamen's life is given in his autobiography, *Radiant science, dark politics* (above, n.12). Further reflections can be found in Kamen, M.D. (1989) Onward into a fabulous half-century. *Photosynthesis Research*, **21**, 139–44.

14 Kamen, *Radiant science, dark politics* (above, n.12).

15 Historical accounts of this era of research in photosynthesis are given in the following papers. Calvin, M. (1989) Forty years of photosynthesis and related activities. *Photosynthesis Research*, **21**, 3–16; Calvin, M. (1992) *Following the trail of light: a scientific odyssey*. Oxford University Press, New York; Calvin, M. and Benson, A.A. (1948) The path of carbon in photosynthesis. *Science*, **107**, 476–80; Benson, A.A. (2002) Following the path of carbon in photosynthesis: a personal story. *Photosynthesis Research*, **73**, 29–49; Benson, A.A. (2002) Paving the path. *Annual Review of Plant Biology*, **53**, 1–25.

16 A comprehensive review of the history of photosynthesis research is given in Govinjee and Krogman, D. (2004) Discoveries of oxygenic photosynthesis (1727–2003): a perspective. *Photosynthesis Research*, **80**, 15–57.

17 The text of Nobel lectures given by laureates is available at: http://nobelprize.org/. For a complete listing of Nobel Prize winners associated with research in photosynthesis, see Govindjee and Krogman, D.W. (2002) A list of personal perspectives with selected quotations, along with lists of tributes, historical notes, Nobel and Kettering awards related to photosynthesis. *Photosynthesis Research*, **73**, 11–20. The drama of Cockcroft and Walton's triumph of ingenuity over adversity is admirably brought to life in Cathart, B. (2004) *The fly in the cathedral*. Penguin, London.

18 See Fuller, R.C. (1999) Forty years of microbial photosynthesis research: where it came from and where it led to. *Photosynthesis Research*, **62**, 1–29.

19 Seaborg, G.T. and Benson, A.A. (1998) Melvin Calvin. *Biographical Memoirs of the National Academy of Sciences, USA,* **76**, 3–21.

20 This point, and the strange fallout of Calvin and Benson, is documented in Fuller, Forty years of microbial photosynthesis research (above, n.18).

21 Calvin and Benson's team discovered the photosynthetic carbon reduction cycle, now named the Calvin–Benson cycle, the process by which the majority of photosynthetic organisms reduce carbon dioxide to form the three-carbon compound phosphoglycerate (or PGA for short) and, eventually, sugars such as hexose, fructose, glucose, and sucrose. This initial step in the Calvin–Benson cycle is catalysed by the enzyme Rubisco. See the following note for further details.

22 The name Rubisco stands for ribulose-1,5-*bis*phosphate carboxylase-oxygenase. As the name implies, the enzyme catalyses both carboxylase and oxygenase reactions. The carboxylase reaction is the chemical attachment of carbon dioxide to a five-carbon acceptor molecule, ribulose 1,5 *bis*phosphate (RuBP), to form a six-carbon intermediate compound that is cleaved into two three-carbon molecules of 3-phosphoglycerate. The oxygenase reaction has the opposite effect and occurs when oxygen instead of carbon dioxide combines with RuBP, denying the possibility of its carboxylation and leading instead to the formation of a two-carbon compound, phosphoglycolate and the eventual loss of carbon dioxide during the regeneration of the RuBP molecule. This is a process known as photorespiration.

23 Medawar, P.B. (1968) *The art of the possible.* Methuen, London.

24 The Hawaiian sugar plantation work was published in Kortschak, H.P., Hartt, C.E., and Burr, G.O. (1965) Carbon dioxide fixation in sugar cane leaves. *Plant Physiology,* **40**, 209–13.

25 Karpilov, Y.S. (1960) The distribution of radioactive carbon 14 among the products of photosynthesis in maize. *Proceedings of the Kazan Agricultural Institute,* **41**, 15–24 [in Russian]. A young Australian scientist also found four-carbon sugars after working with a particular species of salt-marsh plant (*Atriplex spongiosa*): see Osmond, C.B. (1967) Carboxylation during photosynthesis in *Atriplex. Biochimica et Biophysica Acta,* **141**, 197–9.

26 Both Hatch and Slack have extensively documented the historical details leading up to the discovery of C_4 photosynthesis. See Hatch, M.D. (1997) Resolving C_4 photosynthesis: trials, tribulations and other unpublished stories. *Australian Journal of Plant Physiology*, **24**, 413–22; Hatch, M.D. (2002) C_4 photosynthesis: discovery and resolution. *Photosynthesis Research*, **73**, 251–6. See also note 27.

27 The biochemical details are that C_4 plants contain a special enzyme, phosphoenolpyruvate carboxylase (PEP-C), that converts inorganic carbon dioxide into a four-carbon organic acid called oxaloacetic acid (OAA). PEP-C is the attachment enzyme and the only enzyme common to all C_4 plants. Most C_4 plants then convert OAA into malate before pumping it into the bundle sheath cells where it is catalytically converted to sugar by Rubisco after carbon dioxide is enzymically stripped off. C_4 plants have evolved families of 'detachment' enzymes for this purpose going by the acronyms NAD-ME, NADP-ME, and PEP-CK.

28 Osmond, C.B. (1971) The absence of photorespiration in C_4 plants: real or apparent? In *Photosynthesis and photorespiration* (eds M.D. Hatch, C.B. Osmond, and R.O Slayter), pp. 472–82. Academic Press, San Diego.

29 For an accessible account of the mechanisms of C_3 and C_4 photosynthesis, see Walker, D.A. (1993) *Energy, plants and man*. Oxygraphics, Brighton.

30 Hatch, M.D. (1992) I can't believe my luck. *Photosynthesis Research*, **33**, 1–14.

31 Haberlandt, G. (1884) *Physiologische Pflanzenantomie*. Engelman, Leipzig.

32 It has recently been revealed that two obscure species of C_4 plants (*Bienertia cycloptera* and *Borszczowia aralocaspica*) in the salty semi-deserts of Central Asia live without Kranz anatomy. Both species arose from C_3 ancestors and represent a radical departure from Kranz-type C_4 plants because they spatially compartmentalize the carbon dioxide-rich greenhouses not within different parts of the leaf but within different parts of the cell. See Voznesenskaya, E.V., Franceschi, V.R., Killrats, O. *et al.* (2001) Kranz anatomy is not essential for terrestrial C_4 plant photosynthesis. *Nature*, **414**, 543–6. A comparative review of the two approaches to achieving C_4

photosynthesis is given in Edwards, G.E., Franceschi, V.R., and Voznesenskaya, E.V. (2004) Single-cell C_4 photosynthesis versus the dual-cell (Kranz) paradigm. *Annual Review of Plant Biology*, **55**, 173–96.

33 Rye, R., Kuo, P.H., and Holland, H.D. (1995) Atmospheric carbon dioxide concentrations before 2.2 billion years ago. *Nature*, **378**, 603–75; Bekker, A., Holland, H.D., Wang, P.L. *et al.* (2004) Dating the rise of atmospheric oxygen. *Nature*, **427**, 117–20.

34 The Russian botanist Constantin Mereschkowsky (1855–1921) was the first to argue for the so-called endosymbiotic origin of the chloroplast, in a landmark paper: Mereschkowsky, C. (1905) Über Natur und Ursprung der Chromatophoren im Pflanzenreiche. *Biologischen Centralblatt* **25**, 593–604. He based his argument on earlier observations of the phenomenon of symbiosis and on observations showing that the organelles have the capacity to reproduce themselves even when separated from the nucleus. Although widely dismissed at the time, the hypothesis has gained critical support from later structural, genetic, and biochemical studies: see Margulis, L. (1970) *The origin of eukaryotic cells*. Yale University Press, New Haven. See review by Raven, J.A. and Allen, J.F. (2003) Genomics and chloroplast evolution: what did cyanobacteria do for plants? *Genome Biology*, **4**, article 209.

35 Griffiths, H. (2006) Designs on Rubisco. *Nature*, **441**, 940–1.

36 The idea that the evolution of the C_4 photosynthetic pathway is linked to low carbon dioxide was advanced in Ehleringer, J.R., Sage, R.F., Flanagan, L.B., and Pearcy, R.W. (1991) Climate change and the evolution of C_4 photosynthesis. See also Ehleringer, J.R., Cerling, T.E., and Hellicker, B.R. (1997) C_4 photosynthesis, atmospheric CO_2 and climate. *Oecologia*, **112**, 285–99.

37 Pagani, M., Zachos, J.C., Freeman, K.H. *et al.* (2005) Marked decline in atmospheric carbon dioxide concentrations during the Paleogene. *Science*, **309**, 600–3.

38 Lloyd, J. and Farquhar, G.D. (1994) ^{13}C discrimination during CO_2 assimilation by the terrestrial biosphere. *Oecologia*, **99**, 201–15; Still, C.J., Berry, J.A., Collatz, G.J., and DeFries, R.S. (2003) Global distribution of C_3 and C_4 vegetation: carbon cycle implications. *Global Biogeochemical Cycles*, **17**, doi: 10.1029/2001GB001807.

39 Details of the different C_4 species and their distribution between families are given by Sage, R.F. (2001) Environmental and evolutionary preconditions for the origin and diversification of the C_4 photosynthetic syndrome. *Plant Biology*, **3**, 202–13.

40 For a review of the place of C_4 plants in our cultural evolution, see van der Merwe, N.J. and Tschauner, H. (1999) C_4 plants and the development of human societies. In C_4 *plant biology* (eds R.F. Sage and R.K. Monson), pp. 509–49. Academic Press, San Diego.

41 Nambudiri, E.M.V., Tidwell, W.D., Smith, B.N., and Hebbert, N.P. (1978) A C_4 plant from the Pliocene. *Nature*, **276**, 816–17.

42 Dugas, D.P. and Retallack, G.J. (1993) Middle Miocene fossil grasses from Fort Ternan, Kenya. *Journal of Palaeontology*, **67**, 113–28.

43 Thomasson, J.R., Nelson, M.E., and Zakrzewski, J. (1986) A fossil grass (Gramineae) from the Miocene with Kranz anatomy. *Science*, **233**, 876–8.

44 The stochastic 'molecular clock' is explained by assuming most changes to amino acid sequences and nucleotides were neutral mutations. See Kimura, M. (1983) *The neutral theory of molecular evolution*. Cambridge University Press.

45 Jacobs, B.F., Kingston, J.D., and Jacobs, L.L. (1999) The origin of grass-dominated ecosystems. *Annals of the Missouri Botanical Garden*, **86**, 590–643; Kellogg, E.A. (2001) Evolutionary history of grasses. *Plant Physiology*, **125**, 1198–205.

46 Prasad, V., Stromberg, C.A.E., Alimohammadian, H., and Sahni, A. (2005) Dinosaur coprolites and the early evolution of grasses and grazers. *Science*, **310**, 1177–80. See also the commentary: Piperno, D.R. and Dieter-Sues, H. (2005) Dinosaurs dined on grass. *Science*, **310**, 1126–8.

47 Kellogg, E.A. (2000) The grasses: a case study in macroevolution. *Annual Review of Ecology and Systematics*, **31**, 217–38.

48 Extensive documentation of the global C_4 plant expansion 6–8 million years ago, based on the isotopic analysis of fossilized teeth and soil, was reported in Cerling, T.E., Harris, J.M., MacFadden, B.J. *et al.* (1997) Global vegetation change through the Miocene/Pliocene boundary. *Nature*, **389**, 153–8. Some of the ideas in the paper were challenged: see Köhler, M., Moyà-Solà, S., and Agusti, J. (1998) Miocene/Pliocene shift: one step or several? *Nature*, **393**, 126, and

the authors' response: Cerling, T.E., Harris, J.M., MacFadden, B.J. *et al.* (1998) Reply. *Nature*, **393**, 127.

49 In fact some of the records also come from isotopic analyses of fossil soils (palaeosols) but the principle is the same. The isotopic composition of soil organic matter reflects that of the vegetation type from which it is formed. As it decomposes, the carbon dioxide slowly forms soil carbonates over thousands of years, which are preserved as fossils and carry an isotopic composition that reflects that of the decaying vegetation.

50 Cerling, T.E. (1999) Palaeorecords of C_4 plants and ecosystems. In *C_4 plant biology* (eds R.F. Sage and R.K. Monson), pp. 445–69. Academic Press, San Diego.

51 I should point out that these records are an unreliable means of pinpointing the origins of C_4 plants, because they reflect changes in plant community composition. As the molecular clocks tell us, C_4 plants originated long before they became ecologically dominant. However, the oldest isotopic evidence for a tropical C_4 flora does, at 15.3 million years old, pre-date the oldest fossil leaf fragments; see Kingston, J.D., Marino, B.D., and Hill, A. (1994) Isotopic evidence for Neogene hominid palaeoenvironments in the Kenya Rift Valley. *Science*, **264**, 955–9. It is also worth noting that a faint isotopic signal of C_4 plants has been detected in the fossil soils of the southern Great Plants in North America dating to 23 million years ago. This is close to the date of origin (23–35 million years ago) indicated by the molecular clocks; see Fox, D.L. and Koch, P.L. (2003) Tertiary history of C_4 biomass in the Great Plains, USA. *Geology*, **31**, 809–12.

52 See Ehleringer *et al.*, Climate change and the evolution of C_4 photosynthesis, Ehleringer *et al.*, C_4 photosynthesis, atmospheric CO_2 and climate (above, n.36), and Cerling, *et al.*, Global vegetation change through the Miocene/Pliocene boundary (above, n.48).

53 See the references cited in the previous note.

54 Street-Perrott, F.A., Huang, Y., Perrott, G. *et al.* (1997) Impact of lower atmospheric carbon dioxide on tropical mountain ecosystems. *Science*, **278**, 1422–6. See also the commentary: Farquhar, G.D. (1997) Carbon dioxide and vegetation. *Science*, **278**, 1411.

55 Cole, D.R. and Monger, H.C. (1994) Influence of atmospheric CO_2 on the decline of C_4 plants during the last deglaciation. *Nature*, **368**, 533–6.

56 Pagani, M., Freeman, K.H., and Arthur, M.A. (1999) Late Miocene atmospheric CO_2 concentrations and the expansion of C_4 grasses. *Science*, **285**, 876–9.

57 Carbon dioxide reconstructions based on stomata were reported by Royer, D.L., Wing, S.L., Beerling, D.J. *et al.* (2001) Palaeobotanical evidence for near present-day levels of atmospheric CO_2 during part of the Tertiary. *Science*, **292**, 2310–13, and those from the boron isotopes by Pearson, P.N. and Palmer, M.R. (2000) Atmospheric carbon dioxide concentrations over the past 60 million years. *Nature*, **406**, 695–9.

58 Osborne, C.P. and Beerling, D.J. (2006) Nature's green revolution: the remarkable evolutionary rise of C_4 plants. *Philosophical Transactions of the Royal Society*, **B361**, 173–94.

59 An update is given by Antoine, P.O., Shah, S.M.I., Cheema, I.U. *et al.* (2004) New remains of the baluchithere *Paraceratherium bugiense* (Pilgrim 1910) from the late/latest Oligocene of the Bugti Hills, Balochistan, Pakistan. *Journal of Asian Earth Sciences*, **24**, 71–7.

60 Quade, J., Carter, J.M.L., Ojha, T.P. *et al.* (1995) Late Miocene environmental change in Nepal and the Northern Indian subcontinent: stable isotope evidence from paleosols. *Geological Society of America Bulletin*, **107**, 1381–97.

61 Zhisheng, A., Kutzbach, J.E., Prell, W.L., and Porter, S.C. (2001) Evolution of Asian monsoons and phased uplift of the Himalaya plateau since the Late Miocene times. *Nature*, **411**, 62–6.

62 Reconstructions of the palaeo-elevation of Tibet show no change over the past 15 million years, suggesting uplift was complete by this time. Spicer, R.A., Harris, N.B., Widdowson, M. *et al.* (2003) Constant elevation of southern Tibet over the past 15 million years. *Nature*, **421**, 622–4; Currie, B.S., Rowley, D.B., and Tabor, N.J. (2005) Middle Miocene paleoaltimetry of southern Tibet: implications for the role of mantle thickening and delamination of the Himalayan orogen. *Geology*, **33**, 181–4.

63 The idea that fire and carbon dioxide are important determinants of tree and grass cover was originally proposed in Bond, W.J. and Midgley, G.F. (2000) A proposed CO_2-controlled mechanism of woody plant invasion in grasslands and savannas. *Global Change Biology*, **6**, 865–9. It later received strong support from 'experiments' with a model simulating the dynamics of vegetation change in

South Africa under different carbon dioxide concentrations; see Bond, W.J., Midgley, G.F., and Woodward, F.I. (2003) The importance of low atmospheric CO_2 and fire in promoting the spread of grasslands and savannas. *Global Change Biology*, **9**, 973–82.

64 See Bond, W.J., Woodward, F.I., and Midgley, G.F. (2005) The global distribution of ecosystems in a world without fire. *New Phytologist*, **165**, 525–38. See also the commentary on that paper by Bowman, D. (2005) Understanding a flammable planet: climate, fire and the global vegetation patterns. *New Phytologist*, **165**, 341–5. The role of fire in shaping global vegetation patterns is further discussed in the following two papers: Bond, W.J. (2005) Large parts of the world are brown or black: a different view on the 'green world' hypothesis. *Journal of Vegetation Science*, **16**, 261–6; Bond, W.J. and Keeley, J.E. (2005) Fire as a global 'herbivore': the ecology and evolution of flammable ecosystems. *Trends in Ecology and Evolution*, **29**, 387–94.

65 Hughes, F., Vitousek, P.M., and Tunison, T. (1991) Alien grass invasion and fire in the seasonal submontane zone of Hawai'i. *Ecology*, **72**, 743–7; D'Antonio, C.M. and Vitousek, P.M. (1992) Biological invasions by exotic grasses, the grass/fire cycle, the global change. *Annual Review of Ecology and Systematics*, **23**, 63–87.

66 Sage, R.F. (2003) *Quo vadis?* An ecophysiological perspective on global change and the future of C_4 plants. *Photosynthesis Research*, **77**, 209–25.

67 Keeley, J.E. and Rundell, P.W. (2005) Fire and the Miocene expansion of C_4 grasslands. *Ecology Letters*, **8**, 683–90.

68 Koren, I., Kaufman, Y.J., Remer, L.A., and Martins, J.V. (2004) Measurements of the effect of Amazon smoke on inhibition of cloud formation. *Science*, **303**, 1342–5.

69 Andreae, M.O., Rosenfield, D., Artaxo, P. *et al.* (2004) Smoking rain clouds over the Amazon. *Science*, **303**, 1337–42. For a commentary on the significance of these new findings about smoke for climate and those described in Koren *et al.*, Measurements of the effect of Amazon smoke on inhibition of cloud formation (above, n.68), see Graf, H.F. (2004) The complex interaction of aerosols and clouds. *Science*, **303**, 1309–11.

70 The realization that wildfire caused an intensification of the 1988 drought is reported by Liu, Y. (2005) Enhancement of the 1988

northern U.S. drought due to wildfires. *Geophysical Research Letters*, 32, doi: 10.1029/2005GL02241. It represents a significant departure from the belief that it was generated by changes in the conditions of the surface oceans. The origin of the drought in these more conventional terms is considered by Trenberth, K.E., Branstator, G.W., and Arkin, P.A. (1988) Origins of the 1988 American drought. *Science*, 242, 1640–5.

71 Beerling, D.J. and Osborne, C.P. (2006) The origin of the savanna biome. *Global Change Biology*, 12, 2023–31.

72 Even this complex web of feedbacks fails to adequately explain the rise of C_4 plants in the southern Great Plains in North America. Koch and colleagues (Fox and Koch, Tertiary history of C_4 biomass in the Great Plains, USA (above, n.51)) inferred the rise of C_4 plants from the isotopic composition of fossil soils rather than fossil teeth, arguing that soils are formed from the plants themselves whereas teeth reflect the vagaries of diet and selective feeding by the grazing/browsing mammals. Surprisingly, the version of history revealed by fossil soils dates the increase in C_4 grassland biomass in the Great Plains at 2.5 million years ago. This age for the rise of C_4 plants to modern levels in the region is several million years later than the fossil horse teeth, which suggests C_4-dominated diets arrived in what is now Texas around 6.5 million years ago. Why the difference? No one is sure, but part of the answer seems to be that herbivore diet is not always tightly coupled to C_4 plant abundance; some species of horses prefer to graze C_3 or C_4 plants, and others still show no preference. C_4 grasses in the Great Plains did not prosper because of changing fire regimes or carbon dioxide levels, requiring instead intensified grazing pressure to establish and maintain their dominance. For more details, see Fox, D.L. and Koch, P.L. (2004) Carbon and oxygen isotope variability in Neogene paleosol carbonates: constraints on the evolution of the C_4-grasslands of the Great Plains, USA. *Palaeogeography, Palaeoclimatology, Palaeoecology*, 207, 305–29.

73 The relevance of these charcoal datasets from the Pacific Ocean was pointed out by Keeley, J.E. and Rundel, P.W. (2003) Evolution of CAM and C_4 carbon-concentrating mechanisms. *International Journal of Plant Science*, 164 (3 Suppl.), S55–S77. The datasets themselves were

published in Herring, J.R. (1985) Charcoal fluxes into sediments of the North Pacific Ocean: Cenozoic records of burning. In *The carbon cycle and atmospheric CO₂: natural variations from Archean to present* (eds E.T. Sundquist and W.S. Broecker), pp. 419–42. American Geophysical Union, Washington.

74 See Davies, P. (1992) *The mind of God: science and the search for ultimate meaning*. Penguin Science, London.

75 Hoyle, F. (1983) *The intelligent universe*. Michael Joseph, London.

76 Note also that the evolutionary intermediates between a C_3 and a C_4 plant must each be stable and function in their own right, with each step conferring some ecological advantage to the owner.

77 C_4 photosynthesis proliferated in three families of monocots, Poacea (grasses), Cyperaceae (sedges), and Hydrocharitacea, and the numerous families of Eudicots, including Chenopodiaceae, Brassicaceae (cabbages), and Euphorbiaceae. See Sage, R.F. (2004) The evolution of C_4 photosynthesis. *New Phytologist*, **161**, 341–70.

78 Possession of the biochemical machinery for C_4 photosynthesis is only part of the story; making a C_4 plant also requires anatomical modifications for Kranz anatomy. If Kranz anatomy was controlled by some sort of genetic switch analogous to those discovered for other anatomical shifts in plants, its evolution might be reasonably straightforward. The appearance of teosinte, the ancestral form of maize, for example, was transformed to that of domesticated maize through changes to a single gene. Restoring the gene returns modern maize to its ancestral, teosinte-like appearance; see Doebley, J., Stec, A., and Hubbard, I. (1997) The evolution of apical dominance in maize. *Nature*, **386**, 485–8. The exotic plant spikerush (*Eleocharis vivipara*) from the swamps of Florida leads a double life, hinting at the existence of a hidden master switch for Kranz anatomy. When submerged, the leaves of spikerush function like those of a C_3 plant; when emergent its stems develop Kranz anatomy and full-blown C_4 photosynthesis; see Ueno, O. (1996) Structural characterization of photosynthetic cells in an amphibious sedge, *Eleocharis vivipara*, in relation to C_3 and C_4 metabolism. *Planta*, **199**, 382–93; Ueno, O. (1996) Immunocytochemical localization of enzymes involved in the C_3 and C_4 pathways in the photosynthetic cells of an amphibious sedge, *Eleocharis vivipara*. *Planta*, **199**, 394–403.

79 The original paper reported these effects in celery and tobacco plants, and hinted that they occurred in a much wider range of plants. See Hibberd, J.M. and Quick, W.P. (2002) Characteristics of C_4 photosynthesis in stems and petioles of flowering plants. *Nature*, **415**, 451–4. See also the illuminating commentary: Raven, J.A. (2002) Evolutionary options. *Nature*, **415**, 375–6.

80 Osborne and Beerling, Nature's green revolution (above, n.58).

81 An atmospheric composition with low carbon dioxide and high oxygen content can cause particularly high rates of photorespiration in C_3 plants, markedly diminishing their productivity. Evolving a type of C_4 photosynthetic pathway would have largely solved this problem.

82 The search was conducted by systematically screening the stable carbon isotope composition of Carboniferous and Permian plants occurring where theory suggested that C_4 photosynthesis would have been most advantageous (see Osborne and Beerling, Nature's green revolution (above, n.58)).

83 United Nations Population Division (2000) *World population prospects: the 2000 revision*. United Nations, Department of Economics and Social Affairs, New York.

84 Two obvious alternatives are to engineer rice with more Rubisco or better Rubisco. Both, however, have drawbacks. More Rubisco means more nitrogen has to be supplied to the crop but there are limits to the amount of nitrogen that can be taken up. Nitrogen fertilizer use is already heavy and that not taken up by the plant goes into the atmosphere or out into the water. The 'better Rubisco' option might work, but there are limits to its performance involving the biochemical trade-off between its specificity for carbon dioxide and rate of catalysis; see, for example, Zhu, X.-G., Portis, A.R., and Long, S.P. (2004) Would transformation of C_3 crops with foreign Rubisco increase productivity? A computational analysis from kinetic properties to canopy photosynthesis. *Plant, Cell and Environment*, **27**, 155–65.

85 Matusoka, M., Furbank, R.T., Fukayama, H., and Miyao, M. (2001) Molecular engineering of C_4 photosynthesis. *Annual Review of Plant Physiology and Plant Molecular Biology*, **52**, 297–314; Sheehy, J.E., Mitchell, P.L., and Hardy, B. (2000) *Redesigning rice photosynthesis to increase*

yield. Elsevier, Amsterdam; Mitchell, P.L. and Sheehy, J.E. (2006) Supercharging rice photosynthesis to increase yield. *New Phytologist*, **171**, 689–92. See also a general commentary on this issue by Surridge, C. (2002) The rice squad. *Nature*, **416**, 576–8.

86 Foley, J.A., DeFries, R., Asner, G.P. *et al.* (2005) Global consequences of land use. *Science*, **309**, 570–4.

87 Cochrane, M.A., Alencar, A., Schulze, M.D. *et al.* (1999) Positive feedbacks in the fire dynamic of closed canopy tropical forests. *Science*, **284**, 1832–5.

CHAPTER 9

Through a glass darkly

1 See Chaloner, W.G. and Creber, G.T. (1990) Do fossil plants give a climatic signal? *Journal of the Geological Society*, **147**, 343–50.

2 Andrews, H.N. (1980) *The fossil hunters: in search of ancient plants.* Cornell University Press.

3 Seward, A.C. (1892) *Fossils plants as tests of climate.* Clay, London.

4 For a review see Chaloner and Creber, Do fossil plants give a climatic signal? (above, n.1).

5 Bailey, I.W. and Sinnott, E.W. (1915) A botanical index of Cretaceous and Tertiary climate. *Science*, **41**, 831–4; Bailey, I.W. and Sinnott, E.W. (1916) The climatic distribution of certain types of angiosperm leaves. *American Journal of Botany*, **3**, 24–39.

6 Wolfe, J.A. (1979) A method for obtaining climatic parameters from leaf assemblages. *U.S. Geological Survey Bulletin*, **2040**, 1–71.

7 See Brown, V.K. and Lawton, J.H. (1991) Herbivory and the evolution of leaf size and shape. *Philosophical Transactions of the Royal Society*, **B333**, 265–72; Rivero-Lynch, A.P., Brown, V.K., and Lawton, J.H. (1996) The impact of leaf shape on the feeding preference of insect herbivores: experimental and field studies with *Capsella* and *Phyllotreta*. *Philosophical Transactions of the Royal Society*, **B351**, 1671–7.

8 Feild, T.S., Sage, T.L., Czerniak, C., and Iles, J.D. (2005) Hydathodal leaf teeth of *Chloranthus japonicus* (Chloranthaceae) prevent guttation-induced flooding of the mesophyll. *Plant, Cell and Environment*, **28**, 1179–90.

9 Royer, D.L. and Wilf, P. (2006) Why do toothed leaves correlate with cold climates? Gas exchange at leaf margins provides new insights into a classic paleotemperature proxy. *International Journal of Plant Sciences*, **167**, 11–18.

10 Tsiantis, M. and Hay, A. (2003) Comparative plant development: the time of the leaf? *Nature Genetics*, **4**, 169–80; Piazza, P., Jasinski, S., and Tsiantis, M. (2005) Evolution of leaf developmental mechanisms. *New Phytologist*, **167**, 693–710; Byrne, M.E. (2005) Networks in leaf development. *Current Opinion in Plant Biology*, **8**, 59–66; Hay, A. and Tsiantis, M. (2006) The genetic basis for differences in leaf form between *Arabidopsis thaliana* and its wild relative *Cardamine hirsute*. *Nature Genetics, Advance online publication*, doi: 10.1038/ng1835.

11 The original observation was reported in Lutze, J.L., Roden, J.S., Holly, C.J. *et al.* (1998) Elevated atmospheric [CO_2] promotes frost damage in evergreen tree seedlings. *Plant, Cell and Environment*, **21**, 631–5. More details are given in Barker, D.H., Loveys, B.R., Egerton, J.J.G. *et al.* (2005) CO_2 enrichment predisposes foliage of a eucalypt to freezing injury and reduces spring growth. *Plant, Cell and Environment*, **28**, 1506–15. The same effect was also found for ginkgo: see Terry, A.C., Quick, W.P., and Beerling, D.J. (2000) Long-term growth of Ginkgo with CO_2 enrichment increases ice nucleation temperatures and limits recovery of the photosynthetic system from freezing. *Plant Physiology*, **124**, 183–90.

12 Kullman, L. (1998) Tree-limits and montane forests in the Swedish Scandes: sensitive biomonitors of climate change and variability. *Ambio*, **27**, 312–21.

13 Loveys, B.R., Egerton, J.J.G., and Ball, M.C. (2006) Higher daytime leaf temperatures contribute to lower freeze tolerance under elevated CO_2. *Plant, Cell and Environment*, **29**, 1077–86.

14 Royer, D.L., Osborne, C.P., and Beerling, D.J. (2002) High CO_2 increases the freezing sensitivity of plants: implications for paleoclimate reconstructions from fossil plants. *Geology*, **30**, 963–66.

15 Fricke, H.C and Wing, S.L. (2004) Oxygen isotope and paleobotanical estimates of temperature and ^{18}O-latitude gradients over North America during the early Eocene. *American Journal of Science*, **304**, 612–35.

16 *Archaeopteris* stomatal data are given in Osborne, C.P., Beerling, D.J., Lomax, B.H., and Chaloner, W.G. (2004) Biophysical constraints on the origin of leaves inferred from the fossil record. *Proceedings of the*

National Academy of Sciences, USA, **101**, 10360–2. See also Beerling, D. J., Osborne, C.P., and Chaloner, W.G. (2001) Evolution of leaf-form in land plants linked to atmospheric CO_2 decline in the Late Palaeozoic Era. *Nature*, **410**, 352–4.

17 See Harrison, C.J., Corley, S.B., Moylan, E.C. *et al.* (2005) Independent recruitment of a conserved developmental mechanism during leaf evolution. *Nature*, **434**, 509–14; Beerling, D.J. and Fleming, A. (2007) Zimmermann's telome theory of megaphyll leaf evolution: a molecular and cellular critique. *Current Opinion in Plant Biology*, **10**, 1–9.

18 Gray, J.E., Holroyd, G.H., van der Lee, F.M. *et al.* (2000) The *HIC* signaling pathway links CO_2 perception to stomatal development. *Nature*, **408**, 713–16.

19 The case for mosses was made largely on theoretical grounds: see White, J.W.C., Figge, R.A., Ciais, P. *et al.* (1994) A high-resolution record of atmospheric CO_2 content from carbon isotopes in peat. *Nature*, **367**, 153–6. On both experimental and theoretical grounds, we suggested that liverworts were a better bet: see Fletcher, B.J., Beerling, D.J., Brentnall, S.J., and Royer, D.L. (2005) Bryophytes as recorders of ancient CO_2 levels: experimental evidence and a Cretaceous case study. *Global Biogeochemical Cycles*, **6**, doi: 10.1029/2005GB002495; Fletcher, B.J., Brentnall, S.J., Quick, W.P., and Beerling, D.J. (2006) BRYOCARB: a process-based model of thallose liverwort carbon isotope fractionation in response to CO_2, O_2, light and temperature. *Geochimica Cosmochimica Acta*, **70**, 5676–91.

20 Berner, R.A. and Canfield, D.E. (1989) A new model of atmospheric oxygen over time. *American Journal of Science*, **289**, 333–61.

21 Beerling, D.J., Lake, J.A., Berner, R.A. *et al.* (2002) Carbon isotope evidence implying high O_2/CO_2 ratios in the Permo-Carboniferous atmosphere. *Geochimica et Cosmochimica Acta*, **66**, 3757–67.

22 Koti, S., Reddy, K.R., Reddy, V.R. *et al.* (2005) Interactive effects of carbon dioxide, temperature, and ultraviolet-B radiation on soybean (*Glycine max* L.) flower and pollen morphology, pollen production, germination, and tube lengths. *Journal of Experimental Botany*, **56**, 725–36.

23 Estimates kindly provided by R.A. Berner (Yale University) and calculated by removing the effects of land plants in his GEOCARB model [personal communication].

24 Kleidon, A., Fraedrich, K., and Heimann, M. (2000) A green planet versus a desert world: estimating the maximum effect of vegetation on the land surface climate. *Climatic Change*, **44**, 471–93.

25 Otto-Bliesner, B.L. and Upchurch, G.R. (1997) Vegetation induced warming of high latitude regions during the Late Cretaceous period. *Nature*, **395**, 804–7; DeConto, R.M., Brady, E.C., Bergengren, J., and Hay, W.W. (2000) Late Cretaceous climate, vegetation, and ocean interactions. In *Warm climates in Earth history* (eds B.T. Huber, K.G. MacLeod, and S.L. Wing), pp. 275–96. Cambridge University Press.

26 Retallack, G.J. (2001) Cenozoic expansion of grasslands and climatic cooling. *Journal of Geology*, **109**, 407–26.

27 Hoffmann, W.A. and Jackson, R.B. (2000) Vegetation-climate feedbacks in the conversion of tropical savanna to grassland. *Journal of Climate*, **13**, 1593–602; Hoffmann, W.A., Schroeder, W., and Jackson, R.B. (2002) Positive feedbacks of fire, climate and vegetation and the conversion of tropical savanna. *Geophysical Research Letters*, **29**, doi:10.1029/2002GL015424.

28 For reviews, see Stebbins, G.L. (1981) Coevolution of grasses and herbivores. *Annals of Missouri Botanical Gardens*, **68**, 75–86; Janis, C.M., Damuth, J., and Theodor, J.M. (2000) Miocene ungulates and terrestrial primary productivity: where have all the browsers gone? *Proceedings of the National Academy of Sciences, USA*, **97**, 7899–904; MacFadden, B.J. (2000) Cenozoic mammalian herbivores from the Americas: reconstructing ancient diets and terrestrial communities. *Annual Review of Ecology and Systematics*, **31**, 31–59; Strömberg, C.A.E. (2006) Evolution of hypsodonty in equids: testing the hypothesis of adaptation. *Paleobiology*, **32**, 236–58.

29 Falkowski, P.G., Katz, M.E., Knoll, A.H. *et al.* (2004) The evolution of modern Eukaryotic phytoplankton. *Science*, **305**, 354–60.

30 Hetherington, A.M. and Woodward, F.I. (2003) The role of stomata in sensing and driving environmental change. *Nature*, **424**, 901–8.

31 The programme transcript is available at: http://www.bbc.co.uk/sn/tvradio/programmes/horizon/dimming_trans.shtml

32 Aerosols are abundant in the air we breathe and range in size from sub-microscopic to the almost visible. Sulfur compounds, in the form of droplets of sulfuric acid and ammonium sulfate, are the most important from a climatic point of view. In the pre-industrial

era when the air was cleaner, most of the sulfate came from the oceans. Today the main source is from sulfur dioxide produced by industrial processes, and it outstrips the natural sources of sulfur by a factor of three to one.

33 Ramanathan, V., Crutzen, P.J., Kiehl, J.T., and Rosenfeld, D. (2001) Aerosols, climate, and the hydrological cycle. *Science*, **294**, 2119–24.

34 *Sunday Times*, 9 January 2005, 'Culture' section, Critic's choice, p. 76.

35 One major reason for the cleaner atmosphere is thought to be the collapse of Communist economies in the late 1980s, substantially decreasing the amount of pollutants released. Another is greater investment in clean-air technologies in Europe and North America that reduce aerosol emissions and polluting gases from vehicles and smokestacks. We should not run away with the idea, however, that all parts of the world are enjoying cleaner air. In some highly polluted regions, like India, the burning of fossil fuels and wildfires create vast smog clouds that darken the sky for long periods each year, and extend an influence to more remote locations; see Venkataraman, C., Habib, G., Eiguren-Fernandez, A. *et al.* (2005) Residential biofuels in South Asia: Carbonaceous aerosol emissions and climate impacts. *Science*, **307**, 1454–6. The northern islands of the seemingly idyllic Maldives, for example, sit in a stream of dirty air descending from India, which forms a 3-km-thick layer that cuts down sunlight by up to 15%; see Satheesh, S.K. and Ramanathan, V. (2000) Large differences in tropical forcing at the top of the atmosphere and Earth's surface. *Nature*, **405**, 60–3.

36 Wild, M., Gilgen, H., Roesch, A. *et al.* (2005) From dimming to brightening: decadal changes in solar radiation at Earth's surface. *Science*, **308**, 847–50; Pinker, R.T., Zhang, B., and Dutton, E.G. (2005) Do satellites detect trends in surface solar radiation? *Science*, **308**, 850–4. For an important commentary setting this paper in context, see Charlson, R.J., Valero, P.J., and Seinfield, J.H. (2005) In search of balance. *Science*, **308**, 806–7. Charlson and colleagues soberly point out that uncertainties between different methods of measuring changes in the reflectivity (albedo) of our planet are 'as large or larger' than the enhanced greenhouse effect.

37 Andreae, M.O., Jones, C.D., and Cox, P.M. (2005) Strong present-day aerosol cooling implies a hot future. *Nature*, **435**, 1187–90.

38 Stanhill, G. and Shabtai, C. (2001) Global dimming: a review of the evidence for a widespread and significant reduction in global radiation with discussion of its probable causes and possible agricultural consequences. *Agricultural and Forest Meteorology*, **107**, 255–78.

39 Pan evaporation is, as the name suggests, simply a measurement of the amount of water evaporating from a pan. It is easily determined: a pan filled with water is topped up with a known volume of water every morning at the same time to the level it was at that time on the previous morning. All over the world, heroic researchers have been doggedly carrying out this rather mundane task for decades, day in, day out.

40 Roderick, M.L. and Farquhar, G.D. (2004) Changes in Australian pan evaporation from 1970 to 2002. *International Journal of Climatology*, **24**, 1077–90; Roderick, M.L. and Farquhar, G.D. (2005) Changes in New Zealand pan evaporation since the 1970s. *International Journal of Climatology*, **25**, 2031–9.

41 One group (Peterson, T.C., Golubev, V.S., and Groisman, P. Y. (1995) Evaporation losing its strength. *Nature*, **377**, 687–8) argued that decreasing pan evaporation was related to increased cloudiness, which in turn decreased the diurnal temperature variation (daily maximum minus minimum temperature). The other (Brutsaert, W. and Parlange, M.B. (1998) Hydrologic cycle explains the evaporation paradox. *Nature*, **396**, 30) argued that as the air above the pan becomes humidified, it weakens the driving force for further evaporation.

42 Roderick, M.L. and Farquhar, G.D. (2002) The cause of pan evaporation over the past 50 years. *Science*, **298**, 1410–11. For a commentary, see: Ohmura, A. and Wild, M. (2002) Is the hydrological cycle accelerating? *Science*, **298**, 1345–6.

43 Tyndall, J. (1865) *Heat considered as a mode of motion*. Second edn, with additions and illustrations. Longman Green, London.

44 Travis, D.J. Carleton, A.M., and Lauristen, R.G. (2002) Contrails reduce daily temperature range. *Nature*, **418**, 601.

45 Easterling, D.R., Horton, B., Jones, P.D. *et al.* (1997) Maximum and minimum temperature trends for the globe. *Science*, **277**, 364–7.

46 Popper, K.R. (1963) *Conjectures and refutations: the growth of scientific knowledge*. Routledge and Kegan Paul, London.

370 NOTES TO CHAPTER 9

47 This notion is nicely illustrated by what has become known as the 'faint young Sun paradox'. We have already learned of the British astrophysicist Fred Hoyle's amazement that the values of various cosmological constants seem beautifully attuned to allow the emergence of life (Chapter 8). Back in the 1950s, Hoyle and other astronomers developed theoretical models for how stars evolve over time, showing that as they burn their core density increases, accelerating the fusion reactions, to produce more energy and increased luminosity. Accordingly, when Earth formed 4.5 billion years ago, our Sun is calculated to have been around 30% less bright than it is now. A dimmer Sun means a cooler Earth, much cooler, with temperatures remaining below freezing until the Sun got brighter around 2 billion years ago; see Sagan, C. and Mullen, G. (1972) Earth and Mars: evolution of atmospheres and surface temperatures. *Science*, **177**, 52–6. Nothing wrong with that, you might think, but with the publication of indisputable evidence for liquid water on Earth dating to over 4 billion years ago, up went the cry of 'hold on a minute'. Attempts to resolve this particular paradox spawned nearly 30 years of research in laboratories around the world: an atmosphere unusually rich in the greenhouse gas methane is currently the favoured explanation for keeping the young Earth warm; see Kasting, J.F. and Catling, D. (2003) Evolution of a habitable planet. *Annual Reviews of Astronomy and Astrophysics*, **41**, 429–63.

48 Lovelock, J. (1979) *Gaia: A new look at life on Earth*. Oxford University Press.

49 See Cloud, P. (1972) A working model of primitive Earth. *American Journal of Science*, **272**, 537–48; Holland, H.D. (1984) *The chemical evolution of the atmosphere and the oceans*. Princeton Series in Geochemistry. Princeton University Press; Walker, J.C.G. (1974) Stability of atmospheric oxygen. *American Journal of Science*, **274**, 193–214; Berner, R.A. and Canfield, D.E. (1989) A new model of atmospheric oxygen over time. *American Journal of Science*, **289**, 333–61; Garrels, R.M., Lerman, A., and Mackenzie, F.T. (1976) Controls of atmospheric O_2 and CO_2: past, present and future. *American Scientist*, **64**, 306–15.

50 Graham, J.B., Dudley, R., Aguilar, N.M., and Gans, C. (1995) Implications of the late Palaeozoic oxygen pulse for physiology and evolution. *Nature*, **375**, 117–20.

INDEX

Titles in the *Oxford Landmark Science* series